American Mathematical Society

TRANSLATIONS

Series 2 • Volume 230

Selected Papers on Analysis and Differential Equations

American Mathematical Society
Providence, Rhode Island

2000 *Mathematics Subject Classification.* Primary 35–06, 37–06, 43–06.

Library of Congress Cataloging-in-Publication Data

Selected papers on analysis and differential equations.
p. cm. — (American Mathematical Society translations ; series 2, v. 230)
Volume includes English translation of ten expository articles published in the Japanese journal Sūgaku.
Includes bibliographical references.
ISBN 978-0-8218-4881-4 (alk. paper)
1. Mathematical analysis. 2. Differential equations. I. American Mathematical Society.

QA299.6.S425 2010
515—dc22

2010022951

10 9 8 7 6 5 4 3 2 1 15 14 13 12 11 10

Contents

Preface

The volume includes English translations of ten expository articles published in the Japanese journal "Sugaku". These articles are devoted to various topics in partial differential equations, harmonic analyis, and operator algebras.

In the article "Asymptotic Analysis of Nonlinear Dispersive Operators" the author, Kenji Nakanishi, presents recent results in the use of nonperturbative methods in the study of solutions to dispersive nonlinear partial differential equations. The three main topics considered in the article are the global analysis of the scattering problem, asymptotic analysis of the interaction between solitary and dispersive waves, and asymptotic analysis of singular limits for solutions of nonlinear dispersive equations.

Another expository article in this volume devoted to asymptotic analysis of solutions of nonlinear partial differential equations is "Asymptotics of Nonlinear Dispersive-type Evolution Equations" by Nakao Hayashi, where asymptotic behavior of solutions to various classical integrable equations, such as the modified Korteweg–de Vries equation, the cubic nonlinear Schrödinger equation, and the cubic nonlinear Klein–Gordon equation are studied. The author concentrates on solutions corresponding to initial data that do not contain a soliton component. In this case, asymptotic properties of solutions are different from properties of the corresponding linear equations, and to study these properties, new mathematical tools are introduced and described.

The article "Heun's Differential Equation", by Kouchi Takemura, is devoted to Heun's equation, which is an ordinary linear differential equation on the Riemann sphere with four singular points of Fuchsian type. As such, it is a natural generalization of the Gauss hypergeometric equation, which has three Fuchsian singular points. Heun's equation, and the ordinary differential equations of the corresponding confluent type, appear in several areas of physics, including analysis of black holes, crystal transition, and fluid dynamics. From the point of view of mathematics, it is related to Painlevé equations, Teichmüller spaces, and so on. The article presents fundamental facts about Heun's equation and the Gauss hypergeometric equation and outlines possible directions for further study. It also contains recent results related to finite-gap integration of Heun's equation.

"Scattering Theory and Inverse Problems", by Hiroshi Isozaki, is an expository article on two topics: the quantum 3-body problem and the inverse boundary value problem. The main themes of the article are the eigenfunction expansion theorem for the continuous spectrum of 3-body quantum systems and applications of hyperbolic geometry to the inverse boundary value problem.

"Nondoubling Measure and Harmonic Analysis", by Yasuo Komori, is a survey article devoted to recent developments in harmonic analysis of irregular metric spaces. Starting with the Painlevé problem about removable sets in $\mathbb{C}$, the author presents various results related to the Calderón–Zygmund theory, the theory of David and Journé, and the Fefferman–Stein theorem, as well as generalizations of these results to irregular sets obtained by Nazarov, Treil, Volberg, and Tulsa. The common unifying theme in the exposition is the notion of a nondoubling measure on a metric space X, i.e., a measure μ such that the measure of a ball of radius $2r$ centered at a point x cannot be linearly estimated in terms of the measure of the ball of radius r centered at x (uniformly in r and x).

The theory of reproducing kernels started in 1920s with two papers introducing what is now known as the Szegö kernel and the Bergman kernel. Since then, the theory has developed into a large and deep area of complex analysis by many outstanding mathematicians. In the article "Theory of Reproducing Kernels: Applications to Approximate Solutions of Bounded Linear Operator Equations on Hilbert Spaces", the author, Saburou Saitoh, surveys recent results of the theory of reproducing kernels. In addition, he shows how this theory can be used in numerical analysis, in particular, in the Tikhonov regularization procedure for obtaining approximate solutions for bounded linear operator equations in Hilbert spaces.

In the article "An Approach to Superrigidity and Fixed Point Theorems via Harmonic Maps", the authors, Hiroyasu Izeki and Shin Nayatani, emphasize the importance of harmonic analysis in the study of various rigidity properties of lattices in semisimple Lie groups.

In the article "Rational Semigroups, Random Complex Dymamics, and Singular Functions on the Complex Plane", Hiroki Sumi summarizes recent developments in the study of the dynamics of semigroups, of holomorphic maps, of the Riemann sphere, and of random complex dynamics. A random dynamics $F_1, F_2, \ldots$ on a space X is given by a probablity distribution $P = (p_1, \ldots, p_k)$ on a finite set of transformations $\{f_1, \ldots, f_k\}$ of X, and on each step i the transformation F_i is chosen to be equal to f_k with probability p_k, independently of previous choices. In the real case of transformations of the real line, one can consider the probability $T_{+\infty}(x)$ that a random orbit $F_1(x), F_2(F_1(x)), \ldots$, starting at a point $x \in \mathbb{R}$ tends to infinity. It turns out that already in very simple cases ($k = 2$, f_1 and f_2 linear) the function $T_{+\infty}(x)$ and other similarly defined functions turn out to be equal to well-known singular functions (e.g., the Cantor devil's staircase function). One of the main goals of the article is to consider complex analogs of the construction using random complex dynamics on the Riemann sphere. In this case a complex analog of singular functions is a continous function on $\mathbb{C}$ that is constant outside of certain Julia sets of a semigroup generated by several polynomial or rational maps.

The article "Salem Polynomials and the Bimeromorphic Automorphism Group of a Hyper-käler Manifold", by Keiji Oguiso, surveys basic facts about Salem polynomials, $K3$ surfaces, and hyper-kähler manifolds. The author explains a construction of C. McMullen, relating some infinite order automorphisms of certain $K3$ surfaces with Salem numbers (algebraic integers a such that Galois conjugates of a are a, $1/a$, and complex numbers of absolute value 1) and their minimal polynomials (called Salem polynomials). The main goal of the author is to show how McMullen's construction is related to various results from lattice theory, analytic number theory, and complex geometry.

In the paper "Tensor Categories in Operator Algebras", Shigeru Yamagami introduces the notion of a C^*-category, defined as a category admitting a faithful embedding in the category of all Hilbert spaces. Then the author defines the notion of a C^*-tensor category and explains properties of such categories, as well as applications of C^*-tensor categories to Tannaka–Krein-type duality theorems. The article also contains an extensive list of references with a very useful reference guide.

Amer. Math. Soc. Transl.
(2) Vol. **230**, 2010

Asymptotic Analysis of Nonlinear Dispersive Equations

Kenji Nakanishi

ABSTRACT. There has been great progress in the study of partial differential equations and their solutions describing nonlinear waves, from the viewpoint of the relation between the dispersive and the nonlinear structures. In recent years, the progress has come from stepping from perturbation of the linear theories into explorations for truly nonlinear properties.

We will discuss particularly those methods for investigating the solutions given their asymptotic behavior for time and coefficients, mostly concentrating on the following three topics: (1) global analysis for the scattering problem in the energy class, (2) asymptotic analysis of the interaction between solitary and dispersive waves, (3) asymptotic analysis of singular limits for nonlinear dispersive equations.

1. What is a nonlinear dispersive wave?

1.1. Objectives. The nonlinear dispersive equations, as considered here, are those partial differential equations describing space-time dynamics of waves in various phenomena which are dominated by *dispersion*, which extends waves beyond their spatial locality, and by *nonlinearity*, which arises as the interaction between superposed wave components. The described physical phenomena often appear in some extremal situations where the energy loss can be neglected, such as in high energy (laser, plasma), in low temperature (superfluid, superconductor), in macro scales (water waves, cosmic theory), and in micro scales (quantum theory). Typical examples of the equations are the nonlinear Schrödinger equation (NLS):

$$\begin{aligned} &2iu_t - \Delta u + \lambda |u|^p u = 0, \quad u(t,x) : \mathbb{R}^{1+d} \to \mathbb{C}, \\ &\lambda \in \mathbb{R},\ p > 0, \quad \Delta = \partial_{x_1}^2 + \cdots + \partial_{x_d}^2, \end{aligned} \tag{1.1}$$

the nonlinear Klein-Gordon equation (NLKG):

$$u_{tt} - \Delta u + u + \lambda |u|^p u = 0, \quad u(t,x) : \mathbb{R}^{1+d} \to \mathbb{C},\ \lambda \in \mathbb{R},\ p > 0, \tag{1.2}$$

the KdV equation, the KP equation, the Benjamin-Ono equation, and so on. Also, the nonlinear wave equation in the narrow sense (NLW), namely the equation derived from NLKG by removing the term u, has a dispersive effect in the angular

This article originally appeared in Japanese in Sūgaku **59** (2007), 337–352.
2000 *Mathematics Subject Classification.* Primary 35Q55; Secondary 35B40, 35L70.

directions in two or higher spatial dimensions.[1] These examples have all different linear parts. If we also take account of the difference in the nonlinearity, and systems of equations, then there are numerous different equations established as physical models, and also studied mathematically.

A feature of the phenomena described by those equations is that various space-time changes arise without any supply or loss of energy or information, such as self-focusing, collapse, dispersion, and formation of stable shapes for the waves. They are coming from the balance between the nonlinearity and the dispersion, each having very different natures from the other. It is our goal to reveal that process rigorously from the equations. Here our motivation is not in direct applications to the physics or the engineering, but rather in exploring mathematically unknown relations between the functions and the equations, and in establishing analytical methods for that. Actually the corresponding physical models and theories are not necessarily in their final forms, and hence we should use their theoretical requirements and experimental knowledge just as clues to advance our mathematical analysis. Therefore most of our problems, and even some of the equations, are set from mathematical viewpoints in order to seek the roots of the phenomena. Nevertheless it is difficult to treat the nonlinearity and the dispersion together mathematically, so we do not yet have any dominant theory or framework, but various trials with corresponding individual techniques for the time being.

In this survey, we will discuss mainly such trials in recent years including those in progress, in particular those investigating the relation of nonlinearity and dispersion via the asymptotic behavior of solutions with respect to some parameters.

By the way, Sugaku already has three good surveys on the research in this field [**76, 58, 77**]. For the developments up to the 1990's and for the physical backgrounds, I would recommend them rather than this article. Also, the current state of the art can be browsed together with some explanations at the web site http://tosio.math.toronto.edu/wiki/index.php/Main_Page (Dispersive PDE wiki).

1.2. Nonlinearity and dispersion. The dispersion process of waves is similar to the diffusion process in the sense that the spatial distribution is spreading out, but an essential difference is that the former is for the distribution of vibrations, while the latter is for that of positions. A notable mathematical consequence is that for the latter we often have the comparison principle, while for the former it is not available.

Another decisive difference is that the dispersion is a time reversible process whereas the diffusion is time irreversible. In other words, in the diffusive phenomena everything tends to spread out, but in the dispersive phenomena some parts may be gathering while some others are spreading out. Hence even when we investigate states well after the wave has dispersed, we still need in general to trace concentrated states before getting to the dispersed one.

[1] In some exclusive classifications for partial differential equations, the hyperbolic equations such as the Klein-Gordon equation are excluded from the dispersive equations. However, when we talk about nonlinear dispersive equations, the term is intended not for classification but just for grouping. Moreover, the most typical examples from each side, NLS and NLKG, have the most close relationships among the nonlinear dispersive equations, both in the physical models and in the mathematical studies, and I do not know of any research which found significance in separating them. Therefore in this article "nonlinear dispersive equation" denotes the broad class of wave equations which contain both nonlinearity and dispersion.

The difference is more significant in the treatment of nonlinearity. If the equation is linear, we can decompose the solution into infinitely many solutions, each of which evolves independently of the others, so that we can regard them as "particles". But with general nonlinear interaction, such a decomposition is impossible (the completely integrable systems are the special cases where it is possible in some sense). On the other hand, for nonlinear ordinary differential equations, it is not theoretically so difficult to trace the time evolution. But it is spatially mixed up as soon as we have dispersive waves. It is not easy to approximate it by the ODE, because the wave evolution is essentially infinite dimensional in the sense of functional analysis, unlike that of the heat kernel.

In short, nonlinearity and dispersion are not only competing with each other physically by extending and contracting amplitudes and distributions of waves, but they have essentially opposing structures from the mathematical viewpoint. This is the main cause for the various phenomena, and also for the typical difficulties and intricacies as nonlinear partial differential equations.

1.3. Typical behavior of solutions. There are the following three prototypes concerning the dynamical behavior of solutions for nonlinear dispersive equations, depending on the balance between dispersion and nonlinearity.

(1) Scattering: Once the dispersion becomes completely dominant, the wave keeps spreading out in space, decaying the amplitude, and as a consequence, weakening the nonlinear interaction more and more. We call such states dispersive waves, or scattering states, in the sense that the nonlinear interaction is essentially finished. The process leading there is called scattering.

(2) Blowup: On the contrary in the case of nonlinearity with spatially local amplification, it happens that the nonlinear interaction dominates the dispersion, and eventually the local amplitude diverges in finite time. Such a singularity in finite time is called blowup.

(3) Solitary waves: When the nonlinearity and dispersion balance, there arise some solutions which do not disperse nor intensify, preserving their shapes of amplitudes. Among them, those with spatially localized shapes are called solitary waves, those moving in fixed speeds are called traveling waves, and those staying still are called standing waves.

For example, in the case of NLS with the focusing nonlinearity, namely for $\lambda < 0$, the solitary waves are of the following form:

$$\begin{aligned} &u(t,x) = e^{-it\omega + i\frac{|v|^2 t}{2} - ivx + i\theta} \varphi_\omega(x - c - vt), \\ &\omega > 0,\ \theta \in \mathbb{R},\ c, v \in \mathbb{R}^d,\ \varphi_\omega : \mathbb{R}^d \to \mathbb{C}, \end{aligned} \tag{1.3}$$

which is a family of solutions with finite-dimensional continuous parameters that reflect the invariance of the equation. The blow-up solutions have much more variety, and their precise behavior is not yet clear. Just as a special example in the case of NLS with $p = 4/d$,

$$u(t,x) = t^{-d/2} e^{i\omega/t} e^{|x|^2/(4it)} \overline{\varphi_\omega}(x/t), \tag{1.4}$$

with the same φ_ω as above is a solution which blows up as $t \to 0$, concentrating to the origin, with the L^2-norm in x preserved. This is not quite typical behavior for blow-up solutions. On the other hand, the dispersive wave solutions are typically characterized as follows. Let $U(t) = e^{-it\Delta/2}$ be the solution operator for the linear

Schrödinger equation. Then every dispersive solution $u(t)$ should satisfy

$$U(-t)u(t) \overset{t\to\infty}{\to} \exists\varphi \text{ in } H^s(\mathbb{R}^d), \tag{1.5}$$

where H^s denotes the Sobolev space with square integrable s-th derivatives. Since it is asymptotic to the linear solution $U(t)\varphi$, it is easier to capture in analytical ways. There are other types of solutions, such as the breathers for the modified KdV equation, but there is no result which completely classifies the general solutions, except for the cases of the completely integrable systems and those where all the solutions become dispersive waves.

2. Basic ideas

The main problem of analysis for nonlinear dispersive equations is how we estimate the solutions by combining various inequalities for unknown functions. Then it is important that those inequalities are eventually taken over by some quantities or function spaces for the original unknown functions to be estimated; then we say "the estimate is closed". From the logic by itself, this might look like a vicious circle, but when it gains quantitative arguments, we can derive estimates for the unknown functions from known quantities. In most cases, the main tools are inequalities for solutions of the linearized equations, since only a few estimates are directly applicable to nonlinear equations. The perturbation method from the linear theory is to estimate the main parts of the solutions and/or the unknown functions in nonlinear terms by approximating them with the linear solutions. When we try to close the estimate in this way, we call the case "critical" if the nonlinear term exactly balances with the linear principal part, "subcritical" if it is weaker and hence we have some room for analysis, and "supercritical" if the nonlinear term is stronger and we cannot thus close the estimate.[2] Of course, we can get different cases for the same equation depending on the problem and the setting. There are many important problems in the supercritical case, such as the global solvability of the Einstein equation and the (three-dimensional) Navier-Stokes equation, but for now detailed analysis is available only for the subcritical and the critical cases for nonlinear dispersive equations. Those positive results in the supercritical case are mostly by nonconstructive proofs using compactness, and thus we have rather poor information so far.

Before getting into individual problems, let me point out three tendencies widely observed in the technical developments for the estimation.

(1) From pointwise estimates to estimates in sum: The transition from estimates at each point, to estimates in integrals, and further to estimates in discrete sums, can be seen for example in the choice of function spaces, where it appears in the shift from C^k to the Sobolev spaces and to the Besov spaces. In this case the operation measuring the smoothness of functions is changing from a pointwise limit to the limit in integrals and then to the discrete limit. It is natural to estimate nonlocal properties of dispersive waves in integral forms, while the reason for discretization is that we want to decompose wave components, even under nonlinearity, for those with weaker interactions. For example in linear dispersive estimates, the use of space-time integrals in the Strichartz estimate (3.3), instead of the decay

[2]It is customary to classify these cases for specific problems by some characteristic exponents, but the largeness/smallness of them does not necessarily correspond to the supercritical/subcritical cases.

estimate of the L^p-norm at each time, enabled us to deal with much more general nonlinearity and solutions. For a more sophisticated example, Tataru [**67**] obtained a very detailed bilinear estimate for solutions of the wave equation by decomposing them into countably many pieces according to their regularities and directions of propagation, and by applying mixed L^p-norms in different directions to each piece. This was later used by Tao [**68**] as the fundamental estimate for proving global solvability for the wave maps in the critical Sobolev space. While these estimates are applied to the perturbations from the linear equations, the estimate (3.5) below is a rare example which occurs in countably many decompositions and is also directly applicable to nonlinear solutions.

(2) From variable separation to mixed space-time: It is often useful to separate space and time in order to regard the equation as an evolution in function spaces of the spatial variables, when dealing with nonlinearity, and when using conserved quantities in topological arguments. But we can get better dispersive estimates by viewing the properties of the waves in the space-time variables. Space-time integrals make it easier to mix the space-time variables, and typical examples are Klainerman's global Sobolev inequality [**37**] and Bourgain's Fourier restriction norm [**6**]. They investigate the linear dispersion respectively from covariant derivatives in the physical space-time, and from the characteristic surfaces in the Fourier space-time, exploiting the Sobolev spaces defined thereby to obtain nonlinear estimates. The former gives pointwise estimates and is mainly used to get space-time decay, while the latter gives closed estimates in terms of the Sobolev spaces and is useful particularly for treating low regularity solutions. (3.8) mentioned below is another example, which is directly applicable to nonlinear equations, with some similarities in the form and the idea to the global Sobolev inequality.

(3) From linear to nonlinear estimates: Since those arguments by Klainerman and by Bourgain improve the estimates only from the side of linear dispersion by plugging them into the nonlinearity, their mutually complementary relations, such as energy conservation laws, are not reflected. But for the supercritical problems, it can be shown by scaling or an iteration argument that we cannot close the estimate only with the linear approximation. Also in principle, we need essentially nonlinear estimates in those problem settings where we have no small factor in the assumptions, such as in time, coefficients, or initial data. There are mainly two ways to get such estimates. One is directly to estimate the solution for the original nonlinear equation; typical examples are the energy conservation law and the Morawetz estimate (3.4) mentioned later. The information thereby obtained is usually weak, and it is often extremely difficult to reinforce it. The other is somehow to extract the parts in the solution or in the equation where the nonlinearity is the most effective, and to take account of the nonlinearity only there, treating the rest by linear perturbation. If it works, then we can get rather strong estimates. But it often becomes necessary to have small factors for the extraction, and also, this argument is difficult unless the part affected by the nonlinearity is localized explicitly in any way.

For the actual problems, we combine these arguments. In the following, we will discuss these ideas in the contexts of three concrete problems.

3. Energy scattering and global dispersion

Our ultimate goal is to determine (classify) for all solutions what space-time behavior arises from the intertwining of nonlinearity and dispersion, but it is in general very difficult to start our analysis without any restriction on the initial data, and so there are only a few complete results available. In this section, we describe such an example for NLS and NLKG, the problem being to show scattering states for all finite energy solutions. This is almost the only model case so far where we can analyze the competition between nonlinearity and dispersion completely globally both in space-time and in function spaces, though we need a restriction on the nonlinearity $\lambda \geq 0$ and $4/d < p \leq 4/(d-2)$.

First, to explain the difficulty coming from the generality of the initial data, let us compare it with a case for NLS where the data is restricted. The total energy of the solution $u(t,x)$ of NLS is in the form

$$E(u) = \int_{\mathbb{R}^d} |u|^2 + |\nabla u|^2 + \frac{2\lambda}{p+2}|u|^{p+2}dx, \quad \nabla = (\partial_{x_1}, \dots, \partial_{x_d}), \tag{3.1}$$

which is a quantity invariant in time (conserved quantity). We call those solutions for which it is finite the finite energy solutions. This conservation law gives us a priori boundedness in the Sobolev space $H^1(\mathbb{R}^d)$ under appropriate conditions (for example $\lambda \geq 0$), and so it can exclude the possibility of blowup, but it does not contain any information concerning dispersion of the solution (for example, the solitary waves, which are typical nondispersive solutions, have H^1 bounded, as well as most other norms). However, if we assume that at the initial state, $xu(0,x)$ is in L^2, then we have the following differential equation for the pseudo-conformal energy:

$$\frac{d}{dt}\int_{\mathbb{R}^d} |ixu + t\nabla u|^2 + \frac{2\lambda}{p+2}t^2|u|^{p+2}dx = \int_{\mathbb{R}^d} \lambda\frac{8-2np}{p+2}t|u|^{p+2}dx, \tag{3.2}$$

and the boundedness of its time integral implies decay ($O(|t|^{-2})$) of the nonlinear energy for appropriate p and λ. This means that the wave of u gathers at the spatial origin at $t=0$, and immediately disperses after $t=0$. In short, even though dispersive equations give time-reversible evolutions, assuming concentration to the spatial origin for the initial state enables us to guarantee a priori concentration of the solution itself to the space-time origin and its decay at infinity. (Refer to [**12, 18, 57**] for scattering results based on this fact. However the lower bound of p in this case is also an open question.)

For general finite energy solutions, we can at least imagine situations such that several concentrating waves are superposed after translations. In fact for linear dispersive waves, we can derive finiteness of superposition, and moreover, global dispersion in the form of space-time integrals, from finiteness of the energy. In the case of the linear Schrödinger equation, this is expressed in the Strichartz estimate (cf. [**36**]) using the space-time mixed norms

$$\begin{gathered} \|u(t,x)\|_{L^p_t(L^q_x)} \leq C_p\|u(0,x)\|_{L^2_x}, \\ \frac{2}{p} + \frac{d}{q} = \frac{d}{2}, \; 2 \leq p,q \leq \infty, \; (p,q) \neq (2,\infty). \end{gathered} \tag{3.3}$$

By the dual of the Strichartz estimate, we can control the nonlinear term as a perturbation from the linear equation for the NLS with appropriate p ($4/d \leq p \leq 4/(d-2)$) if the above norm is globally finite for u and ∇u, and thus we can obtain the scattering state by the contraction mapping argument globally in space-time, assuming that the initial state has sufficiently small energy.

However, when the energy becomes greater than a certain amount, the total sum of all nonlinear interactions between wave components controlled by the above left-hand side becomes much greater than their own space-time integrals, and thus the attempt to close the estimate in this way completely fails. Hence we need additional information which implies the finiteness of nonlinear interactions in the whole space-time, for which we currently have essentially only one method, that is, a family of space-time a priori estimates called the Morawetz type. In the case of NLS and NLKG in three dimensions or higher with repulsive nonlinearity ($\lambda > 0$), it follows that [**48**]

$$(3.4)\quad \iint_{\mathbb{R}^{1+d}} \frac{|\nabla_\theta u|^2}{r} + \frac{\lambda |u|^{p+2}}{r} dxdt \le CE(u), \quad r = |x|,\ \theta = \frac{x}{r},\ \nabla_\theta = \nabla - \theta \partial_r.$$

For dimensions higher than three, we can add another term, $\iint r^{-3}|u|^2 dxdt$.

By the integral with weight $1/r$, the above estimate can exclude nondispersive solutions such as solitary waves with constant speed, or generally those with positive lower bound on the nonlinear energy in the region $r \le C|t|$. However since the weight becomes weaker as it goes away from the origin, which is opposite to that for the pseudo-conformal energy, the estimate is far from closed for the nonlinear terms. So we need some argument to derive stronger estimates such as the Strichartz type, by using the finite propagation of energy for spatial translation, combining it with the Morawetz-type estimates, and thus classifying various possible patterns in the wave dispersion and concentration.

The first of such arguments was by Morawetz-Strauss [**49**], which decomposes energy concentration in the time direction by using the time decay of the linear solution operator in the nonlinear terms in the Duhamel formula. It was later extended by Brenner, Ginibre and Velo [**9, 20**] to the whole finite energy solutions, by improving the linear and nonlinear estimates. These methods have some restrictions in their applicability to the equations, such as integrability of the time decay of the linear solutions, but Bourgain [**7**] improved it by the ideas of space-time resolution of energy concentration, and the energy induction.

In the case of NLW with the Sobolev critical exponent ($d \ge 3$, $p = 4/(d-2)$, $\lambda \ge 0$), we can give some a priori estimate for the space-time concentration of energy. Precisely, the following infinite sum is bounded by the energy [**52**]:

$$(3.5)\quad \sup_{j \in \mathbb{Z}} \sup_{2^j < |t| < 2^{j+1}} \int_{|x|<C|t|} |u|^{2^*} dx, \quad 2^* = \frac{2d}{d-2} : \text{Sobolev critical exponent,}$$

where quadratic terms from the linear part were omitted to avoid complications. This expresses quantitatively that only a finite number of dyadic slices in t of the light cone contain concentrated energy above any fixed amount, and it can be regarded as a quantitative version of the argument for nonconcentration of energy in [**23, 62**]. Since concentrated states aligned in spatial directions can be similarly estimated by the finite propagation property, we can directly count the number of energy concentration points in space-time by further combining them with the Morawetz estimate, without using the energy induction or any resolution of the solution [**52**] (although the global dispersive estimates had been obtained before that by a nonconstructive argument [**2**]). The power of the above estimate (3.5) can be seen by comparing it with another proof for the global dispersive estimate given recently by Tao [**69**]. However, (3.5) exploits the behavior of energy specific to NLW, and so similar estimates are not available for NLS or NLKG. On the other hand, for the global dispersion of the three-dimensional NLS ($d = 3$, $\lambda \geq 0$), Colliander-Keel-Staffilani-Takaoka-Tao [**14**] proved what they call the interaction Morawetz estimate

$$\int_{\mathbb{R}^{1+3}} |u|^4 dxdt \leq C\|u\|^4_{L^\infty_t(H^{1/2}_x)}. \tag{3.6}$$

This gives a drastically simplified proof for the scattering of the Sobolev subcritical NLS, and it also played an important role for their result in the critical case. Its derivation can be regarded as realizing the combination of the Morawetz estimate and the propagation estimate in the form of convolution, leading to the a priori estimate. But it also depends on the special properties of the three-dimensional NLS, and so similar estimates are not available for NLKG or in higher dimensions (but estimates in the form of a double integral are available[3] [**60**]).

As can be seen in the above concentration estimate (3.5), a special feature of energy critical equations is that the nonlinear term and the linear term have the same strength (in the sense of energy), regardless of the space-time scale. Because of this, time-local problems and time-global problems are essentially equivalent, in particular in the case where the equation is scaling-invariant (such as NLS and NLW), and moreover, the instantaneous nonlinear interaction in the concentrated state can be approximated by the scattering operator. On the other hand, the energy size can bring an essential difference to the solutions, which is different from the subcritical cases. For these reasons, the energy critical problems have more importance than simply technical borderlines. As for the NLKG with the critical exponent, the scattering problem was solved by combining Bourgain's idea and the time-resolution argument for the lower frequency and thereby showing concentration in higher frequencies [**52**]. The case of NLS was even harder due to the lack of strict finite propagation. Colliander-Keel-Staffilani-Takaoka-Tao [**15**] finally solved it by the idea of applying the Morawetz-type estimates after optimizing Bourgain's energy resolution argument both in space-time and in frequency. In higher dimensions, the low critical power causes an additional problem in the perturbation argument for the energy resolution, which was overcome by using the Strichartz-type norm with unusual exponents [**52**] (that is, for NLKG, and the same for NLS [**71**]).

More recently, the scattering results have been obtained for the two-dimensional Sobolev critical case, i.e. for exponentially growing nonlinearity [**29**], and for the NLS with the L^2 critical exponent $p = 4/d$ under spherical symmetry [**72**]. In the former case, the critical threshold appears in the energy size itself, where even the time-local perturbative estimates are dependent on the strength of the energy concentration. In that sense the nonlinearity is even stronger than in the higher-dimensional critical case. As for the latter case, the equation has an additional invariance for the pseudo-conformal transform, and moreover the optimal dispersion by the Strichartz estimate has the same scaling as the nonlinear term, which further enhances the global nature of the problem.

Thus the research for the energy scattering problems is still actively in progress, but the bottleneck has been always that the Morawetz estimates at the starting

---

[3] Note added in translation: These estimates were recently extended to all spatial dimensions for NLS [**78, 80**].

point do not have so much flexibility. Concerning the dispersion of the equations, the original Morawetz estimate in three dimensions was directly extended to the higher dimensions, but there was an essential difficulty in extending to the lower dimensions. That is due to the fact that the original Morawetz estimate essentially uses the spatial variables only, and so it is influenced by the Laplace equation rather than the properties of waves. More precisely, in order to make the Morawetz estimate with positive definiteness and some weight function, we need a function $q(x)$ satisfying

$$q(x) \geq 0, \quad -\Delta q(x) \geq 0 \quad (\text{in} \mathcal{D}'(\mathbb{R}^d)), \tag{3.7}$$

which is impossible in one or two dimensions (cf. [**55**]).

The above problem was resolved by modifying the estimate into a form essentially of space-time, taking account of the dispersive property of the waves, and by replacing the positive definiteness for the error part with integrability due to time decay. In the case of NLS, we have in all dimensions [**53**]

$$\int_{\mathbb{R}^{1+d}} \frac{|ixu + t\nabla u|^2}{(t^2 + r^2)^{3/2}} + \frac{|u|^{p+2}}{(t^2 + r^2)^{1/2}} dx dt \leq CE(u). \tag{3.8}$$

We note that although the left-hand side is similar to the pseudo-conformal energy, it cannot be derived as its weighted version since they originate from different terms.

For NLKG, we have a similar estimate where the quadratic part is replaced with $|xu_t + t\nabla u|^2$, and they transfer to each other continuously in the nonrelativistic limit. However, the estimate for the one-dimensional NLKG is more delicate because the corresponding wave equation has traveling waves and the Morawetz-type estimates cannot in principle distinguish NLKG from NLW. In this case, we can eliminate the possibility of traveling waves for NLKG by using the L^2 finiteness on the light cone surface, thus obtaining global dispersive estimates inside the light cone [**53, 55**].

For the condition on the nonlinear term, some monotone repulsiveness is required such as a single power (more precisely, we need that the nonlinear energy density is in the form $V(|u|^2)|u|^2$ with a nonnegative and nondecreasing V). Previously, a coercive condition had been required because the nonlinear term was used in the Morawetz-type estimates. It has been improved to the above condition by using the linear part (the quadratic term) instead [**51, 54, 55**]. However to get nonscattering by the existence of solitary waves for example requires that the potential function takes a negative value. Hence the gap between these conditions, namely the nonmonotonic case, remains unsolved. We remark that the idea using the linear part is inherited in the interaction Morawetz estimate (3.6), where it gives a completely different form from the nonlinear term, and so is more essential regardless of the assumptions on the nonlinearity.

Removing the monotone condition of the nonlinearity seems an important step in order to proceed to the case where we have nonscattering states as well. A promising approach has recently been given by Kenig-Merle [**30, 31**], who obtained scattering (and blow-up) results for the attractive nonlinearity (i.e. with the opposite sign), under some variational restrictions on the energy quantity. Also, it is an important problem for the future to extend the method to the equations with derivatives, such as the wave maps, the derivative NLS and the Maxwell-Klein-Gordon system.

4. Asymptotic behavior and space-time decay around solitary waves

As we mentioned above, we do not have any method so far to analyze all general solutions asymptotically in the case where nonscattering solutions exist. Hence a practical method to investigate the global interaction between nondispersive and dispersive waves is to analyze those solutions consisting of explicit nondispersive solutions and small dispersive perturbations. For example for the minimal energy solitary waves (the ground states) for NLS (1.3), it is known under some condition on the nonlinearity ($\lambda < 0$, $p < 4/d$) that those solutions starting sufficiently close to a solitary wave in the energy space stay close to the solitary waves modulo the freedom of translations due to the invariance of the equation (the orbital stability) [**11**]. Since the argument is based only on the energy control by variational arguments, it tells us nothing about the dispersive property of the error part or the asymptotic behavior of the interaction with the main part. However it is generally expected that the error part goes away by the dispersion, the interaction with the main part decays in time, and thus only the solitary wave stays locally in space.

Such results were obtained in the case that the initial perturbation is spatially localized by using the idea evolving the parameters for the solitary waves such that the solitary wave part becomes orthogonal to the dispersive part, and also using the dispersive estimate for the linearized operators around the solitary waves, for the NLS with a potential

$$2iu_t - \Delta u + Vu + \lambda |u|^p u = 0, \quad V : \mathbb{R}^d \to \mathbb{R}, \tag{4.1}$$

with small solitary waves bifurcating from the linear ground state [**65, 74**], and for the NLS with large solitary waves which are essentially nonlinear (assuming some properties on the spectra of the linearized operators) [**10, 16**]. In these arguments, the assumption of spatial localization on the dispersive part is necessary to get fast time decay for the interaction with the main part, which does not decay in time.

However since we can anticipate that nonlinear estimates such as the variational method will become more important when we deal with more general solutions, it is not desirable to have the gap on the initial assumption from the variational result. Also, we cannot hope for fast time decay of the dispersive component in general when it is interacting with many different localized components. This problem was resolved in the case of small solitary waves by the idea of replacing the estimates on the dispersive component as well as the solitary wave parameters with space-time integrals and applying the endpoint Strichartz estimate (the case $p = 2$ in (3.3)) to the linear part of the equation for the dispersive part and to the quadratic part of the parameter equation [**24**]. This result was further extended recently to lower dimensions [**46, 47**] by replacing the endpoint Strichartz estimate with $L^2_{t,x}$ weighted estimates. However, in these results in the energy space, the asymptotic behaviors of the solitary wave parameters are not completely classified due to the slow decay of the dispersive part. Also the problem for the large solitary waves remains open because of a difficulty in the error estimate for the dispersive part corresponding to the parameter evolution.

In the above problem, the fast spatial decay at infinity of the solitary waves is an essential factor for the interaction with the dispersive part to decay fast in time. Hence otherwise we get a new difficulty, namely when we have nondispersive components with weaker spatial decay. As a typical example, NLS has plane wave solutions in the form $ae^{i\omega t+i\xi x}$. They have no decay and hence the standard analysis in L^2 or the Sobolev space H^s does not apply to the question of how those solutions close to them evolve in time. But it is not artificial as a physical model, considering that almost all nonlinear dispersive equations are obtained by some scaling limit in the whole space from dynamics originally localized in space. In the case of the cubic repulsive NLS ($\lambda = 1$, $p = 2$), after removing the phase part by using the invariance, the equation is reduced to

$$2i\psi_t - \Delta\psi = (1 - |\psi|^2)\psi, \quad \psi = 1 + u, \quad \lim_{|x|\to\infty} u(t,x) = 0, \tag{4.2}$$

which is also called the Gross-Pitaevskii equation in some contexts such as superfluid and nonlinear optics. Its global solvability has recently been obtained under the more general boundary condition $|\psi| \to 1$ (and in three dimensions or lower), for the finite energy solutions [**17**], and even for a larger class which contains static vortexes with infinite energy [**5**]. For those ψ as above, the interaction between the perturbation part u and the nondecaying main part 1 appears as a linear and quadratic modification from NLS in the equation for u:

$$2iu_t - \Delta u + 2u_1 = -2u_1u - |u|^2 - |u|^2u, \quad u = u_1 + iu_2, \ (u_1, u_2) \in \mathbb{R}^2. \tag{4.3}$$

Here the linear term $2u_1$ is of course effective at the time infinity and hence gives a modification from the free Schrödinger equation for the asymptotic behavior of u, but since it is a linear term with a constant coefficient, it does not cause any problem by itself. The problem is coming from the quadratic terms. The low degree implies weak decay by the dispersion, which causes difficulty especially in lower dimensions where the dispersion itself is weaker. Another problem is that we get a singularity $|\xi|^{-1}$ at the Fourier origin when we insert the linearized solution into those quadratic terms, reflecting the fact that they come from the interaction with the constant function 1. The singularity is also more problematic for lower dimensions.

In particular in two spatial dimensions, asymptotic behavior of the solutions has not been revealed completely even for the model NLS without the latter problem

$$2iu_t - \Delta u = \lambda_1 u^2 + \lambda_2 |u|^2 + \lambda_3 \overline{u}^2, \tag{4.4}$$

even with localized small data at the initial time or the final time $t \to \infty$ (cf. [**28**]). The difficulty changes depending on whether the quadratic nonlinear terms cause resonance. Here the resonance means technically the case (or the component) where the phase oscillation vanishes in the space-time inner product of the nonlinear terms and the linear solutions. More precisely, the problem appears when the space-time derivatives of the phase vanish in view of the stationary phase method. In the case of (4.4), the space and the time derivatives actually vanish simultaneously for the term $|u|^2$, which is present also in the Gross-Pitaevskii equation, and moreover the singularity appears exactly at the Fourier origin (cf. [**63, 64**]).

Because of these difficulties, the scattering problem had appeared very difficult especially in two spatial dimensions. However, by using the special structure of the nonlinear term we can remove the singularity in the resonance component. More precisely, the quadratic transform which changes the energy of u into the usual quadratic term and a quartic term only:

$$u \mapsto z = H^{-1}((2 - \Delta)u_1 + |u|^2) + iu_2, \quad H = \sqrt{-\Delta(2 - \Delta)}, \tag{4.5}$$

gives an equation for the new unknown z, where all quadratic terms gain decay at the Fourier origin, and thus enables analysis by bilinear space-time oscillatory integrals. Consequently, we can construct dispersive solutions with an asymptotic formula[4] [**26**]

$$z(t) \sim z^0(t) + \frac{i}{2}\int_\infty^t e^{i\frac{H}{2}(t-s)}|\sqrt{-\Delta(2-\Delta)^{-1}}z^0(s)|^2 ds, \quad z^0 = e^{i\frac{H}{2}t}\varphi. \tag{4.6}$$

Here the second term generally has singularity $|\xi|^{-1}$ at the Fourier origin and so does not belong to L^2. The presence of quadratic terms in the asymptotic formula implies that the nonlinear interaction cannot be neglected even at the time infinity.

Such a transform is valid only for the special nonlinearity of the Gross-Pitaevskii equation, and we do not expect similar things for the NLS with general quadratic nonlinearities. However, a similar nonlinear cancellation has been observed in the Schrödinger maps [**21**]. These are examples where the nonlinearity in the physical models has better properties than mathematically simpler models, and similarly for the null forms in the wave equations [**38**].

By the way, the quadratic term in (4.6) is different from that in (4.5), and essentially the same as that obtained by Hayashi-Naumkin [**27**] for those NLS (4.4) with the special coefficients of the nonlinearity such that it is orthogonal to the principal part in $\mathbb{C}$ (for example u_1^2). As for the higher dimensions, the scattering problem is relatively easier [**25, 26**], but it is an open question if small energy solutions all scatter in three dimensions, where the estimate for $t \to \infty$ is a supercritical problem (on the other hand, in two dimensions there are traveling waves with arbitrarily small energy).

In the models such as for superfluids, it will be interesting to investigate further the evolution of solutions with traveling wave components (especially those with vortexes). If we formulate it as in the above equation for u, we get a problem similar to stability of the solitary waves, but it has another difficulty that the traveling wave also has slow spatial decay ($O(|x|^{1-d})$) [**22**]. For example, it is not even in L^2 in two dimensions. However, it is intriguing that the singular leading term of the dispersive solutions has the same order, of which we may take advantage for further analysis. As another example of slow decay, Krieger-Schlag [**33**] studied asymptotic behavior around the standing waves with algebraic decay order for the NLW with the attractive Sobolev critical nonlinearity.

For the future development in asymptotic analysis at time infinity for low-order nonlinearity, it is desired to treat interactions with nondispersive waves such as standing waves, besides other important problems such as the analysis of completely resonating components such as $|u|^2$. However, the current techniques for low-order terms are strongly dependent on the oscillation of the fundamental solution, and therefore it is not easy to extend them to the case where it is not explicit. As a related result, Krieger-Schlag [**34**] developed an argument to use the generalized Fourier transform for the linearized operator in the case of the one-dimensional L^2 critical NLS. But the analysis for larger classes of solutions than those close to the solitary waves appears to be difficult in the direction of making estimates finer. On the other hand, Tao [**70**] studied general bounded solutions which are not necessarily dispersive, by combining perturbative estimates in the frequency resolution and topological arguments in the energy.

[4]Note added in translation: This formula misses a term, the correct one is given in [**79**].

5. Singular limits and control over nonlinear resonance

In all the above problems, either the problem of nonlinearity or that of dispersion always becomes dominant in their balance. Our true goal is to consider the situation where the balance translates between them, but it seems extremely difficult to capture such phenomena in general from the evolution of unknown functions, as long as we rely on perturbative arguments. Hence as an experimental argument, if we add a parameter in the equation to change the balance, then we may be able to trace transitions of the solutions along the parameter. In this section, we discuss the problems concerned with transition between nonlinear dispersive equations by singular limits where the highest terms vanish in the parameter limits.

Although this may appear to be an artificial setting of problems, it is as important as other fundamental questions such as unique existence of solutions, because it is related to the meanings of the equation and the solutions themselves, taking into account that almost all nonlinear dispersive equations are derived from other equations in some limit of parameters. In fact there has been a lot of research in the singular limit problems deriving nonlinear dispersive equations [**75, 61, 1, 59, 32, 13**]. But it is rather recent that they are studied from the viewpoint of nonlinearity and dispersion, which enabled us to treat drastically larger classes of convergent solutions, leading to in some cases strong convergence for finite energy solutions.

In these problems, the central question is about uniform estimates on the solutions with respect to the parameters, while it is rather easy to derive convergence from them. Also uniform estimates are more important for the purpose of investigating the balance between nonlinearity and dispersion. An example of the problems is the nonrelativistic limit from NLKG, the Maxwell-Klein-Gordon and the Maxwell-Dirac systems which include the light speed as a parameter, to NLS or the Schrödinger-Poisson system. In these cases, we can argue similarly to the case without the parameter by retrieving the estimates in function spaces which appropriately depend on the parameter [**40, 41, 43, 3, 4**]. However we need some more tricks for nonperturbative aspects of the problem, such as the space-time global transfer in the limit, which was treated through a bypass equation [**56, 42**]. In particular, the limit from NLKG to NLS with the Sobolev critical exponent still remains unsolved.

However, depending on how the parameter is involved, those estimates available for fixed equations may not be uniform for the parameter even locally in time. For example, let us take the subsonic limit $\alpha \to \infty$ from the Zakharov system

$$\begin{aligned} &2iu_t - \Delta u = -nu, \quad u : \mathbb{R}^{1+3} \to \mathbb{C}, \\ &\alpha^{-2} n_{tt} - \Delta n = \Delta |u|^2, \quad n : \mathbb{R}^{1+3} \to \mathbb{R}, \end{aligned} \tag{5.1}$$

to NLS ($\lambda = -1$, $p = 2$). If we fix the parameter at a finite value (or at the infinity), then we can solve the equations locally in time by using Bourgain's bilinear estimate for Fourier restriction norms and the Strichartz estimates [**8, 19**] (they cannot be solved globally because of the presence of blow-up solutions). However, it can be shown by the iteration method that it is impossible to obtain a uniform estimate in the parameter for this problem by means of linear perturbations in the Sobolev spaces in general, not only by the above specific arguments [**44**]. This nonuniformity is due to the fact that if we insert the linear approximation into the quadratic terms, then at a certain frequency component (around 2α) we get resonance (cancellation of oscillation), for which we cannot compensate the amplifying effect of the parameter by using oscillatory integrals. This problem is not observed for fixed parameters, where this resonant frequency is bounded. Such phenomena are not specific to the Zakharov system and can happen in general for singular limits of nonlinear dispersive equations.

Nevertheless we can prove that the actual solutions are uniformly bounded in the energy space H^1 with respect to the parameter α by taking advantage of the energy conservation [**45**]. More precisely, we derive estimates on a time interval which is independent of the parameter by exploiting the fact that the nonlinear part of the energy is subcritical, and hence controllable by interpolation between the energy and the uniform smallness of the perturbation in some negative Sobolev space. Although this argument is concerned essentially with the special frequency of resonance, it can be regarded as capturing the balance between the nonlinearity and the dispersion beyond simple perturbation.

In the above argument, it is key to have uniform boundedness of the conserved energy, but such a thing is not always available in limit problems. For example, if we further step back the approximations of the equation by replacing the Schrödinger part of (5.1) with the Klein-Gordon equation

$$c^{-2}u_{tt} - \Delta u + c^2 u = -nu, \quad u : \mathbb{R}^{1+3} \to \mathbb{R}^3, \tag{5.2}$$

and consider the high-frequency and subsonic limit $(c, \alpha) \to \infty$, then by the high-frequency limit the energy becomes unbounded (which has the same form as the mass energy mc^2 in the nonrelativistic limit). Nevertheless, we can obtain uniform estimates, by combining some approximate energy conservation where we remove high frequency in time, nonresonant bilinear estimates on the error terms, the Strichartz estimates to control the Fourier restriction norms needed for the bilinear estimates, and the interpolation estimate on the Strichartz norms by the Fourier restriction norms and the energy [**45**]. In this argument, the Fourier restriction norms lose one regularity due to the resonance phenomenon, and then the interpolation between it and the energy norm takes $1/2$ regularity from the Strichartz norms, which are, fascinatingly, the maximums allowed for those estimates to be carried out. We remark that the above interpolation inequality is a precise version of that used first by Koch-Tzvetkov [**39**] for the Benjamin-Ono equation, which is further applied to various problems such as the Maxwell-Schrödinger system and the Schrödinger maps [**50, 35**].

The above argument strongly depends on the nonlinear energy, reflecting the supercritical character of the problem, and therefore it is difficult to derive more information than the strong convergence in the energy space. By another argument, if we assume further boundedness of the initial data in H^s $(s > 3/2)$, then we can derive uniform boundedness of the solutions in the same space, as well as uniform estimates on the Strichartz norms and the Fourier restriction norms as for the linear solutions except for the resonant frequency [**44**]. The proof exploits the bilinear estimate and the Strichartz estimate for the solution after removing the resonant frequency, and for the resonant frequency they are replaced with a localized energy used as the higher-order Sobolev norms. The restriction $s > 3/2$ is needed in order to treat those error terms which appear as soon as we make the energy higher orders (or localize it). Even though this might be a technical problem, the gap can be possibly fundamental because it is a question beyond the perturbation methods. It is an open problem together with whether the resonant frequency actually behaves differently from the linear solutions. As a similar phenomenon, Takaoka-Tsutsumi [**66**] studied the initial value problem for the modified KdV equation in the case where the nonlinear flow differs essentially from the linear one (due to very low regularity of the initial data) and succeeded in describing the difference explicitly in the frequency space.

These arguments work when resonance is restricted to some specific frequencies, but it is not always the case in other limit problems. Even in the Zakharov system, if we take account of another physical parameter and thus consider the electrostatic and subsonic limit, then there appear two separate resonance frequencies, and so we have to consider the possibility of energy transition between them. In particular the lower resonant frequency causes a problem in the higher energy estimate for the higher resonance frequency, and consequently we cannot get such strong estimates as in the previous case. It is desirable to develop methods applicable to those cases where the resonant components are not localized to specific frequencies. Also, the above limit problems treat only the part close to the final approximating equation NLS, and we are still far from getting back in this way to the Maxwell-Euler system, which is the starting point in the derivation of the Zakharov system (for a WKB approximation argument, see [**73**]).

6. For the future

Although this is certainly oversimplification, the main advances in the research of nonlinear dispersive equations in the 1980s were the methods separately using linear dispersive estimates and nonlinear estimates, and in the 1990s the analysis was by nonlinear estimates in multilinear forms with built-in linear dispersion. In this century, the main stream of the research is heading towards essentially nonlinear properties beyond the perturbation methods.

Here the difficult question common to various cases is how we can digest and reinforce limited information of essentially nonlinear characters and join them with more precise linear analysis. The procedure of improving linear estimates and thereby analyzing the nonlinear parts is in principle possible step by step. But in most cases it leads to too many complications to proceed before getting to the essential problems, and it is often clear that there are logical limitations before the goals. Therefore it is more important to descend from the nonlinear estimates to the linear ones, but for that task we have to rely so far too much on trial and error and on intuition for individual problems.

Although I am not yet convinced that this direction is possible, the situation will drastically change if a method is developed which can deal with those transforms and decompositions used in various steps of nonlinear treatments for the solutions, the function spaces and the equations, in a systematic way not restricted to some specific problems. I conclude this survey with a hopeful expectation that right now may be the best appropriate time for such an endeavor, with the worldwide increasing number of researchers, thereby accelerating accumulating individual results in nonlinear dispersive equations.

References

[1] H. Added and S. Added, *Equations of Langmuir turbulence and nonlinear Schrödinger equation: Smoothness and approximation.* J. Funct. Anal. **79** (1988), no. 1, 183–210. MR950090 (89h:35273)

[2] H. Bahouri and P. Gérard, *High frequency approximation of solutions to critical nonlinear wave equations.* Amer. J. Math. **121** (1999), no. 1, 131–175. MR1705001 (2000i:35123)

[3] P. Bechouche, N. J. Mauser and S. Selberg, *Nonrelativistic limit of Klein-Gordon-Maxwell to Schrödinger-Poisson.* Amer. J. Math. **126** (2004), no. 1, 31–64. MR2033563 (2004m:35256)

[4] P. Bechouche, N. J. Mauser and S. Selberg, *On the asymptotic analysis of the Dirac-Maxwell system in the nonrelativistic limit.* J. Hyperbolic Differ. Equ. **2** (2005), no. 1, 129–182. MR2134957 (2007b:35275)

[5] F. Bethuel and D. Smets, *A remark on the Cauchy Problem for the 2D Gross-Pitaevskii equation with nonzero degree at infinity.* Differential Integral Equations, **20** (2007), no. 3, 325–338. MR2293989 (2008g:35177)

[6] J. Bourgain, *Fourier transform restriction phenomena for certain lattice subsets and applications to nonlinear evolution equations, I. Schrödinger equations, II. The KdV equation.* Geom. Funct. Anal. **3** (1993), 107–156, 209–262. MR1209299 (95d:35160a); MR1215780 (95d:35160b)

[7] J. Bourgain, *Scattering in the energy space and below for 3D NLS.* J. Anal. Math. **75** (1998), 267–297. MR1655835 (99m:35180)

[8] J. Bourgain and J. Colliander, *On wellposedness of the Zakharov system.* Int. Math. Res. Not. (1996), no. 11, 515–546. MR1405972 (97h:35206)

[9] P. Brenner, *On space-time means and everywhere defined scattering operators for nonlinear Klein-Gordon equations.* Math. Z. **186** (1984), 383–391. MR744828 (85h:35183)

[10] V. S. Buslaev and G. S. Perelman, *Scattering for the nonlinear Schrödinger equations: States close to a solitary wave.* St. Petersburg Math. J. **4** (1993), 1111–1142. MR1199635 (94b:35256)

[11] T. Cazenave and P. L. Lions, *Orbital stability of standing waves for some nonlinear Schrödinger equations.* Comm. Math. Phys. 85 (1982), no. 4, 549–561. MR677997 (84i:81015)

[12] T. Cazenave and F. B. Weissler, *Rapidly decaying solutions of the nonlinear Schrödinger equation.* Comm. Math. Phys. **147** (1992), no. 1, 75–100. MR1171761 (93d:35150)

[13] T. Colin, G. Ebrard, G. Gallice and B. Texier, *Justification of the Zakharov model from Klein-Gordon–wave systems.* Comm. Partial Differential Equations **29** (2004), no. 9-10, 1365–1401. MR2103840 (2005k:35372)

[14] J. Colliander, M. Keel, G. Staffilani, H. Takaoka and T. Tao, *Global existence and scattering for rough solutions of a nonlinear Schrödinger equation on* $\mathbb{R}^3$. Comm. Pure Appl. Math. **57** (2004), no. 8, 987–1014. MR2053757 (2005b:35257)

[15] J. Colliander, M. Keel, G. Staffilani, H. Takaoka and T. Tao, *Global well-posedness and scattering for the energy-critical nonlinear Schrödinger equation in* $\mathbb{R}^3$. Ann. of Math. (2) **167** (2008), 767–865. MR2415387 (2009f:35315)

[16] S. Cuccagna, *Stabilization of solutions to nonlinear Schrödinger equations.* Comm. Pure Appl. Math. **54** (2001), no. 9, 1110–1145. MR1835384 (2002g:35193)

[17] P. Gérard, *The Cauchy problem for the Gross-Pitaevskii equation.* Ann. Inst. H. Poincaré Anal. Non Linéaire **23** (2006), no. 5, 765–779. MR2259616 (2007g:35230)

[18] J. Ginibre, T. Ozawa and G. Velo, *On the existence of the wave operators for a class of nonlinear Schrödinger equations.* Ann. Inst. H. Poincaré Phys. Théor. **60** (1994), no. 2, 211–239. MR1270296 (95c:35232)

[19] J. Ginibre, Y. Tsutsumi, and G. Velo, *On the Cauchy problem for the Zakharov system.* J. Funct. Anal. **151** (1997), no. 2, 384–436. MR1491547 (2000c:35220)

[20] J. Ginibre and G. Velo, *Scattering theory in the energy space for a class of nonlinear Schrödinger equations.* J. Math. Pures Appl. **64** (1985), 363–401. MR839728 (87i:35171)

[21] J. Ginibre and G. Velo, *Long range scattering for the modified Schrödinger map in two space dimensions.* J. Differential Equations **245** (2008), 3570–3614. MR2462694 (2010c:35178).

[22] P. Gravejat, *Asymptotics for the travelling waves in the Gross-Pitaevskii equation.* Asymptot. Anal. **45** (2005), 227–299. MR2191764 (2007k:35460)

[23] M. Grillakis, *Regularity and asymptotic behavior of the wave equation with a critical nonlinearity.* Ann. of Math. (2) **132** (1990), 485–509. MR1078267 (92c:35080)

[24] S. Gustafson, K. Nakanishi and T. Tsai, *Asymptotic stability and completeness in the energy space for nonlinear Schrödinger equations with small solitary waves.* Int. Math. Res. Not. (2004) no. 66, 3559–3584. MR2101699 (2005g:35268)

[25] S. Gustafson, K. Nakanishi and T.-P. Tsai, *Scattering theory for the Gross-Pitaevskii equation.* Math. Res. Lett. **13** (2006), no. 2, 273–285. MR2231117 (2007f:35267)

[26] S. Gustafson, K. Nakanishi and T. Tsai, *Global dispersive solutions for the Gross-Pitaevskii equation in two and three dimensions.* Ann. Henri Poincaré **8** (2007), 1303–1331. MR2360438 (2008m:35335)

[27] N. Hayashi and P. I. Naumkin. *Asymptotic in time of solutions to nonlinear Schrödinger equations in 2D.* Funkcial. Ekvac. **49** (2006), no. 3, 415–425. MR2297946 (2007m:35250)

[28] N. Hayashi, P. I. Naumkin, A. Shimomura and S. Tonegawa, *Modified wave operators for nonlinear Schrödinger equations in one and two dimensions.* Electron. J. Differential Equations 2004, No. 62, 16 pp. MR2047418 (2004m:35247)

[29] S. Ibrahim, M. Majdob, K. Nakanishi and N. Masmoudi, *Energy scattering for the 2D critical wave equation.* Preprint, arXiv:0806.3150v1 [math.AP].

[30] C. Kenig and F. Merle, *Global well-posedness, scattering and blow-up for the energy-critical, focusing, non-linear Schrödinger equation in the radial case.* Invent. Math. **166** (2006), no. 3, 645–675. MR2257393 (2007g:35232)

[31] C. Kenig and F. Merle, *Global well-posedness, scattering and blow-up for the energy critical focusing non-linear wave equation.* Acta. Math. **201** (2008), 147–212. MR2461508

[32] C. Kenig, G. Ponce and L. Vega, *On the Zakharov and Zakharov-Schulman systems.* J. Funct. Anal. **127** (1995), no. 1, 204–234. MR1308623 (96a:35197)

[33] J. Krieger and W. Schlag, *On the focusing critical semi-linear wave equation.* Amer. J. Math. **129** (2007), no. 3, 843–912 MR2325106 (2009f:35231)

[34] J. Krieger and W. Schlag, *Non-generic blow-up solutions for the critical focusing NLS in 1-D.* J. Euro. Math. Soc. **11** (2009), 1–125. MR2471133 (2009m:35488)

[35] J. Kato, *Existence and uniqueness of the solution to the modified Schrödinger map.* Math. Res. Lett. **12** (2005), no. 2-3, 171–186. MR2150874 (2006b:35302)

[36] M. Keel and T. Tao, *Endpoint Strichartz estimates*, Amer. J. Math. **120** (1998), no. 5, 955–980. MR1646048 (2000d:35018)

[37] S. Klainerman, *Global existence of small amplitude solutions to nonlinear Klein-Gordon equations in four space-time dimensions*, Comm. Pure Appl. Math. **38** (1985), 631–641. MR803252 (87e:35080)

[38] S. Klainerman and M. Machedon, *Smoothing estimates for null forms and applications*, Int. Math. Res. Not. (1994), no. 9, 383–390. MR1301438 (95i:58174)

[39] H. Koch and N. Tzvetkov, *On the local well-posedness of the Benjamin-Ono equation in* $H^s(\mathbb{R})$, Int. Math. Res. Not. (2003), no. 26, 1449–1464. MR1976047 (2004b:35284)

[40] S. Machihara, K. Nakanishi and T. Ozawa, *Nonrelativistic limit in the energy space for nonlinear Klein-Gordon equations*, Math. Ann. **322** (2002), no. 3, 603–621. MR1895710 (2003b:35199)

[41] S. Machihara, K. Nakanishi and T. Ozawa, *Small global solutions and the nonrelativistic limit for the nonlinear Dirac equation.* Rev. Mat. Iberoamericana **19** (2003), no. 1, 179–194. MR1993419 (2005h:35293)

[42] N. Masmoudi and K. Nakanishi, *From nonlinear Klein-Gordon equation to a system of coupled nonlinear Schrödinger equations.* Math. Ann. **324** (2002), no. 2, 359–389. MR1933862 (2004c:35288)

[43] N. Masmoudi and K. Nakanishi, *Nonrelativistic limit from Maxwell-Klein-Gordon and Maxwell-Dirac to Poisson-Schrödinger.* Int. Math. Res. Not. (2003), no. 13, 697–734. MR1949296 (2004b:81042)

[44] N. Masmoudi and K. Nakanishi, *From the Klein-Gordon-Zakharov system to the nonlinear Schrödinger equation*, J. Hyperbolic Differ. Equ. **2** (2005), no. 4, 975–1008. MR2195989 (2007e:35262)

[45] N. Masmoudi and K. Nakanishi, *Energy convergence for singular limits of Zakharov type systems.* Invent. Math. **172** (2008), no. 3, 535–584. MR2393080 (2009f:35298)

[46] T. Mizumachi, *Asymptotic stability of small solitons to 1D NLS with potential.* To appear in J. Math. Kyoto Univ.

[47] T. Mizumachi, *Asymptotic stability of small solitons for 2D Nonlinear Schrödinger equations with potential.* Preprint, arXiv:math/0609323.

[48] C. S. Morawetz, *Time decay for the nonlinear Klein-Gordon equations.* Proc. Roy. Soc. Ser. A **306** (1968) 291–296. MR0234136 (38:2455)

[49] C. S. Morawetz and W. A. Strauss, *Decay and scattering of solutions of a nonlinear relativistic wave equation.* Comm. Pure Appl. Math. **25** (1972), 1–31. MR0303097 (46:2239)

[50] M. Nakamura and T. Wada, *Global existence and uniqueness of solutions to the Maxwell-Schrödinger equations.* Comm. Math. Phys. **276** (2007), no. 2, 315–339. MR2346392 (2008m:81046)

[51] K. Nakanishi, *Unique global existence and asymptotic behaviour of solutions for wave equations with non-coercive critical nonlinearity.* Comm. Partial Differential Equations **24** (1999), no. 1-2, 185–221. MR1671997 (2000f:35104)

[52] K. Nakanishi, *Scattering theory for the nonlinear Klein-Gordon equation with Sobolev critical power.* Int. Math. Res. Not. (1999), no. 1, 31–60. MR1666973 (2000a:35174)

[53] K. Nakanishi, *Energy scattering for nonlinear Klein-Gordon and Schrödinger equations in spatial dimensions* 1 *and* 2. J. Funct. Anal. **169** (1999), no. 1, 201–225. MR1726753 (2000m:35141)

[54] K. Nakanishi, *Energy scattering for Hartree equations.* Math. Res. Lett. **6** (1999), no. 1, 107–118. MR1682697 (2000d:35173)

[55] K. Nakanishi, *Remarks on the energy scattering for nonlinear Klein-Gordon and Schrödinger equations.* Tohoku Math. J. (2) **53** (2001), no. 2, 285–303. MR1829982 (2002e:35220)

[56] K. Nakanishi, *Nonrelativistic limit of scattering theory for nonlinear Klein-Gordon equations.* J. Differential Equations **180** (2002), no. 2, 453–470. MR1894180 (2003e:35202)

[57] K. Nakanishi and T. Ozawa, *Remarks on scattering for nonlinear Schrödinger equations.* NoDEA Nonlinear Differential Equations Appl. **9** (2002), no. 1, 45–68. MR1891695 (2003a:35177)

[58] T. Ozawa, *Scattering theory for nonlinear Schr6dinger equations.* Sugaku **50** (1998), no. 4, 337–349. MR1690627 (2000b:35192)

[59] T. Ozawa and Y. Tsutsumi, *The nonlinear Schrödinger limit and the initial layer of the Zakharov equations.* Differential Integral Equations **5** (1992), no. 4, 721–745. MR1167491 (93d:76079)

[60] E. Ryckman and M. Visan, *Global well-posedness and scattering for the defocusing energy-critical nonlinear Schrödinger equation in* $\mathbb{R}^{1+4}$. Amer. J. Math. **129** (2007), no. 1, 1–60. MR2288737 (2007k:35474)

[61] S. Schochet and M. Weinstein, *The nonlinear Schrödinger limit of the Zakharov equations governing Langmuir turbulence.* Comm. Math. Phys. **106** (1986), no. 4, 569–580. MR0860310 (87j:35227)

[62] J. Shatah and M. Struwe, *Well-posedness in the energy space for semilinear wave equations with critical growth.* Int. Math. Res. Not. (1994), no. 7, 303–309. MR1283026 (95e:35132)

[63] A. Shimomura, *Nonexistence of asymptotically free solutions for quadratic nonlinear Schrödinger equations in two space dimensions.* Differential Integral Equations **18** (2005), no. 3, 325–335. MR2122723 (2005k:35384)

[64] A. Shimomura and Y. Tsutsumi, *Nonexistence of scattering states for some quadratic nonlinear Schrödinger equations in two space dimensions.* Differential Integral Equations **19** (2006), no. 9, 1047–1060. MR2262096 (2007k:35475)

[65] A. Soffer and M.I. Weinstein, *Multichannel nonlinear scattering theory for nonintegrable equations I, II.* Comm. Math. Phys. **133** (1990), 119–146; J. Diff. Eqns. **98** (1992), 376–390. MR1071238 (91h:35303)

[66] H. Takaoka and Y. Tsutsumi, *Well-posedness of the Cauchy problem for the modified KdV equation with periodic boundary condition.* Int. Math. Res. Not. (2004), no. 56, 3009–3040. MR2097834 (2006e:35295)

[67] D. Tataru, *On global existence and scattering for the wave maps equation.* Amer. J. Math. **123** (2001), no. 1, 37–77. MR1827277 (2002c:58045)

[68] T. Tao, *Global regularity of wave maps. II. Small energy in two dimensions.* Comm. Math. Phys. **224** (2001), no. 2, 443–544. MR1869874 (2002h:58052)

[69] T. Tao, *Spacetime bounds for the energy-critical nonlinear wave equation in three spatial dimensions.* Dyn. Partial Differ. Equ. **3** (2006), no. 1, 93–110. MR2227039 (2007c:35116)

[70] T. Tao, *A (concentration-)compact attractor for high-dimensional non-linear Schrödinger equations.* Dyn. Partial Differ. Equ. **4** (2007), no. 1, 1–53. MR2304091 (2007k:35479)

[71] T. Tao and M. Visan, *Stability of energy-critical nonlinear Schrödinger equations in high dimensions.* Electron. J. Differential Equations (2005), No. 118, 28 pp. MR2174550 (2006e:35307)

[72] T. Tao, M. Visan and X. Zhang, *Global well-posedness and scattering for the mass-critical nonlinear Schrödinger equation for radial data in high dimensions.* Duke Math. J. **140** (2007), 165–202. MR2355070 (2010a:35249)

[73] B. Texier, *WKB asymptotics for the Euler-Maxwell equations.* Asymptot. Anal. **42** (2005), 211–250. MR2138794 (2006a:35288)

[74] T.-P. Tsai and H.-T. Yau, *Classification of asymptotic profiles for nonlinear Schrödinger equations with small initial data.* Adv. Theor. Math. Phys. **6** (2002), no.1, 107–139. MR1992875 (2004m:35254)

[75] M. Tsutsumi, *Nonrelativistic approximation of nonlinear Klein-Gordon equations in two space dimensions.* Nonlinear Anal. **8** (1984), no. 6, 637–643. MR746721 (85h:35211)

[76] M. Tsutsumi, *Nonlinear Schrödinger equations.* Sugaku **47** (1995), no. 1, 18–37. MR1362516

[77] Y. Tsutsumi, *Global existence and blow-up of solutions for nonlinear wave equations.* Sugaku **53** (2001), no. 2, 139–156. MR1839958 (2002c:35189)

Added in translation

[78] J. Colliander, M. Grillakis and N. Tzirakis, *Tensor products and correlation estimates with applications to nonlinear Schrödinger equations.* Preprint, arXiv:0807.0871v2 [math.AP]. Comm. Pure. Appl. Math. **62** (2009), 920–968. MR2527809 (2010c:35175)

[79] S. Gustafson, K. Nakanishi and T.-P. Tsai, *Scattering theory for the Gross-Pitaevskii equation in three dimensions.* Preprint, arXiv:0803.3208v1 [math.AP]. Commun. Contemp. Math. **11** (2009), 657–707. MR2559713

[80] F. Planchon and L. Vega, *Bilinear virial identities and applications.* Ann. Sci. École Norm. Sup. (4) **42** (2009), 261–290. MR2518079 (2010b:35441)

Graduate School of Sciences, Kyoto University, Kyoto University, Kyoto 606-8502, Japan

Translated by KENJI NAKANISHI

Amer. Math. Soc. Transl.
(2) Vol. **230**, 2010

Asymptotics of Nonlinear Dispersive-type Evolution Equations

Nakao Hayashi

1. Introduction

1.1. Asymptotics of solutions to linear equations. The nonlinear Schrödinger equation, the Korteweg-de Vries equation, the Benjamin-Ono equation are well known as nonlinear dispersive equations which are used in various natural sciences to describe natural phenomena.

Our purpose in this expository article is to consider the asymptotic behavior of solutions to the modified Korteweg-de Vries equation, the cubic nonlinear Schrödinger equation and the cubic nonlinear Klein-Gordon equation which is considered as a relativistic version of the Schrödinger equation. We clarify the relationship between the oscillating property of the solutions to linear equations that correspond to these nonlinear problems and to that of the nonlinear terms.

These nonlinear equations are integrable, and it is known that soliton solutions exist by the inverse scattering method. However here we concentrate our attention on the initial data or the final data which do not have soliton solutions (for papers concerning soliton solutions, see [**1**]and [**66**]). In this case, the asymptotics of solutions to nonlinear problems are different from that of linear problems since the difference between the frequency of cubic nonlinear terms and that of solutions to linear problems occurs as resonant phenomena. Namely, it is impossible to find solutions of nonlinear problems in the neighborhood of those of linear problems. Therefore we need another method. To make this fact clear mathematically, we investigate the properties of solutions to linear problems and apply the results to nonlinear problems.

The Schrödinger equation is written as

$$\text{(1)} \qquad \begin{cases} 2u_t + i\partial_x^2 u = 0, & (t,x) \in \mathbf{R} \times \mathbf{R}, \\ u(0,x) = \phi(x), & x \in \mathbf{R}, \end{cases}$$

and the solution $u_{\mathrm{S}}(t)$ is represented as $u_{\mathrm{S}}(t) = U_{\mathrm{S}}(t)\phi$, where $U_{\mathrm{S}}(t) = \mathcal{F}^{-1}\exp\left(it|\xi|^2/2\right)\mathcal{F}$ is the free Schrödinger evolution group, and $\mathcal{F}$, $\mathcal{F}^{-1}$ are the Fourier

This article originally appeared in Japanese in Sūgaku **60** (1) (2008), 1–22.
2000 *Mathematics Subject Classification.* Primary 35Q40; Secondary 35Q53.

transform and the inverse Fourier transform, respectively. It is known that the main term of $u_{\mathrm{S}}(t)$ which is considered as an approximate solution is given by

$$i^{1/2}t^{-1/2}e^{-ix^2/(2t)}\phi\left(-x/t\right), \tag{2}$$

where $i^2 = -1$ (see Section 2 for the derivation of (2)).

In the case of the Airy equation

$$\begin{cases} 3u_t + \left(-\partial_x^2\right)\partial_x u = 0, & (t,x) \in \mathbf{R}\times\mathbf{R}, \\ u(0,x) = \phi(x), & x \in \mathbf{R}, \end{cases} \tag{3}$$

which is the linear problem of the Korteweg-de Vries equation, if we let $U_{\mathrm{A}}(t) = \mathcal{F}^{-1}\exp\left(it\left|\xi\right|^2\xi/3\right)\mathcal{F}$ be the evolution group of the Airy equation, then the solution $u_{\mathrm{A}}(t)$ is written as $u_{\mathrm{A}}(t) = U_{\mathrm{A}}\left(t\right)\phi$ and an approximate solution $v_\phi\left(t,q\right)$ is described as

$$v_\phi\left(t,q\right) = \begin{cases} t^{-1/2}\left|q\right|^{-1/2} 2\mathrm{Re}C_1 e^{2it|q|^3/3}\widehat{\phi}(q), & x > 0, \\ 0, & x \le 0 \end{cases} \tag{4}$$

under the conditions such that ϕ is real and satisfies suitable symmetry conditions, where $q = \left(x/t\right)^{1/2}, C_1 = \left(2i\right)^{1/2}$, and $\mathrm{Re}f$ is the real part of f (see Section 3 for an approximate solution). We note here that if the data ϕ do not satisfy the symmetry condition stated in Section 3, then $v_\phi\left(t,q\right)$ is not an approximate solution.

Finally, we introduce the Klein-Gordon equation

$$\begin{cases} u_{tt} + \left(1-\partial_x^2\right)u = 0, & (t,x) \in \mathbf{R}\times\mathbf{R}, \\ u(0) = u_0, \quad u_t(0) = u_1, & x \in \mathbf{R}. \end{cases} \tag{5}$$

This is written as

$$\left(\partial_t + i\sqrt{1-\partial_x^2}\right)\left(\partial_t - i\sqrt{1-\partial_x^2}\right)u = 0;$$

hence if the data are real, then it is sufficient to consider the problem

$$\left(\partial_t + i\sqrt{1-\partial_x^2}\right)v = 0, \quad v(0) = f \tag{6}$$

since the solution $v_{KG}(t)$ of equation (6) is written as $v_{KG}(t) = U_{\mathrm{KG}}(t)f$ if we let $U_{\mathrm{KG}}(t) = \mathcal{F}^{-1}\exp\left(it\sqrt{1+\xi^2}\right)\mathcal{F}$ be the Klein-Gordon evolution group, and if we let $u(t)$ be a solution of equation (5), then we have $u(t) = 2\mathrm{Re}v_{KG}(t)$. It is shown that

$$v_f\left(t\right) = \begin{cases} (it)^{-1/2}\left(1-\eta^2\right)^{-3/4}e^{it\sqrt{1-\eta^2}}\widehat{f}\left(\eta/\left(1-\eta^2\right)^{1/2}\right), & |\eta| \le 1, \\ 0, \quad |\eta| \ge 1, \quad \eta = x/t \end{cases} \tag{7}$$

is an approximate solution of equation (6) (see Section 4). From the above facts we find that the frequencies of solutions of these equations are $e^{it\eta^2/2}$, $e^{2it|\eta|^{3/2}/3}$, $e^{it\sqrt{1-\eta^2}}$, $\eta = x/t$, respectively. Thus these equations are considered as wave equations.

1.2. Wave operator and modified wave operator. We explain the effects of frequencies to nonlinear problems and study the scattering problem. First, we briefly explain the definition of the wave operator. For a given function u_+, we assume that there exists a unique solution u of nonlinear problems satisfying the asymptotics

$$\lim_{t\to\infty}\left\|u(t) - U(t)u_+\right\|_X = 0, \tag{8}$$

where $U(t)u_+$ is the solution of linear problems with the initial data u_+ and $\|\cdot\|_X$ is the norm of the Banach space X. We can define the map $W_+ : u_+ \mapsto u(t)$ and call it the wave operator. We also call u_+ the final state (or the final value) since it is the value of $U(-t)u(t)$ at infinity. In general, as it is impossible to show existence of the wave operator, we have to modify the setting of the problem. We will do it as follows: Let us take a function u_+ from a suitable function space and define a function $f(u_+, t)$ by u_+. Then we try to find a unique solution of nonlinear problems under the asymptotic condition

$$\lim_{t\to\infty} \|u(t) - U(t)f(u_+, t)\|_X = 0 \tag{9}$$

instead of the condition (8). Namely, the problem is solved if we can define the function f satisfying the asymptotic condition (9) by taking the structure of the nonlinear terms into consideration. If we can give the positive answer, we can define the map $MW_+ : u_+ \mapsto u(t)$ instead of the wave operator. We call MW_+ the modified wave operator since we modified the final states. We explain the reason why we must modify the framework of the problem. Assume that the solutions of the nonlinear problems are obtained by the perturbation of linear problems. Then the asymptotic behavior of the nonlinear terms is determined by putting solutions of linear problems into nonlinear terms. Though the amplitudes of the nonlinear terms are different from those of solutions of linear problems, the frequency of the nonlinear term exists and coincides with that of solutions to linear problems in our cases. We call the nonlinear term the resonant term. If we do not use some transformation to remove the resonant term, we find that the L^2-norm of the nonlinear terms considered in this paper decays in time of order t^{-1} and is not integrable in time. This is the reason why we do not treat the problems in the usual setting of scattering theory and we need modify the final states. We finally note that the modified Korteweg-de Vries equation and the cubic nonlinear Klein-Gordon equation are different from the cubic Schrödinger equation since the nonlinear terms of the previous two cases include nonresonant cubic terms contrary to the last case and it can be shown that these terms are stable through an integration by parts.

1.3. Scattering operator and modified scattering operator. In this subsection we explain the scattering problem briefly. The scattering problem is to study the domain and the range of the wave operator (the modified wave operator). As stated in the previous subsection, we can define the wave operator $W_+ : u_+ \mapsto u(0)$. However we do not state anything about the domain and the range of it. Now we assume that there exists a Banach space X such that $W_+ : X \to X$. In other words, for any $u_+ \in X$, we assume that there exists a unique solution $u(t) \in C([0,\infty); X)$ of the nonlinear problem such that

$$\lim_{t\to\infty} \|u(t) - U(t)u_+\|_X = 0, \tag{10}$$

and we consider the initial value problem with the data $u(0)$ which are determined by the solution $u(t)$ in the time interval $-\infty < t \leq 0$. If the initial value problem has a unique global-in-time if solution $u(t) \in C((-\infty, 0]; X)$ and if we can find a unique $u_- \in X$ from the solution $u(t)$ satisfying

$$\lim_{t\to-\infty} \|U(-t)u(t) - u_-\|_X = 0,$$

then we can define the inverse wave operator $W_-^{-1} : u(0) \in X \mapsto u_- \in X$. From this operator we can define $S = W_-^{-1}W_+ : X \to X$. We call the operator the scattering operator. Similarly, we define the modified scattering operator by $MS = (MW_-^{-1})MW_+$. Even if we can prove the existence of the wave operator (the modified wave operator) and the asymptotic behavior of solutions to the initial value problem, the existence of the scattering operator (the modified scattering operator) is not clear. If we restrict our attention to problems on nonlinear Schrödinger equations or nonlinear Klein-Gordon equations with super-critical nonlinearities, there are a lot of works on the scattering theory since the time decay of the nonlinear terms is faster than the time decay of solutions to linear problems (for examples, see [**5**],[**6**],[**12**],[**13**], [**38**],[**49**],[**56**], [**57**],[**63**],[**67**]). However, there are a few results on the existence of modified scattering operators in the case that the nonlinearities are critical ones.

From Section 2 to Section 4, we consider the asymptotic behavior of solutions to linear problems. In Section 5, Section 6 and Section 7, we survey results on existence of the modified wave operators. Section 8 is devoted to the study of the domain and range of the modified wave operators for nonlinear Schrödinger equations with critical nonlinearities and as products, we show the scattering operators are well defined. In the last section we consider the related problems and open problems.

Before closing this section, we introduce the usual weighted Sobolev space

$$H_r^{m,s} = \{f \in L^r \ : \ \|f\|_{m,s,r} = \|(1-\partial_x^2)^{m/2}(1+|x|^2)^{s/2} f\|_r < \infty\}, \|f\|_r = \|f\|_{L^r}.$$

For simplicity, we use

$$H^{m,s} = H_2^{m,s}, \|f\|_{m,s} = \|f\|_{m,s,2}, H^m = H_2^{m,0}, \|f\| = \|f\|_2.$$

2. Schrödinger type equations

In this section we consider the Schrödinger type equations

$$\left\{ \begin{array}{ll} \rho u_t - i\left(-\partial_x^2\right)^{\rho/2} u = 0, & (t,x) \in \mathbf{R}\times\mathbf{R}, \\ u(0,x) = \phi(x), & x \in \mathbf{R} \end{array} \right. \tag{11}$$

and investigate asymptotic behavior of solutions, where $\rho \geq 2$, and we define $\left(-\partial_x^2\right)^{\rho/2} = \mathcal{F}^{-1}|\xi|^\rho \mathcal{F}$. For simplicity, we use the notation

$$\|v\|_{\mathbf{Z}^\alpha} \equiv \left\|\{\xi\}^{-\alpha} v(\xi)\right\|_\infty + \left\|\{\xi\}^{1-\alpha} v'(\xi)\right\|_\infty, \{\xi\} = |\xi|/\langle\xi\rangle, \langle\xi\rangle = (1+|\xi|^2)^{1/2}.$$

When $\rho = 2$, it is called the usual Schrödinger equation.

The main result in this section is the following asymptotics of solutions to linear equations obtained in [**29**].

THEOREM 2.1. *Let $U(t) = \mathcal{F}^{-1}\exp\left(it|\xi|^\rho/\rho\right)\mathcal{F}$, where $\rho \geq 2$. Then the solution of equation (11) is written as $U(t)\phi$. For this solution we have the estimate*

$$\|U(t)\phi\|_\infty \leq Ct^{-(1+\alpha)/\rho}\left\|\left(-\partial_x^2\right)^{-\alpha/2}\phi\right\|_1 \tag{12}$$

for any $t > 0$ and $\alpha \in [0, \rho/2 - 1]$. Moreover we have the asymptotics

$$U(t)\phi = u_\phi(t,q) + R_1(t,x), \tag{13}$$

where $q = -\left(|x|/t\right)^{1/(\rho-1)}(x/|x|)$, the main term is represented as

$$u_\phi(t,q) = \left(i(\rho-1)\right)^{-1/2} t^{-1/2}|q|^{1-\rho/2} e^{-i(1-1/\rho)|q|^\rho t}\widehat{\phi}(q),$$

and the remainder term satisfies $\|R_1(t)\|_\infty \le C|t|^{-\alpha/\rho-1/\rho}\left\|\widehat{\phi}\right\|_{\mathbf{Z}^\alpha}$ *when* $0<\alpha<\rho-1$ *and* $\|R_1(t)\| \le C|t|^{-\alpha/\rho-1/(2\rho)}\left\|\widehat{\phi}\right\|_{\mathbf{Z}^\alpha}$ *when* $0<\alpha<\rho/2-1/2$.

The estimate (12) of Theorem 2.1 is obtained in [**3**] when $\rho=4$ and in [**19**] when $\rho=3$. In the case of $\rho>3$, it is shown in [**16**]. Therefore the estimate (12) is a well-known result. The important point of the above theorem is the estimate of the remainder term $R_1(t)$ which says that the L^2-norm of $R_1(t)$ decays in time faster than $t^{-(\rho-1)/(2\rho)}$ under suitable conditions on ϕ,α. We need a suitable time decay estimate of the remainder term to get global solutions in time to nonlinear Schrödinger type equations with critical nonlinearities. This fact is due to our proof in [**29**] depending on the space time estimates of solutions, which is the so-called Stricharz estimate. However for the usual nonlinear Schrödinger equations ($\rho=2$) we can use the special operator $J=x+it\partial_x$ to get the time decay and to avoid the use of the space time estimates. For details, see Section 8.

3. Airy type equation

In this section we state the asymptotic formula of solutions to the following Airy type equation:

$$\begin{cases} \rho u_t-\left(-\partial_x^2\right)^{(\rho-1)/2}\partial_x u=0, & (t,x)\in\mathbf{R}\times\mathbf{R},\\ u(0,x)=\phi(x), & x\in\mathbf{R}.\end{cases} \tag{14}$$

In the case of $\rho=3$, it is called the ordinary Airy equation. We recall some properties of the solutions. The asymptotic behavior of the solutions of the Airy type equation in the region $x>0$ is different from that in $x<0$. In $x<0$, the asymptotics of solutions for equation (14) is similar to that of the Schrödinger type equation, and in $x>0$, it is known that the solutions of the equation (14) decay rapidly at infinity. Equation (14) represents the linear part of the Benjamin-Ono equation if $\rho=2$ ([**4**], [**52**]), the Korteweg-de Vries equation if $\rho=3$ ([**41**]), and the Kawahara equation if $\rho=5$ ([**40**]). The following theorem plays an important roll in studying these nonlinear equations. In Section 6, we consider the case $\rho=3$. Hence we write the free evolution operator as $U_\rho(t)$ with ρ.

THEOREM 3.1 ([**30**]). *Let* $U_\rho(t)=\mathcal{F}^{-1}\exp\left(it|\xi|^{\rho-1}\xi/\rho\right)\mathcal{F}$ *and* $\rho\ge 2$. *Then the solution of equation* (14) *is given by* $U_\rho(t)\phi$. *For this solution we have the estimate*

$$\|U_\rho(t)\phi\|_\infty \le Ct^{-(1+\alpha)/\rho}\left\|\left(-\partial_x^2\right)^{-\alpha/2}\phi\right\|_1 \tag{15}$$

for any $t>0$ *if* $\alpha\in[0,\rho/2-1]$. *Furthermore, we have the asymptotic formula*

$$U_\rho(t)\phi=t^{-1/2}\operatorname{Re}\left(FE\widehat{\phi}(\chi)\right)+O\left(\left\|\widehat{\phi}\right\|_{\mathbf{Z}^\alpha}t^{-\alpha/\rho-(1/\rho)(1-1/p)}\right), \tag{16}$$

in the sense of the $L^p(2\le p\le\infty)$ *norm if* $0<\alpha<(\rho-1)(1-1/p)$, *where*

$$F(\chi)=\sqrt{4/i(\rho-1)}\theta(x)|\chi|^{1-\rho/2}, E=e^{-i(1-1/\rho)t|\chi|^{\rho-1}\chi}$$

and $\chi=|x/t|^{1/(\rho-1)}$, $\theta(x)=1$ *when* $x<0$ *and* $\theta(x)=0$ *when* $x\ge 0$.

The proof of Theorem 3.1 is the same as that of the asymptotics of the free evolution group $\mathcal{F}^{-1}\exp\left(it|\xi|^\rho/\rho\right)\mathcal{F}$ except for the proof of the domain $\{x>0\}$.

In this domain the free evolution group $\mathcal{F}^{-1}\exp\left(it\,|\xi|^{\rho-1}\xi/\rho\right)\mathcal{F}$ decays rapidly at infinity since $|\xi|^{\rho-1}\xi$ is an odd function and the symbol is smooth. Under the condition that the Fourier transform of ϕ vanishes at the origin, we find that the main term disappears on the domain $\{x > 0\}$. From this fact, the first term of the right-hand side of the asymptotic formula (16) depends on $\{x < 0\}$. For the same reason as stated in the previous section, the important point is that the L^2-norm of the remainder term decays in time faster than $t^{1/(2\rho)-1/2}$ for $\alpha > \rho/2-1$, $\rho \geq 2$ to apply the nonlinear problem. Finally, we consider the relation between the vanishing condition on $\widehat{\phi}$ at the origin and the usual weighted Sobolev space. For $0 \leq j \leq k = [\alpha]$, if $\widehat{\phi}^{(j)}(0) = 0$ and $\phi \in H^{0,k+2}$, then

$$\left\|\widehat{\phi}\right\|_{\mathbf{Z}^\alpha} \leq C\left\|\widehat{\phi}\right\|_\infty + C\left\|\widehat{\phi}^{(k+1)}\right\|_\infty \leq C\left\|\widehat{\phi}\right\|_{k+2,0} = C\,\|\phi\|_{0,k+2},$$

where $[\alpha]$ is the largest integer less than α. In particular, in the case of $\rho = 3$, we can take $1/2 < \alpha < 1$; hence if $\widehat{\phi}(0) = 0$ and $\phi \in H^{0,2}$, then the condition of the theorem is satisfied since we have $\left\|\widehat{\phi}\right\|_{\mathbf{Z}^\alpha} \leq C\,\|\phi\|_{0,2}$.

4. Klein-Gordon equation

In this section, we consider the asymptotic formula of the free evolution group $U(t) = \mathcal{F}^{-1}e^{-it\langle\xi\rangle}\mathcal{F}$ which corresponds to solutions to the linear Klein-Gordon equation. For simplicity, we use the notation

$$\|v\|_{\mathbf{Z}_p} = \|\widehat{v}\|_{0,2,\infty} + \|\partial_\xi \widehat{v}\|_{0,3-3/(2p),p}$$

for $2 \leq p \leq \infty$. We now state the main result in this section (obtained in [**31**]).

THEOREM 4.1. *For any $t > 0$, we have the estimate*

$$\|U(t)\,v\|_\infty \leq Ct^{-1/2}\,\|v\|_{3/2,0,1}. \tag{17}$$

Moreover if $v \in \mathbf{Z}_p$, then the asymptotic formula

$$U(t)\,v = \mathcal{D}_t BM\mathcal{F}v\,(x/\langle ix\rangle) + \begin{cases} O\left(\|v\|_{\mathbf{Z}_p} t^{-1+1/r+1/(2p)}\right), & 2 \leq p < \infty, \\ O\left(\|v\|_{\mathbf{Z}_\infty} t^{-1+1/r}\log t\right), & p = \infty \end{cases} \tag{18}$$

holds in the L^r sense for any $t \geq 1$, and $2 \leq r \leq \infty$, where $\mathcal{D}_t\phi = (it)^{-1/2}\phi(x/t)$, $M = e^{-it\langle ix\rangle}$, $B = \theta(x)\langle ix\rangle^{-3/2}$, $\langle ix\rangle = \sqrt{1-x^2}$, $\theta(x) = 1$ for $|x| < 1$ and $\theta(x) = 0$ for $|x| \geq 1$.

The $L^\infty - L^1$ estimate (17) is proved in [**44**]. The important estimate is that of the remainder term. If we let $p = 2$ in the formula (18), and if we replace v by $U(-t)\,v$, then we find that

$$\begin{aligned} &\|v - \mathcal{D}_t BM\mathcal{F}U(-t)\,v\,(x/\langle ix\rangle)\|_r \\ \leq\ & Ct^{1/r-3/4}\left(\|\mathcal{F}U(-t)\,v\|_{0,2,\infty} + \|xU(-t)\,v\|_{9/4,0}\right). \end{aligned}$$

From the properties of $\mathcal{D}_t, B, M$, we also find that

$$\|\mathcal{D}_t BM\mathcal{F}U(-t)\,v\,(x/\langle ix\rangle)\|_r \leq Ct^{1/r-1/2}\,\|\mathcal{F}U(-t)\,v\|_{0,3/2,r}.$$

Therefore the a priori estimates $\|\mathcal{F}U(-t)v\|_{\infty}$ and $\|xU(-t)v\|$ are important to obtain global solutions to the initial value problem of the cubic nonlinear Klein-Gordon equation. We remark that these a priori estimates are used to study the initial value problem for nonlinear Schrödinger equations with critical nonlinearities. This is natural since nonlinear Klein-Gordon equations are a relativistic version of nonlinear Schrödinger equations.

5. Nonlinear Schrödinger type equations

In this section we apply the result of Section 2 to nonlinear Schrödinger type equations

$$\begin{cases} i\rho\partial_t u + \left(-\partial_x^2\right)^{\rho/2} u = \rho\lambda\,|u|^{p-1}\,u, & (t,x)\in\mathbf{R}\times\mathbf{R}, \\ \lim_{t\to\infty} U(-t)\,u(t) = u_+, & x\in\mathbf{R}, \end{cases} \tag{19}$$

where $p \geq 1$ and $U(t)$ is the same one as that defined in Section 2. If $p > 3$, we can find a unique solution satisfying the final value condition as above. In the theorem below, we assume $p < 5$ for simple descriptions. However this condition is not essential. In the case of $p \leq 3$ we cannot find a solution under the final value condition as above. We need to modify the framework of the problem as stated in Section 1. More precisely, we state it in the theorem.

THEOREM 5.1 ([**29**]). *We assume that the final value $u_+ \in L^2$ satisfies the condition $\|\hat{u}_+\|_{\mathbf{Z}^\alpha} + \|u_+\| < \infty$ for $\alpha > \rho/2 - 1$, where $\mathbf{Z}^\alpha$ is the same one as that defined in Section 2.*

We first state the result for the super-critical nonlinearities.

(1) Let $3 < p < 5$. Then there exists a $T > 1$ and a unique solution $u(t) \in C\left([T,\infty);L^2\right)$ of equation (19). Furthermore there exist positive constants c_1, c_2 such that the asymptotic formula

$$c_2 t^{-(p-3)/2} \leq \|u(t) - U(t)u_+\| \leq c_1 t^{-(p-3)/2}$$

holds. We next state the case of the critical nonlinearity.

(2) Let $\lambda \in \mathbf{R}, p = 3$ and $\|\hat{u}_+\|_{\mathbf{Z}^\alpha} + \|u_+\|$ be sufficiently small. Then equation (19) has a unique solution $u(t) \in C\left([T,\infty);L^2\right)$, and we have the asymptotic formula

$$\|u(t) - u_w(t)\| + \left(\int_t^\infty \|u(\tau) - u_w(\tau)\|_\infty^{2\rho}\,d\tau\right)^{1/(2\rho)} \leq Ct^{-b}$$

for $(\rho-1)/(2\rho) < b < 1/2$, where

$$u_w(t,q) = C_1 t^{-1/2}\,|q|^{(2-\rho)/2}\,e^{i(1-1/\rho)t|q|^\rho}\widehat{u}_+(q)e^{i\lambda|C_1|^2|q|^{2-\rho}|\widehat{u}_+(q)|^2\log t},$$

$q = -\left(|x|/t\right)^{1/(\rho-1)}(x/|x|)$, and $C_1 = (i(\rho-1))^{1/2}$. The next result is concerned with nonlinear terms which act as dissipative terms.

(3) Let $p = 3$ and $\operatorname{Im}\lambda < 0$. Then there exists a $T > 1$ and a unique solution $u(t) \in C\left([T,\infty);L^2\right)$ of equation (19). We also have the asymptotic formula

$$\|u(t) - u_W(t)\| + \left(\int_t^\infty \|u(\tau) - u_W(\tau)\|_\infty^{2\rho}\,d\tau\right)^{1/(2\rho)} \leq Ct^{-b}$$

for $(\rho-1)/(2\rho) < b < 1/2$, where

$$u_W(t,q) = C_1 t^{-1/2}\,|q|^{(2-\rho)/2}\,e^{i(1-1/\rho)t|q|^\rho}\widehat{u_+}(q)e^{i\frac{(\operatorname{Re}\lambda)}{(\operatorname{Im}\lambda)}\varphi(t,q)-\varphi(t,q)}$$

and $e^{-2\varphi(t,q)} = (1 - 2(\operatorname{Im}\lambda) |C_1|^2 |q|^{(2-\rho)} |\widehat{u}_+(q)|^2 \log t)^{-1}$.

From Theorem 5.1, the next corollary follows.

COROLLARY 5.2. *Let* $3 < p < 5$. *Then there exists the wave operator* $W_+ : L^2 \cap \mathbf{Z}^\alpha \to L^2$. *Let* $p = 3$ *and* $\operatorname{Im}\lambda < 0$. *Then there exists the modified wave operator* $MW_+ : L^2 \cap \mathbf{Z}^\alpha \to L^2$. *Furthermore when* $p = 3, \lambda \in \mathbf{R}$, *then there exists the modified wave operator* MW_+ *which maps the neighborhood of the origin of* $L^2 \cap \mathbf{Z}^\alpha$ *to* L^2.

The next theorem implies the nonexistence of the usual wave operator.

THEOREM 5.3 ([**29**]). *Let* $p = 3$, $\operatorname{Im}\lambda < 0$ *or* $p = 3, \lambda \in \mathbf{R}$ *and the final value* u_+ *satisfies* $\|u_+\| + \|\hat{u}_+\|_{\mathbf{Z}^\alpha} < \infty$. *Furthermore we assume that there exists a solution* u *of equation (19) satisfying* $\lim_{t\to\infty} \|u(t) - U(t)u_+\| = 0, \|u(t)\|_\infty \leq Ct^{-1/2}$. *Then* u *is identically zero.*

We now state the previous results on existence of modified wave operators. In [**53**], under the conditions such that $\rho = 2$, $N(u, \overline{u}) = i\lambda |u|^2 u$, $\lambda \in \mathbf{R}$, $u_+ \in H^{3,0} \cap H^{1,2}$, $\|\widehat{u}_+\|_\infty$ is small enough, and suitable modifications of the final value, global-in-time solutions are obtained. Namely, a modified wave operator is constructed first; see also [**54**]. In the case of $\rho = 2$, progress of the problem including the nonlinear terms $u^3, \bar{u}^2 u, \bar{u}^3$ is obtained in papers [**35**], [**62**]. In the higher order case $\rho > 2$, by a method similar to [**53**], the existence of a modified wave operator is shown in [**58**] when $\rho = 4$, $N(u, \overline{u}) = i\lambda |u|^2 u$, $\lambda \in \mathbf{R}$. This result is extended to the case of the dissipative nonlinearity ($\operatorname{Im}\lambda < 0$) in [**59**].

The results in this section are improvements of the previous results [**53**], [**58**], [**59**] and [**61**]. In order to prove the existence of modified scattering operators, we need to consider the initial value problem for the nonlinear Schrödinger type equations with critical nonlinearities:

$$\begin{cases} \rho u_t - i(-\partial_x^2)^{\rho/2} u = i\lambda\rho |u|^2 u, & (t, x) \in \mathbf{R} \times \mathbf{R}, \\ u(0, x) = u_0, & x \in \mathbf{R}. \end{cases} \tag{20}$$

When $\rho = 2$, it is shown that in [**18**], equation (20) has a unique global small solution and its asymptotic behavior in the $L^2 \cap L^\infty$ sense is given under the conditions such that $u_0 \in H^\gamma \cap H^{0,\gamma}$, $\gamma > 1/2$, $\|u_0\|_{\gamma,0} + \|u_0\|_{0,\gamma}$ is small enough. This result implies the existence of an inverse wave operator. However it is not enough to show existence of a modified scattering operator. In the case of the usual nonlinear Schrödinger equation ($\rho = 2$), we can improve the domain of an inverse wave operator. As a product, we have a modified scattering operator. We describe such an improvement in Section 8 in detail. Next we give some open problems. When $\rho > 2$, the asymptotic behavior of solutions to the initial value problem is not known. The nonexistence of the wave operator for the subcritical case $|u|^{p-1} u$, $1 < p < 3$ is an open problem since the previous proofs of nonexistence of a wave operator depend on the sharp time decay of solutions. However if $\lambda < 0, \rho = 2$, and $i\lambda |u|^{p-1} u$, $1 < p < 3$, then in [**2**] we have the nonexistence result since we can prove the sharp $L^q, q \geq 2$ time decay estimate of solutions by the pseudo-conformal conservation law in [**12**]. For general $\rho > 2$ it is an open problem. We also have a conjecture that if $\operatorname{Im}\lambda > 0$, then the blowup of solutions occurs. However this is also an open problem.

5.1. Generalization. In this subsection, we apply the method of paper [**29**] to nonlinear Schrödinger equations of derivative type. We present the result in paper [**37**]. In paper [**37**], the equation

$$iu_t + u_{xx} = 2i(|u|^2 u)_x, \quad (t,x) \in \mathbf{R} \times \mathbf{R} \tag{21}$$

is considered. This equation appears in plasma physics (see [**45**], [**46**] for the derivation of (21)), and it is well known that soliton solutions exist and the inverse scattering method is useful. For an existence result by functional analytic methods, we refer to [**15**] and the references cited therein. Solutions of equation (21) are not asymptotically free (see Theorem 5.5 below). Therefore we cannot treat the scattering problem in the usual framework. The method in [**15**] depends on a nonlinear transformation which translates the original equation to a system of equations without a derivative of an unknown function in nonlinearities. In other words, we can apply the method used in a single equation to this system.

In [**37**], the following theorem is shown.

THEOREM 5.4. *There exists an $\epsilon_1 > 0$ such that if $\|\phi_+\|_{2,1} < \epsilon_1$, then equation (21) has a unique global solution $u \in C(\mathbf{R}; H^2) \cap L^4_{loc}(\mathbf{R}; H^{2,0}_\infty)$ for any $\phi_+ \in H^4 \cap H^{0,4}$. For $1/2 < \alpha < 1$, we have the asymptotics*

$$\lim_{t \to \infty} \left\| u(t) - \exp(iS^+(t)) U(t) \phi_+ \right\|_{2,0} = O(t^{-\alpha}),$$

where $S^+(t,x) = (\log|t|)(x/(2|t|)) \left| \hat{\phi}_+ (x/(2t)) \right|^2.$

From Theorem 5.4, it follows that the modified wave operator $MW_+ : \phi_+ \mapsto u(0)$ is defined as the map from the neighborhood of the origin of $H^4 \cap H^{0,4}$ to H^2.

THEOREM 5.5. *We assume that the solution $u(t) \in C(\mathbf{R}; H^1)$ to equation (21) satisfies*

$$\|u(t)\|_{1,0} \le C(\|u(0)\|_{1,0}), \|u(t)\|_4 = O(|t|^{-1/4}) \tag{22}$$

for $t \to \infty$. Moreover we assume that there exists a $\phi_+ \in H^{0,1}$ such that $\lim_{t\to\infty} \|u(t) - U(t)\phi_+\| = 0$. Then u is identically zero.

By Theorem 5.4 and the energy conservation law it is clear that there exists a unique global solution of equation (21) which satisfies the assumption (22). Therefore Theorem 5.5 means nonexistence of wave operators.

On the other hand, by using the method used in paper [**29**] we can treat the following type of equations:

$$2iu_t + u_{xx} = N(u, u_x), \quad (t,x) \in \mathbf{R} \times \mathbf{R}. \tag{23}$$

The nonlinear term N can be decomposed as $N = N_1 + N_2$, where

$$N_1 = \lambda_1 |u|^2 u + i\lambda_2 |u|^2 u_x + i\lambda_3 u^2 \overline{u}_x + \lambda_4 |u_x|^2 u + \lambda_5 \overline{u} u_x^2 + i\lambda_6 |u_x|^2 u_x$$

satisfies the self-conjugate property (gauge-invariant property); namely, for any $\theta \in \mathbf{R}$ we have $N_1(e^{i\theta}u) = e^{i\theta} N_1(u)$. The nonlinear term

$$N_2 = a_1 u^3 + a_2 u \overline{u}^2 + a_3 \overline{u}^3 + a_4 \overline{u}^2 \overline{u}_x + a_5 \overline{u} \overline{u}_x^2 + a_6 \overline{u}_x^3 + a_7 |u|^2 \overline{u}_x + a_8 u \overline{u}_x^2$$

does not satisfy the self-conjugate property. To use the energy method, we assume that the derivative of the nonlinear term N with respect to u_x must be pure imaginary. By the definition of N_1, N_2, we have

$$\frac{\partial N}{\partial u_x} = i\lambda_2 |u|^2 + \lambda_4 u \overline{u}_x + 2\lambda_5 \overline{u} u_x + 2i\lambda_6 |u_x|^2;$$

hence the conditions will be $\lambda_2, \lambda_6 \in \mathbf{R}, \mathrm{Re}(\lambda_4 + 2\lambda_5) = 0$. Under the conditions, we can apply the energy method to get local-in-time solutions to equation (23). We next consider additional conditions to apply the method in [**29**]. In order to find such conditions, we substitute the approximate solution

$$v_{u_+} = (it)^{-1/2} e^{ix^2/(2t)} \widehat{u}_+ (x/t)$$

of the linear problem into the nonlinear terms. If we do it, we know that each term of N_2 is the nonresonant term by the difference of frequencies. Hence the coefficients a_j can be taken from $\mathbf{C}$ for $j = 1, \ldots, 8$. For the nonlinear term N_1, we have the main term

$$t^{-1} \Lambda (q) v_{u_+}$$

with $q = x/t$ and

$$\Lambda (q) = \lambda_1 + (-\lambda_2 + \lambda_3) q + (\lambda_4 - \lambda_5) q^2 - \lambda_6 q^3,$$

which means that the condition of the coefficients will be $\mathrm{Im}\Lambda (q) \leq 0$. The approximate solution u_w of the nonlinear problem is written as

$$u_w = (1/\sqrt{it}) e^{ix^2/(2t)} \widehat{w} (t, x/t)$$

when $\mathrm{Im}\Lambda (q) = 0$, where

$$\widehat{w} (t, q) = \widehat{u}_+ (t, q) \exp \left(i\Lambda (q) \left| \widehat{u}_+ (q) \right|^2 \log t \right).$$

In the case of $\mathrm{Im}\Lambda (q) < 0$ the approximate solution u_W is given by

$$u_W = (1/\sqrt{it}) e^{(ix^2)/(2t)} \widehat{W} (t, x/t),$$

where

$$\widehat{W} (t, q) = \widehat{u}_+ (t, q) \exp \left(i\Lambda (q) \varphi(t, q) / \mathrm{Im}\Lambda (q) \right),$$
$$e^{-2\varphi(t,q)} = (1 - 2(\mathrm{Im}\Lambda(q)) |\widehat{u}_+(q)|^2 \log t)^{-1}.$$

From the above result, it follows that we have the asymptotic formula of solutions under the conditions of coefficients of nonlinear terms and suitable conditions on u_+. The asymptotic formula implies that solutions of the nonlinear problem decay in time faster the order $(\log t)^{-1/2}$ than those of the linear problem when the nonlinear terms are dissipative ones. Thus we can extend the previous work of the asymptotics of solutions to a wide class of nonlinearities. For coefficients which do not satisfy the above conditions, it is expected that solutions will blow up in finite time. However this is an open problem. For general $\rho > 2$, our method can be applicable to the subcritical nonlinearity $|u|^{p-1}u$, $p < 3$ if p is close to 3 (see [**34**]). However if p is close to 1, then the problem is open. Concerning nonlinear heat equations and nonlinear dissipative wave equations with power nonlinearities, there are a lot of works for the subcritical cases. (For examples, we refer to [**50**], [**51**] and the references cited therein.)

6. Modified Korteweg-de Vries equation

In this section we consider the modified Korteweg-de Vries equation

$$3\partial_t u - \partial_x^3 u = 3\lambda \partial_x u^3, \quad (t, x) \in \mathbf{R}^+ \times \mathbf{R}, \quad \lambda \in \mathbf{R} \tag{24}$$

and present the result obtained in [**30**] on the existence of modified wave operators. As stated in Section 5, there are a few results of the final value problems on nonlinear dispersive equations with nonlinearities including derivatives of unknown functions.

We also describe the asymptotic behavior of solutions to (24) from the results in [**30**].

THEOREM 6.1. *Let the final value u_+ satisfy the conditions such that $\widehat{u}_+(0) = \partial_\xi \widehat{u}_+(0) = 0, u_+ \in H^{2,4}$ and u_+ is a real-valued function. Then there exist a time T and ε such that equation (24) has a unique solution*

$$u \in L^\infty([T,\infty); H^2) \cap C([T,\infty); H^1)$$

if $\|u_+\|_{2,2} \le \varepsilon$. Moreover, the asymptotic formula

$$\|u(t) - U_3(t) w(t)\|_{2,0} \le Ct^{-b}, \quad \widehat{w}(t,\xi) = e^{3i\lambda|\xi|^{-1}\xi|\widehat{u}_+(\xi)|^2 \log t} \widehat{u_+}(\xi)$$

holds, where b is a constant satisfying $1/3 < b < 1/2$.

From the result, the corollary follows.

COROLLARY 6.2. *Let $X = \{u_+ \in H^{2,4} \ : \ \widehat{u}_+(0) = \partial_\xi \widehat{u}_+(0) = 0\}$. Then the modified wave operator MW_+ is defined as a map from the neighborhood of the origin in X to H^2.*

We now explain the asymptotic formula in Theorem 6.1 in detail. The sign of the coefficient in the equation is different from that in Theorem 3.1. Therefore we note that the properties of solutions in the domains $x > 0$ and $x < 0$ change each other. We have the asymptotics

$$\partial_x^j U_3(t) w(t) = \begin{cases} t^{-1/2} |q|^{-1/2+j} 2\mathrm{Re}(i)^j C_1 e^{i(2/3)t|q|^3} \widehat{u_+}(q), & x > 0 \\ 0, \quad x \le 0 \end{cases} + R,$$

where $j = 0, 1, 2$, $q = |x/t|^{1/2}$, $C_1 = (2i)^{1/2}$ and the remainder term satisfies

$$\|R(t)\| \le Ct^{-b} \|u_+\|_{2,3}, \quad 1/3 < b < 1/2.$$

Therefore we obtain

$$\left\|\partial_x^j u(t) - t^{-1/2} |q|^{-1/2+j} 2\mathrm{Re}(i)^j C_1 e^{i(1-1/\rho)t|q|^3} \widehat{u_+}(q)\right\|_{L^2(0,\infty)}$$
$$+ \quad \left\|\partial_x^j u(t)\right\|_{L^2(-\infty,0)} \le Ct^{-b} \|u_+\|_{2,4}.$$

Namely, we note that Theorem 6.1 gives the asymptotic estimate in the domain $x > 0$, but only gives the estimate of time decay of solutions in the domain $x < 0$. The following result gives nonexistence of the usual wave operator.

THEOREM 6.3. *Let u_+ satisfy the conditions in Theorem 6.1, and let u be a solution of equation (24) satisfying the asymptotics $\lim_{t\to\infty} \|u(t) - U_3(t)u_+\| = 0$. Then u is identically zero.*

6.1. Generalization. If we use the method used in paper [**30**], we can handle the following type of equations:

$$3\partial_t u - \partial_x^3 u = 3N(u, u_x), \quad (t,x) \in \mathbf{R} \times \mathbf{R}, \tag{25}$$

where the nonlinear term N is represented as

$$N_1 = \lambda_1 u^3 + \lambda_2 u^2 u_x + \lambda_3 u u_x^2 + \lambda_4 u_x^3$$

and the coefficient of each nonlinear term is a real number. In order to apply the method in this section to the above, we assume that $\lambda_1 \le 0, \lambda_3 \le 0$. In the case of $\lambda_1 = \lambda_3 = 0$, the approximate solution u_w of the nonlinear problem is given by $u_w = U_3(t)w$, where

$$\widehat{w}\,(t,\xi) = \widehat{u}_+\,(\xi)\exp\left(i|\xi|^{-1}(\lambda_2\xi + 3\lambda_4\xi^3)|\widehat{u}_+(\xi)|^2\log t\right).$$

If $\lambda_1 < 0$ or $\lambda_3 < 0$, the approximate solution u_W will be $u_W = U_3(t)W$, where

$$\widehat{W}\,(t,\xi) = \widehat{u}_+\,(\xi)\exp\left(i\Lambda\,(\xi)\varphi(t,\xi)/\mathrm{Im}\Lambda\,(\xi)\right),$$
$$e^{-2\varphi(t,\xi)} = (1 - 2(\mathrm{Im}\Lambda(\xi))|\widehat{u}_+(\xi)|^2\log t)^{-1}$$

and $\Lambda(\xi) = |\xi|^{-1}(3i\lambda_1 + \lambda_2\xi + i\lambda_3\xi^2 + 3\lambda_4\xi^3)$. Thus if the coefficients of the nonlinear terms satisfy the above conditions, and if the final value u_+ satisfies vanishing conditions at the origin, then we can find an asymptotic formula of solutions. For general $\rho > 2$, our method can be applicable. We refer to [**30**] for results on $\rho > 2$. It is expected that solutions will blow up in finite time when $\mathrm{Im}\Lambda(\xi) > 0$. This problem is also an open one.

7. Nonlinear Klein-Gordon equation

In this section we consider global existence of solutions to the nonlinear Klein-Gordon equation

$$u_{tt} + \left(1 - \partial_x^2\right)u = \mu\,|u|^2\,u, \quad (t,x) \in \mathbf{R}\times\mathbf{R} \tag{26}$$

under the final value condition $\lim_{t\longrightarrow\infty}\left(u\,(t) - 2\mathrm{Re}U\,(t)\,w_+\,(t)\right) = 0$, where μ is real, $U\,(t) = \mathcal{F}^{-1}e^{-it\langle\xi\rangle}\mathcal{F}$ is the Klein-Gordon evolution operator and $\widehat{w}_+$ is defined through u_+ by

$$\widehat{w}_+\,(t,\xi) = \widehat{u}_+\,(\xi)\,e^{(3/2)i\mu\langle\xi\rangle^2|\widehat{u}_+(\xi)|^2\log t}.$$

We cannot solve the problem in the neighborhood of the free solution since the cubic nonlinear term is the critical nonlinearity in one space dimension (see [**14**],Theorem 1). In order to find a solution of equation (26) we define the approximate solution as above and try to find a solution in the neighborhood of that. We let u be a solution of equation (26). Then we put

$$\begin{pmatrix}\widetilde{u}\\ \widetilde{v}\end{pmatrix} = \begin{pmatrix}\left(1 + \langle i\partial_x\rangle^{-1}i\partial_t\right)u/2\\ \left(1 - \langle i\partial_x\rangle^{-1}i\partial_t\right)u/2\end{pmatrix}.$$

New variables $\widetilde{u},\widetilde{v}$ satisfy the equations

$$\begin{aligned}\partial_t\widetilde{u} + i\,\langle i\partial_x\rangle\,\widetilde{u} &= -\mu\,\langle i\partial_x\rangle^{-1}|\widetilde{u}+\widetilde{v}|^2\,(\widetilde{u}+\widetilde{v})\,/(2i),\\ \partial_t\widetilde{v} - i\,\langle i\partial_x\rangle\,\widetilde{v} &= \mu\,\langle i\partial_x\rangle^{-1}|\widetilde{u}+\widetilde{v}|^2\,(\widetilde{u}+\widetilde{v})\,/(2i).\end{aligned} \tag{27}$$

If u is a real-valued function, then $\widetilde{v}\,(t) = \overline{\widetilde{u}\,(t)}$, which means that (27) is written as

$$\partial_t\widetilde{u} + i\,\langle i\partial_x\rangle\,\widetilde{u} = -\mu\,\langle i\partial_x\rangle^{-1}\left|\widetilde{u}+\overline{\widetilde{u}}\right|^2\left(\widetilde{u}+\overline{\widetilde{u}}\right)/(2i). \tag{28}$$

From this equation, we find nonlinear terms consist of resonant and nonresonant terms. We are now in a position to state our main result in this section.

THEOREM 7.1 ([**31**]). *Let the final value u_+ be a real-valued function, $u_+ \in H^{5,3}$ and $\|u_+\|_{4,2}$ be sufficiently small. Then equation (28) has a unique global solution $\widetilde{u} \in C\left([1,\infty);H^1\right)$ which satisfies the asymptotic estimate*

$$\left\|\widetilde{u}\,(t) - U\,(t)\,\mathcal{F}^{-1}\widehat{w}_+\right\|_{1,0} \le Ct^{-b},$$

where b is a constant satisfying $1/4 < b < 1/2$.

From the above result, we have the following theorem.

THEOREM 7.2. *Let the final value $u_+ \in H^{5,3}$ be real, and $\|u_+\|_{4,2}$ be small enough. Then equation (26) has a unique global solution $u \in C([1,\infty); H^1)$ satisfying the asymptotic estimate*

$$\left\| u(t) - 2ReU(t)\mathcal{F}^{-1}\widehat{w}_+ \right\|_{1,0} \le Ct^{-b},$$

where b is a constant satisfying $1/4 < b < 1/2$.

By Theorem 7.2, we have

COROLLARY 7.3. *The modified wave operator MW_+ is well defined as a map from the neighborhood of the origin in $H^{5,3}$ to H^1.*

We briefly explain how we find a solution. From Theorem 4.1, the asymptotic formula

$$U(t)f = u_f + R$$

follows, where the main term is represented as

$$u_f = \begin{cases} (it)^{-1/2}\langle i\eta\rangle^{-3/2} e^{it\sqrt{1-\eta^2}}\widehat{f}(\eta/\langle i\eta\rangle), & |\eta| \le 1 \\ 0, \quad |\eta| > 1, \quad \eta = x/t \end{cases}$$

and as the estimate of the remainder term we have $\|R(t)\| \le Ct^{-b}$, where $1/4 < b < 1/2$. As stated in the above we need to modify the final value u_+ since the nonlinear terms contain the critical one. Let us consider how we modify it. We assume that a solution of the nonlinear problem behaves like a solution of the linear problem. Then the main term of the nonlinear term with $\mu = 1$ will be

$$\begin{aligned} & |U(t)u_+|^2 U(t)u_+ \\ \simeq\ & t^{-1}\langle i\eta\rangle^{-3}(it)^{-1/2}\langle i\eta\rangle^{-3/2} e^{it\sqrt{1-\eta^2}}\widehat{u_+}(\eta/\langle i\eta\rangle)\left|\widehat{u_+}(\eta/\langle i\eta\rangle)\right|^2 \\ \simeq\ & U(t)\mathcal{F}^{-1}\langle\xi\rangle^3 |\widehat{u}_+(\xi)|^2 \widehat{u}_+(\xi), \end{aligned}$$

where we used the notation $a \simeq b$ which means that $a - b$ can be regarded as the remainder term for our problem. The L^2-norm of the main term is not integrable in time; hence we need to use the approximate solution to eliminate this term. In what follows, we describe $\widetilde{u}$ by u for simplicity. By a simple calculation, we have by (28),

$$2\left(e^{it\langle\xi\rangle}\widehat{u}\right)_t = i\mu\left(\langle\xi\rangle^{-1} e^{it\langle\xi\rangle}\right)\mathcal{F}\left(3|u|^2 u + 3|u|^2\overline{u} + u^3 + \overline{u}^3\right).$$

If we substitute a solution of the linear equation to the nonlinear term, we find that the first term $|u|^2u$ of the right-hand side is a resonant term by comparing frequencies. To eliminate this term we introduced the approximate solution. Indeed we obtain

$$2(\widehat{w}_+(t,\xi))_t = 3i\mu\langle\xi\rangle^2 t^{-1}|\widehat{u}_+(\xi)|^2\widehat{w}_+(t,\xi)$$

from which it follows that

$$\begin{aligned} & 2\left(e^{it\langle\xi\rangle}\widehat{u} - \widehat{w}_+(t,\xi)\right)_t \\ =\ & i\mu\langle\xi\rangle^{-1} e^{it\langle\xi\rangle}\mathcal{F}\left(|u+\overline{u}|^2(u+\overline{u})\right. \\ & \left. -3t^{-1}\mathcal{F}^{-1}e^{-it\langle\xi\rangle}\langle\xi\rangle^3|\widehat{w}_+(t,\xi)|^2\widehat{w}_+(t,\xi)\right). \end{aligned} \tag{29}$$

The right-hand side of (29) can be rewritten as

$$\begin{aligned}
& |u+\overline{u}|^2 (u+\overline{u}) - 3t^{-1}\mathcal{F}^{-1}e^{-it\langle\xi\rangle}\langle\xi\rangle^3 |\widehat{u}_+(\xi)|^2 \widehat{w}_+(t,\xi) \\
= \; & 3\left(|u|^2 u - t^{-1}U(t)\mathcal{F}^{-1}\langle\xi\rangle^3 |\widehat{u}_+(\xi)|^2 \widehat{w}_+(t,\xi)\right) \\
& + \left(3|u|^2\overline{u} - 3\left|U(t)\mathcal{F}^{-1}\widehat{w}_+\right|^2 \overline{U(t)\mathcal{F}^{-1}\widehat{w}_+}\right) \\
& + \left(u^3 - \left(U(t)\mathcal{F}^{-1}\widehat{w}_+\right)^3\right) + \left(\overline{u}^3 - \overline{\left(U(t)\mathcal{F}^{-1}\widehat{w}_+\right)^3}\right) \\
& + 3\left|U(t)\mathcal{F}^{-1}\widehat{w}_+\right|^2 \overline{U(t)\mathcal{F}^{-1}\widehat{w}_+} + \left(U(t)\mathcal{F}^{-1}\widehat{w}_+\right)^3 \\
& + \overline{\left(U(t)\mathcal{F}^{-1}\widehat{w}_+\right)^3}. \qquad (30)
\end{aligned}$$

By Theorem 4.1 we get

$$\begin{aligned}
u(t,x) &= U(t)\mathcal{F}^{-1}\widehat{w}_+ = u_{\mathcal{F}^{-1}\widehat{w}_+} + R \\
&= \begin{cases} (it)^{-1/2}\langle i\eta\rangle^{-3/2} e^{it\sqrt{1-\eta^2}}\widehat{u}_+(\eta/\langle i\eta\rangle) \\ \times e^{(3/2)i\mu\langle i\eta\rangle^{-2}|\widehat{u}_+(\eta/\langle i\eta\rangle)|^2\log t}, & |\eta|\le 1 \\ 0, & |\eta|>1 \end{cases} + R.
\end{aligned}$$

This asymptotic formula and Theorem 4.1 give that the main term of the first term of the right-hand side of (30),

$$t^{-1}U(t)\mathcal{F}^{-1}\langle\xi\rangle^3 |\widehat{u}_+(\xi)|^2 \widehat{w}_+(t,\xi),$$

will be

$$\left|u_{\mathcal{F}^{-1}\widehat{w}_+(t,\xi)}\right|^2 u_{\mathcal{F}^{-1}\widehat{w}_+(t,\xi)}.$$

We can also prove that the third term of the right-hand side of (30) is the remainder term. Therefore we get the main result since the difference between u and $U(t)\mathcal{F}^{-1}\widehat{w}_+$ is regarded as the remainder term.

Under conditions such that the data are small, smooth enough and have compact support, the initial value problem is studied. In [**14**], the L^∞ time decay estimate of solutions is obtained and in [**8**] or [**65**], the asymptotic behavior of solutions to the nonlinear Klein-Gordon equations with cubic nonlinearities including $|u|^2 u$ is studied (see also [**42**]). Recently, the compactness condition on the data is removed in [**32**]. Concerning the initial value problem, there are a lot of works. However, there are few results on the final value problem except for the case of higher-order nonlinearities. There are some works on the initial value problem for the cubic nonlinear Klein-Gordon equations with nonlinearities including variables $u, u_t, u_x, u_{xx}, u_{tx}$ in [**39**], [**47**], [**48**] under the conditions that the data are small, smooth and decay rapidly in space at infinity and the nonlinearities satisfy the non-resonant condition. They find a solution in the neighborhood of the free solution. Therefore cubic nonlinearities are not necessarily resonant terms.

7.1. Generalization. The method in [**31**] is applicable to the general cubic nonlinear Klein-Gordon equation

$$u_{tt} + \left(1-\partial_x^2\right)u = F(u,u_t,u_x), \qquad (t,x)\in\mathbf{R}\times\mathbf{R},$$

where

$$\begin{aligned}
F(u,u_t,u_x) &= \left(\gamma_1 u^2 + \gamma_2 u_t^2 + \gamma_3 u_x^2 + \gamma_4 u_t u_x\right)u + \left(\gamma_5 u^2 + \gamma_6 u_t^2 + \gamma_7 u_x^2\right)u_t \\
&\quad + \left(\gamma_8 u^2 + \gamma_9 u_t^2 + \gamma_{10} u_x^2\right)u_x.
\end{aligned}$$

Under the assumption that the asymptotic behavior of the solutions to the nonlinear problem is the same as that of the linear problem, we substitute the leading term of the solutions to the linear problem into the nonlinear term. Then we make a transformation to remove resonant terms. We put $\widetilde{u} = \left(1 + \langle i\partial_x \rangle^{-1} i\partial_t\right) u/2$. Then the nonlinear problem becomes

$$2\partial_t \widetilde{u} + 2i \langle i\partial_x \rangle \widetilde{u} = i \langle i\partial_x \rangle^{-1} G(\widetilde{u}, \widetilde{u}_x),$$

where the nonlinear term $G(u, u_x)$ is written by F as follows:

$$\begin{aligned} & G(a, a_x) \\ = \; & F(a + \overline{a}, -i \langle i\partial_x \rangle (a - \overline{a}), (a_x + \overline{a}_x)) \\ = \; & \Big(\gamma_1 (a + \overline{a})^2 - \gamma_2 (\langle i\partial_x \rangle a - \langle i\partial_x \rangle \overline{a})^2 + \gamma_3 (a_x + \overline{a}_x)^2 \\ & \qquad - \gamma_4 i (\langle i\partial_x \rangle a - \langle i\partial_x \rangle \overline{a})(a_x + \overline{a}_x)\Big)(a + \overline{a}) \\ & - i\left(\gamma_5 (a + \overline{a})^2 - \gamma_6 (\langle i\partial_x \rangle a - \langle i\partial_x \rangle \overline{a})^2 + \gamma_7 (a_x + \overline{a}_x)^2\right) \langle i\partial_x \rangle (a - \overline{a}) \\ & + \left(\gamma_8 (a + \overline{a})^2 - \gamma_9 (\langle i\partial_x \rangle a - \langle i\partial_x \rangle \overline{a})^2 + \gamma_{10} (a_x + \overline{a}_x)^2\right)(a_x + \overline{a}_x). \end{aligned}$$

By a simple calculation we find that the leading terms (resonant terms) of G will be

$$\begin{aligned} & G(a, a_x) \\ = \; & 3\gamma_1 |a|^2 a + \gamma_2 \left(2 |\langle i\partial_x \rangle a|^2 a - (\langle i\partial_x \rangle a)^2 \overline{a}\right) + \gamma_3 \left(2 |a_x|^2 a + a_x^2 \overline{a}\right)^2 \\ & - \gamma_4 i \left(\left(\langle i\partial_x \rangle a\right) \overline{a}_x a - \left(\langle i\partial_x \rangle \overline{a}\right) a_x a + \left(\langle i\partial_x \rangle a\right) a_x \overline{a}\right) \\ & - i \Big(2\gamma_5 |a|^2 \langle i\partial_x \rangle a - \gamma_5 a^2 \langle i\partial_x \rangle \overline{a} + 3\gamma_6 |\langle i\partial_x \rangle a|^2 \langle i\partial_x \rangle a \\ & \qquad + 2\gamma_7 |a_x|^2 \langle i\partial_x \rangle a - \gamma_7 \overline{a}_x^2 \langle i\partial_x \rangle \overline{a}\Big) \\ & + \left(2\gamma_8 |a|^2 a_x + \gamma_8 a^2 \overline{a}_x + 2\gamma_9 |\langle i\partial_x \rangle a|^2 a_x - \gamma_9 (\langle i\partial_x \rangle a)^2 \overline{a}_x + 3\gamma_{10} |a_x|^2 a_x\right). \end{aligned}$$

We let the asymptotic behavior of a be the same as that of a solution to the linear problem. Namely, by Theorem 4.1, we regard a as

$$(it)^{-1/2} \langle i\eta \rangle^{-3/2} e^{it\sqrt{1-\eta^2}} \widehat{u}_+ (\eta / \langle i\eta \rangle).$$

If we do so, we can exchange $\langle i\partial_x\rangle a$ and $\partial_x a$ by $\langle i\eta\rangle^{-1} a$ and $i\langle i\eta\rangle^{-1}\eta a$ respectively. Thus we get, if we neglect remainder terms,

$$3\gamma_1 |a|^2 a = 3\gamma_1 \langle i\eta\rangle^{-3} t^{-1} |\widehat{u}_+|^2 a,$$
$$\gamma_2 \left(2 |\langle i\partial_x\rangle a|^2 a - (\langle i\partial_x\rangle a)^2 \overline{a}\right) = \gamma_2 \langle i\eta\rangle^{-5} t^{-1} |\widehat{u}_+|^2 a,$$
$$\gamma_3 \left(2|a_x|^2 a + a_x^2\overline{a}\right)^2 = \gamma_3 \langle i\eta\rangle^{-5} \eta^2 t^{-1} |\widehat{u}_+|^2 a,$$
$$\gamma_4 i\left((\langle i\partial_x\rangle a)\overline{a}_x a - (\langle i\partial_x\rangle \overline{a}) a_x a + (\langle i\partial_x\rangle a) a_x \overline{a}\right) = -\gamma_4 \langle i\eta\rangle^{-5} \eta t^{-1} |\widehat{u}_+|^2 a,$$
$$\begin{aligned} &- i\Big(2\gamma_5 |a|^2 \langle i\partial_x\rangle a - \gamma_5 a^2 \langle i\partial_x\rangle \overline{a} \\ &\quad + 3\gamma_6 |\langle i\partial_x\rangle a|^2 \langle i\partial_x\rangle a + 2\gamma_7 |a_x|^2 \langle i\partial_x\rangle a - \gamma_7 \overline{a}_x^2 \langle i\partial_x\rangle \overline{a}\Big) \\ &= -i\left(\gamma_5 \langle i\eta\rangle^{-1} + 3\gamma_6 \langle i\eta\rangle^{-3} + 3\gamma_7 \langle i\eta\rangle^{-3}\eta^2\right) \langle i\eta\rangle^{-3} t^{-1} |\widehat{u}_+|^2 a \end{aligned}$$

and

$$\begin{aligned} &2\gamma_8 |a|^2 a_x + \gamma_8 a^2 \overline{a}_x + 2\gamma_9 |\langle i\partial_x\rangle a|^2 a_x - \gamma_9 (\langle i\partial_x\rangle a)^2 \overline{a}_x + 3\gamma_{10} |a_x|^2 a_x \\ = \;& i\left(\gamma_8 \langle i\eta\rangle^{-1}\eta + 3\gamma_9 \langle i\eta\rangle^{-3}\eta + 3\gamma_{10}\eta^3\right) \langle i\eta\rangle^{-3} t^{-1} |\widehat{u}_+|^2 a. \end{aligned}$$

The right-hand sides of the above identities are considered as the leading terms. Therefore we find that the leading term is

$$-\Lambda(\eta/\langle i\eta\rangle) t^{-1} |\widehat{u}_+|^2 a,$$

where

$$\begin{aligned} -\Lambda(\xi) = &\left(3\gamma_1 \langle\xi\rangle^3 + \gamma_2 \langle\xi\rangle^5 + \gamma_3\xi^2 \langle\xi\rangle^3 - \gamma_4\xi \langle\xi\rangle^4\right) \\ &+ i\left(-\gamma_5 \langle\xi\rangle^4 - 3\gamma_6 \langle\xi\rangle^6 - 3\gamma_7\xi^2 \langle\xi\rangle^4 + \gamma_8\xi \langle\xi\rangle^3 + 3\gamma_9\xi \langle\xi\rangle^5 + 3\gamma_{10}\xi^3 \langle\xi\rangle^3\right). \end{aligned}$$

From the right-hand side of this identity, we can find the asymptotic behavior of solutions if $-\mathrm{Im}\Lambda(\xi) \geq 0$. In the case of $\mathrm{Im}\Lambda(\xi) = 0$, if we modify the final value as

$$\widehat{w}_+(t,\xi) = \widehat{u}_+(\xi) e^{2^{-1} i\Lambda(\xi)\langle\xi\rangle^{-1} |\widehat{u}_+(\xi)|^2 \log t},$$

the approximate solution is constructed from this. The condition $\mathrm{Im}\Lambda(\xi) = 0$ is called the null condition and was introduced in the paper [**8**]. In the case of

$$\begin{aligned} &-\mathrm{Im}\Lambda(\xi) \\ = \;&\left(-(\gamma_5 + 3\gamma_6) - 3(\gamma_6 + \gamma_7)\xi^2\right) \langle\xi\rangle^4 + \left((\gamma_8 + 3\gamma_9)\xi + 3(\gamma_9 + \gamma_{10})\xi^3\right) \langle\xi\rangle^3 > 0, \end{aligned}$$

we take $\widehat{w}_+(t,\xi)$ as follows:

$$\begin{aligned} \widehat{w}_+(t,\xi) &= \widehat{u}_+(\xi) \exp\left(i\Lambda(\xi)\varphi(t,\xi)/(2\mathrm{Im}\Lambda(\xi))\right), \\ e^{-\varphi} &= (1 - \mathrm{Im}\Lambda(\xi) |\widehat{u}_+(\xi)|^2 \log t)^{-1}. \end{aligned}$$

Then we obtain the asymptotics of solutions through $\widehat{w}_+(t,\xi)$. As stated in Section 5 and Section 6, if $\mathrm{Im}\Lambda(\xi) > 0$, then the problem is still open.

8. Nonlinear Schrödinger equations of resonant type

We reconsider the asymptotic behavior of solutions to nonlinear Schrödinger type equations considered in Section 5. We state that modified wave operators are constructed in a wider class of data than the ones in Section 5. As we use lower-order Sobolev spaces, we can consider the problem in higher space dimensions. However we must restrict our attention to the usual nonlinear Schrödinger equations, namely $\rho = 2$.

Let us consider the existence of modified wave operators for the nonlinear Schrödinger equation with critical nonlinearities

(31) $$2i\partial_t u + \Delta u = 2g_n(u), \quad (t,x) \in \mathbf{R} \times \mathbf{R}^n,$$

where $g_n(u) = \lambda |u|^{2/n} u, \lambda \in \mathbf{R}, 1 \leq n \leq 3$ and $\Delta = \sum_{j=1}^n \partial_{x_j}^2$. We assume that the final state u_+ is in the neighborhood of the origin in the Sobolev space $H^{0,\gamma}$, where γ satisfies $n/2 < \gamma < \min\{2, 1+2/n\}$. Our purpose is to find a solution of (31) satisfying the asymptotic formula

(32) $$\lim_{t\to\infty} \left\| u(t) - (1/(it)^{n/2}) e^{i|x|^2/(2t)} \widehat{u_+}((\cdot)/t) \exp\left(-i\lambda \left|\widehat{u_+}((\cdot)/t)\right|^{2/n} \log t\right) \right\| = 0.$$

We now state our main result in this section.

THEOREM 8.1 ([**28**]). *Let $n/2 < \gamma \leq \min\{2, 1+2/n\}$, the final state u_+ be in $H^{0,\gamma}$ and $\|\widehat{u_+}\|_\infty \leq \varepsilon$ for small $\varepsilon > 0$. Then the integral equation associated with (31) has a unique solution $u \in C([0,\infty); L^2)$ satisfying the asymptotic formula*

$$\left\| u(t) - (1/(it)^{n/2}) e^{i|x|^2/(2t)} \widehat{u_+}((\cdot)/t) \exp\left(-i\lambda \left|\widehat{u_+}((\cdot)/t)\right|^{2/n} \log t\right) \right\| \leq Ct^{-b/2},$$

where $1/2 < b < \gamma$.

From the theorem, we have

COROLLARY 8.2. *Let $n/2 < \gamma$. Then the modified wave operator MW_+ is defined as a map from the neighborhood of the origin in $H^{0,\gamma}$ to L^2.*

In the one space dimensional case, we have the following theorem. If we modify the approximate solution a little bit, similar results are obtained in two or three space dimensions. We refer to [**28**] for it.

THEOREM 8.3 ([**28**]). *Let $1/2 < \gamma \leq 2, n = 1$, the final state u_+ be in $H^{0,\gamma}$ and satisfy $\|\widehat{u_+}\|_\infty \leq \varepsilon$ for small $\varepsilon > 0$. Then the integral equation associated with (31) has a unique solution u such that*

$$u \in C([0,\infty); L^2), |J|^\beta u \in C([0,\infty); L^2), 1/2 < \beta < \gamma \leq 2$$

and

$$\left\| U(-t)\left(u(t) - (1/(it)^{1/2}) e^{(i|x|^2)/(2t)} \widehat{u_+}((\cdot)/t) \exp\left(-i\lambda \left|\widehat{u_+}((\cdot)/t)\right|^2 \log t\right)\right) \right\|_{0,\delta}$$
$$\leq Ct^{-b/2+\delta/2}, 0 \leq \delta < b < \beta, 1/2 < b < 2, t < -1$$

hold, where $|J|^\beta = U(t)|x|^\beta U(-t)$.

From the theorem, we have

COROLLARY 8.4. *Let $1/2 < \gamma < \beta$. Then the modified wave operator MW_+ is defined as a map from the neighborhood of the origin in $H^{0,\gamma}$ to $H^{0,\beta}$.*

We also have the following result obtained by investigating the asymptotic behavior of solutions to the initial value problem.

THEOREM 8.5 ([**28**]). *Let* $1/2 < \beta \leq 2$, *the initial function* $u(0)$ *be in* $H^{0,\beta}$ *and satisfy* $\|u(0)\|_{0,\beta} \leq \varepsilon$ *for small* $\varepsilon > 0$. *Then there exists a unique function* $u_- \in H^{0,\delta}$ *satisfying the asymptotic formula*

$$\left\|(\mathcal{F}U(-t)\,u)\exp\left(i\lambda\int_{-1}^{t}|\mathcal{F}U(-\tau)\,u|^2\,\tau^{-1}d\tau\right)-\widehat{u_-}\right\|_{\delta,0}\leq C\,\|u(0)\|_{0,\beta}^3\,|t|^{-b/2+\delta/2},$$

where $0 \leq \delta < b < \beta, 1/2 < b < 2, t < 0$, $u(t)$ *is a unique solution of equation* (31) *determined by* $u(0)$ *and satisfies*

$$u\in C((-\infty,0]\,;L^2),|J|^{\beta}u\in C((-\infty,0]\,;L^2).$$

Moreover, if $1/2 < \eta < \delta$, *then the asymptotic formula*

$$\begin{aligned}&\left\|U(-t)\left(u(t)-(1/(it)^{1/2})\widehat{u_-}((\cdot)/t)\exp\left(-i\lambda\,|\widehat{u_-}((\cdot)/t)|^2\log|t|\right)\right)\right\|_{0,\eta}\\ \leq\ & C\,\|u(0)\|_{0,\beta}^3\,|t|^{-(\beta-\delta)/2+C\varepsilon}+C\left(\|u_-\|_{0,\delta}+\|u_-\|_{0,\delta}^2\right)|t|^{-(\delta-\eta)/2+C\varepsilon},t<0\end{aligned}$$

holds.

Theorem 8.5 shows existence of the inverse modified wave operator such that

$$MW_-^{-1}:u(0)\in H^{0,\beta}\to u_-\in H^{0,\delta},1/2<\delta<\beta.$$

We combine the above and Theorem 8.3 into the following corollary.

COROLLARY 8.6. *Let* $1/2 < \delta < \gamma$. *Then the modified scattering operator* $MS_+ = (MW_-)^{-1}MW_+$ *is defined as a map from the neighborhood of the origin in* $H^{0,\gamma}$ *to* $H^{0,\delta}$.

We survey the previous works [**7**], [**11**], [**53**]. In Theorem 2 of [**53**], the one space dimensional case is studied and the following result is proved: Let the final state u_+ be in $H^{0,3} \cap H^2$ and $\|\widehat{u_+}\|_\infty$ be sufficiently small. Then equation (31) has a unique solution $u \in C([0,\infty); H^1)$ satisfying

$$\left\|u(t)-(1/(it)^{1/2})e^{i|x|^2/(2t)}\widehat{u_+}((\cdot)/t)\exp\left(-i\lambda\,|\widehat{u_+}((\cdot)/t)|^2\log t\right)\right\|_{1,0}\leq Ct^{-b/2},$$

where $1 < b < 2$. In [**11**], the result is extended to two or three dimensional cases. In paper [**7**], results in [**53**] are improved as follows : Let the final state u_+ satisfy the same conditions as those in [**53**]. Then equation (31) has a unique solution $u \in C([0,\infty); H^{0,1} \cap H^1)$ satisfying

$$\left\|u(t)-(1/(it)^{1/2})e^{i|x|^2/(2t)}\widehat{u_+}((\cdot)/t)\exp\left(-i\lambda\,|\widehat{u_+}((\cdot)/t)|^2\log t\right)\right\|_{1,0}\leq Ct^{-1}(\log t)^3$$

and

$$\begin{aligned}&\left\|U(-t)\left(u(t)-(1/(it)^{1/2})e^{i|x|^2/(2t)}\widehat{u_+}((\cdot)/t)\exp\left(-i\lambda\,|\widehat{u_+}((\cdot)/t)|^2\log t\right)\right)\right\|_{0,1}\\ \leq\ & Ct^{-1}(\log t)^3.\end{aligned}$$

From the last estimate of the above and results on the initial value problem obtained in [**18**] we have the modified scattering operator such that $MS_+ : H^{0,3} \cap H^{1,2} \to L^2$. For a proof of the result, we refer to Corollary 2 in [**7**].

We note here that our results in [**28**] yield constructions of the modified scattering in higher space dimensions and the inverse modified scattering operator. Thus our results are an improvement of [**7**], [**11**], [**53**].

Their method requires smoothness on the Fourier transform of the final state since they use

$$(1/(it)^{1/2})e^{i|x|^2/(2t)}\widehat{u_+}\left((\cdot)/t\right)\exp\left(-i\lambda\left|\widehat{u_+}\left((\cdot)/t\right)\right|^2\log t\right)$$

as the approximate solution. They substitute this one into the linear Schrödinger equation. After that they find a unique solution of the nonlinear problem around the approximate solution. Therefore we differentiate $\widehat{u_+}\left(\cdot/t\right)$ twice, which requires the condition $u_+ \in H^{0,2}$. In [**28**], we use

$$\mathcal{F}^{-1}U\left(t\right)\widehat{u_+}\exp\left(-i\lambda\left|\widehat{u_+}\right|^{2/n}\log t\right)$$

as the approximate solution. The difference between these two approximate solutions is

$$e^{i|x|^2/(2t)}D\mathcal{F}\left(e^{i|x|^2/(2t)}-1\right)\mathcal{F}^{-1}\widehat{u_+}\exp\left(-i\lambda\left|\widehat{u_+}\right|^{2/n}\log t\right),$$

where $D\phi(x) = (2it)^{-1/2}\phi(x/t)$. As stated in this section we have the modified scattering operator for nonlinear Schrödinger equations. However the problem is open for nonlinear Klein-Gordon equations with critical nonlinearities and the modified Korteweg-de Vries equation.

9. Other problems

To study the asymptotic behavior of solutions to initial value problems is important to prove the existence of inverse wave operators or inverse modified ones. For the initial value problem to the derivative nonlinear Schrödinger equation

$$iu_t + u_{xx} + i(|u|^2u)_x = 0, \quad (t,x) \in \mathbf{R}\times\mathbf{R}, \tag{33}$$

existence of the inverse wave operator is shown in [**17**]. This result is extended to the cases with more general nonlinearities in [**22**], [**36**]. In these papers, smoothing effects of solutions to the linear Schrödinger equation are used to overcome the derivative loss in nonlinearities. We do not know if smoothing effects can be applicable to final value problems or not. In [**33**], the nonlinear Schrödinger equation with nonlinearities consisting of a linear combination of $u^3, \bar{u}^3, \bar{u}^2u$ is considered and the asymptotic behavior of solutions is obtained under some suitable conditions on the initial function. The initial value problem for the nonlinear Schrödinger equation with a special quadratic nonlinearity such that

$$iu_t + u_{xx} = \left(\overline{u}_x\right)^2, \quad (t,x) \in \mathbf{R}\times\mathbf{R} \tag{34}$$

is treated in [**26**] and the global existence of solutions and the existence of the inverse modified wave operator are studied. For quadratic nonlinearities, it is expected that the linear part of the equation decays in time faster than the nonlinear terms. This is the difficulty of the above problem. We note that the existence of wave operators for equation (34) is an open problem. The initial value problems for the generalized Korteweg-de Vries equation with super-critical nonlinearities,

$$3u_t + u_{xxx} + (|u|^{\rho-1}u)_x = 0, \quad (t,x) \in \mathbf{R}\times\mathbf{R}, \tag{35}$$

and the modified Korteweg-de Vries equation

$$3u_t + u_{xxx} + (u^3)_x = 0, \quad (t,x) \in \mathbf{R} \times \mathbf{R} \tag{36}$$

are considered in [**19**],[**21**],[**23**] and existence results of inverse wave operators and inverse modified wave operators are shown. If we replace the linear part of the above equations by $u_t + (H\partial_x^2/2)u = 0$, then we call them the generalized Benjamin-Ono type equation and the modified Benjamin-Ono type equation, respectively. For the initial value problems for these equations, we refer to papers [**20**], [**25**].

We have some existence results of modified wave operators from Corollary 5.2 to Corollary 8.4. On the other hand, for the modified scattering operator we have Corollary 8.6 in the case of the usual nonlinear Schrödinger equation only. As stated in Section 1, this problem depends on the domain and range of the modified wave operator. There are no results on the existence of modified scattering operators except for the usual nonlinear Schrödinger equation.

For the initial value problem to quadratic nonlinear Klein-Gordon equations in two space dimensions, we refer to [**10**], [**55**]. For the case of nonlinear Schrödinger equations, we refer to [**9**], [**24**]. Nonlinear dissipative equations including nonlinear heat and nonlinear dissipative wave equations are studied extensively. We refer to [**27**],[**43**] and the references cited therein for them.

10. Acknowledgments

We would like to thank the referees for their careful reading of the first draft written in Japanese and for useful comments.

References

[1] M.J.Ablowitz and H. Segur, *Solitons and the Inverse Scattering Transform*, SIAM, Philadelphia, 1981. MR642018 (84a:35251)

[2] J.E.Barab, *Nonexistence of asymptotically free solutions for a nonlinear Schrödinger equation,* J. Math. Phys., **25** (1984), pp. 3270-3273. MR761850 (86a:35121)

[3] M.Ben-Artzi, H.Koch and J-C. Saut, *Dispersion estimates for fourth order Schrödinger equations,* C.R. Acad. Sci. Paris Sér. I Math. **330** (2000), pp.87-92. MR1745182 (2001a:35149)

[4] T.B.Benjamin, *Internal waves of permanent form in fluids of great depth,* J. Fluid Mech., **29** (1976), pp. 559-592.

[5] P.Brenner, *On space time means and everywhere defined scattering operators for nonlinear Klein-Gordon equations*, Math. Z., **186** (1984), pp. 383-391. MR744828 (85h:35183)

[6] P.Brenner, *On scattering and everywhere defined scattering operators for nonlinear Klein-Gordon equations*, J. Differential Equations, **56** (1985), pp. 310-344. MR780495 (86f:35155)

[7] R.Carles, *Geometric optics and long range scattering for one dimensional nonlinear Schrödinger equations*, Commun. Math. Phys. **220** (2001), pp. 41-67. MR1882399 (2002k:35301)

[8] J.-M. Delort, *Existence globale et comportement asymptotique pour l'équation de Klein-Gordon quasi-linéaire à données petites en dimension* 1, Ann. Sci. École Norm. Sup. (4) **34**, (2001), pp.1-61. MR1833089 (2002h:35194)

[9] J.-M. Delort, *Global solutions for small nonlinear long range perturbations of two dimensional Schrödinger equations*, Mém. Soc. Math. Fr. (N.S.) No. 91 (2002), vi+94 pp. MR1942854 (2004g:35212)

[10] J.-M. Delort, D.Fang and R.Xue, *Global existence of small solutions for quadratic quasilinear Klein-Gordon systems in two space dimensions*, J. Funct. Anal., **211** (2004), pp. 288-323. MR2056833 (2005f:35214)

[11] J.Ginibre and T.Ozawa, *Long range scattering for nonlinear Schrödinger and Hartree equations in space dimension* $n \geq 2$, Commun. Math. Phys. **151** (1993), pp. 619-645. MR1207269 (93m:35168)

[12] J.Ginibre and G.Velo, *On a class of nonlinear Schrödinger equations. I. The Cauchy problem, general case; II. Scattering theory, general case*, J. Funct. Anal., **32** (1979), pp.1-71. MR533219 (82c:35058)

[13] J.Ginibre and G.Velo, *Time deacy of finite energy solutions of the nonlinear Klein-Gordon equation*, Ann. Inst. H. Poincaré Phys. Théor., **41** (1985), pp. 399-442. MR824083 (87g:35208)

[14] V.Georgiev and B.Yordanov, *Asymptotic behaviour of the one dimensional Klein-Gordon equation with a cubic nonlinearity*, preprint, 1996.

[15] N.Hayashi, *The initial value problem for the derivative nonlinear Schrödinger equation in the energy space,* J. Nonlinear Anal. T.M.A., **20** (1993), pp. 823-833. MR1214746 (94c:35007)

[16] N.Hayashi, T.Matos and P.I.Naumkin, *Large time behavior of solutions of higher-order nonlinear dispersive equations of KdV - type with weak nonlinearity*, Differential and Integral Equations. **12**, (1999), pp. 23-40. MR1668529 (99m:35206)

[17] N.Hayashi and P.I.Naumkin, *Asymptotic behavior in time of solutions to the derivative nonlinear Schrödinger equation revisited,* Discrete and Continuous Dynamical Systems, **3** (1997), pp. 383-400. MR1444201 (98e:35155)

[18] N.Hayashi and P.I.Naumkin, *Asymptotics in large time of solutions to nonlinear Schrödinger and Hartree equations,* Amer. J. Math., **120** (1998), pp.369-389. MR1613646 (99f:35190)

[19] N.Hayashi and P.I.Naumkin, *Large time asymptotics of solutions to the generalized Korteweg-de Vries equation,* J. Funct. Anal., **159** (1998), pp. 110-136. MR1654178 (99i:35139)

[20] N.Hayashi and P.I.Naumkin, *Large time asymptotics of solutions to the generalized Benjamin-Ono equation,* Trans. Amer. Math. Soc., **351** (1999), pp. 109-130. MR1491867 (99i:35148)

[21] N.Hayashi and P.I.Naumkin, *Large time behavior of solutions for the modified Korteweg-de Vries equation,* International Mathematics Research Notices, (1999), pp. 395-418. MR1687327 (2000d:35201)

[22] N.Hayashi and P.I.Naumkin, *Large time behavior of solutions for derivative cubic nonlinear Schrödinger equations without a self-conjugate property*, Funkcialaj Ekvacioj **42** (1999), pp. 311-324. MR1718755 (2000h:35149)

[23] N.Hayashi and P.I.Naumkin, *On the modified Korteweg - de Vries equation,* Mathematical Physics, Analysis and Geometry, **4** (2001), pp. 197-227. MR1871426 (2002m:35191)

[24] N.Hayashi and P.I.Naumkin, *Global existence of small solutions to the quadratic nonlinear Schrödinger equations in two space dimensions* SIAM J. Math. Anal., **32** (2001), pp. 1390-1403. MR1856255 (2002g:35194)

[25] N.Hayashi and P.I.Naumkin, *On the reduction of the modified Benjamin - Ono equation to the cubic derivative nonlinear Schrödinger equation,* Discrete Continuous Dynamical Systems, **8** (2002), pp.237-255. MR1877838 (2002k:35292)

[26] N.Hayashi and P.I.Naumkin, *A quadratic nonlinear Schrödinger equation in one space dimension,* J. Differential Equations, **186** (2002), pp. 165-185. MR1941097 (2003j:35298)

[27] N. Hayashi, E.I.Kaikina, P. I. Naumkin and I. A. Shishmarev, *Asymptotics for Dissipative Nonlinear Equations,* Lecture Note in Mathematics, Springer, Berlin, Heidelberg, New York **1884** (2006), 557 pp., ISBN 3-540-32059-8 MR2230524 (2009d:35001)

[28] N.Hayashi and P.I.Naumkin, *Domain and range of the modified wave operator for Schrödinger equations with a critical nonlinearity,* Commun. Math. Phys., **267** (2006), pp. 477-492. MR2249776 (2007h:81226)

[29] N.Hayashi and P.I.Naumkin, *Asymptotic properties of solutions to dispersive equation of Schrödinger type*, J. Math. Soc. Japan. **60** (2008), pp. 631–652. MR2440407 (2009j:35351)

[30] N.Hayashi and P.I.Naumkin, *Final state problem for Korteweg-de Vries type equations*, J. Math. Phys., **47** (2006), 123501, 16 pp. MR2285150 (2007h:35281)

[31] N.Hayashi and P.I.Naumkin, *Final state problem for the cubic nonlinear Klein-Gordon equation,* J. Math. Phys. **50** (2009), no. 103511, 14 pp. MR2572684

[32] N.Hayashi and P.I.Naumkin *The initial value problem for the cubic nonlinear Klein-Gordon equation,* Z. Angew. Math. Phys. **59** (208), 1002–1028. MR2457221 (2010b:35298)

[33] N.Hayashi and P.I.Naumkin, *Nongauge invariant cubic nonlinear Schrödinger equations*, Pacific Journal of Applied Mathematics (PJAM), **1** (2007), pp. 27-43. MR2467127 (2009k:35297)

[34] N.Hayashi and P.I.Naumkin, *Modified wave operators for nonlinear Schrödinger equations with subcritical dissipative nonlinearities*, Inverse Problems and Imaging **1** (2007), No. 2, pp. 391-398. MR2282275 (2008a:35225)

[35] N.Hayashi, P.I.Naumkin, A. Shimomura and S.Tonegawa,*Modified wave operators for nonlinear Schrödinger equations in one and two space dimensions* Electron. J. Differential Equations **2004** (2004), No. 62, pp. 1-16. MR2047418 (2004m:35247)

[36] N.Hayashi, P.I.Naumkin and H.Uchida, *Large time behavior of solutions for derivative cubic nonlinear Schrödinger equations,* Publ. Res. Inst. Math. Sci. **35** (1999), pp. 501-513. MR1710752 (2000j:35246)

[37] N.Hayashi and T.Ozawa, *Modified wave operators for the derivative nonlinear Schrödinger equations* , Math. Annalen, **298** (1994), pp. 557-576. MR1262776 (95f:35240)

[38] N.Hayashi and Y.Tsutsumi, *Remarks on the scattering problem for nonlinear Schrödinger equations,* Differential Equations and Mathematical Physics, Lecture Note in Mathematics, Edited by I.W.Knowles and Y.Saito, Springer-Verlag **1285** (1986), pp. 162-168. MR921265 (89d:35155)

[39] S.Katayama, *A note on global existence of solutions to nonlinear Klein-Gordon equations in one space dimension,* J. Math. Kyoto Univ., **39** (1999), pp. 203-213. MR1709289 (2000e:35193)

[40] T.Kawahara, *Oscillatory solitary waves in dispersive media,* J. Phys. Soc. Japan, **33** (1972), pp. 260-264.

[41] D.J.Korteweg and F.de Vries, *On the change of form of long waves advancing in a rectangular canal, and on a new type of long stationary waves,* Philosophical Magazine, **39** (1895), pp. 422-443.

[42] H.Lindblad and A.Soffer, *A remark on long range scattering for the nonlinear Klein-Gordon equation,* J. Hyperbolic Differential Equations., **2** (2005), no. 1, pp. 77–89. MR2134954 (2006d:35143)

[43] A.Matsumura and K.Nishihara, *Global solutions for nonlinear differential equations (Mathematical analysis for compressive vicious fluid),* Nihonhyouron-Sha (Japanese), ISBN 4-535-78386-1(2004).

[44] B.Marshall, W.Strauss and S.Wainger, *$L^p - L^q$ estimates for the Klein-Gordon equation,* J. Math. Pures Appl., **59**, (1980), pp. 417-440. MR607048 (82j:35133)

[45] W.Mio, T.Ogino, K.Minami and S.Takeda, *Modified nonlinear Schrödinger equation for Alfven waves propagating along the magnetic field in cold plasmas,* J. Phys. Soc. Japan, **41** (1976), pp. 265-271. MR0462141 (57:2116)

[46] E.Mjølhus, *On the modulational instability of hydromagnetic waves parallel to the magnetic field,* J. Plasma Phys., **16** (1976), pp. 321-334.

[47] K.Moriyama, *Normal forms and global existence of solutions to a class of cubic nonlinear Klein-Gordon equations in one space dimension,* Differential and Integral Equations, **10** (1997), pp. 499-520. MR1744859 (2001a:35125)

[48] K.Moriyama, S.Tonegawa and Y.Tsutsumi, *Almost global existence of solutions for the quadratic semilinear Klein-Gordon equation in one space dimension,* Funkcialaj Ekvacioj, **40** (1997), pp. 313-333. MR1480281 (98k:35130)

[49] K.Nakanishi, *Energy scattering for nonlinear Klein-Gordon and Schrödinger equations in spatial dimensions* 1 *and* 2, J. Funct. Anal. **169** (1999), pp. 201-225. MR1726753 (2000m:35141)

[50] K.Nishihara, *Global asymptotics for the damped wave equation with absorption on higher dimensional space,* J. Math. Soc. Japan **58** (2006), no. 3, pp. 805-836. MR2254412 (2007d:35194)

[51] K.Nishihara and H.Zhao, *Decay properties of solutions to the Cauchy problem for the damped wave equations with absorption,* J. Math. Anal. Appl. **313** (2006), no. 2, pp. 598-610. MR2183322 (2006i:35252)

[52] H.Ono, *Algebraic solitary waves in stratified fluids* J. Phys. Soc. Japan, **39** (1975), pp. 1082-1091. MR0398275 (53:2129)

[53] T.Ozawa, *Long range scattering for nonlinear Schrödinger equations in one space dimension,* Commun. Math. Phys., **139** (1991), pp. 479-493. MR1121130 (92j:35172)

[54] T.Ozawa, *Scattering theory for nonlinear Schrödinger equations,* Suugaku (Japanese), **50** (1998), pp. 337-349. MR1690627 (2000b:35192)

[55] T.Ozawa, K.Tsutaya and Y.Tsutsumi, *Global existence and asymptotic behavior of solutions for the Klein-Gordon equations with quadratic nonlinearity in two space dimensions,* Math. Z., **222** (1996), pp. 341-362. MR1400196 (97e:35112)

[56] H.Pecher, *Nonlinear small data scattering for the wave and Klein-Gordon equation*, Math. Z., **185** (1984), pp. 261-270. MR731347 (85h:35165)

[57] H.Pecher, *Low energy scattering for Klein-Gordon equations*, J. Funct. Anal, **63** (1985), pp. 101-122. MR795519 (86j:35143)

[58] J.Segata, *Modified wave operator for the fourth order nonlinear Schrödinger type equation with subcritical nonlinearity*, Math. Methods Appl. Sci., **29** (2006), pp. 1785-1800. MR2253364 (2007e:35265)

[59] J.Segata and A.Shimomura, *Modified wave operator for the fourth order nonlinear Schrödinger type equation with subcritical nonlinearity*, preprint, 2005

[60] J.Shatah, *Normal forms and quadratic nonlinear Klein-Gordon equations,* Commun. Pure Appl. Math., **38** (1985), pp. 685-696. MR803256 (87b:35160)

[61] A.Shimomura, *Asymptotic behavior of solutions for Schrödinger equations with dissipative nonlinearities,* Commun. P.D.E, **31** (2006), pp. 1407-1423. MR2254620 (2007m:35257)

[62] A.Shimomura and S.Tonegawa, *Long-range scattering for nonlinear Schrödinger equations in one and two space dimensions*, Differential Integral Equations **17** (2004), pp. 127-150. MR2035499 (2005a:35258)

[63] W.A.Strauss, *Nonlinear scattering theory at low energy,* J. Funct. Anal., **41** (1981), pp. 110–133. MR614228 (83b:47074a)

[64] H.Sunagawa, *Remarks on the asymptotic behavior of the cubic nonlinear Klein-Gordon equations in one space dimension*, Differential and Integral Equations, **18** (2005), pp. 481-494. MR2136975 (2006d:35169)

[65] H.Sunagawa, *Large time behavior of solutions to the Klein-Gordon equation with nonlinear dissipative terms*, J. Math. Soc. Japan, **58**(2006), pp.379-400. MR2228565 (2007g:35155)

[66] M.Toda, *Nonlinear waves and soliton*, Nihonhyouron-sha, (Japanese), ISBN 4-535-78316-0(2000).

[67] Y.Tsutsumi, *Scattering problem for nonlinear Schrödinger equations*, Ann. Inst. Henri Poincaré (Phys. Théor.) **43** (1985), pp. 321-347. MR824843 (87g:35221)

Department of Mathematics, Graduate School of Science, Osaka University, Osaka, Toyonaka, 560-0043, Japan

E-mail address: nhayashi@math.sci.osaka-u.ac.jp

Translated by NAKAO HAYASHI

Amer. Math. Soc. Transl.
(2) Vol. **230**, 2010

Heun's Differential Equation

Kouichi Takemura

1. Introduction

Heun's differential equation is an ordinary linear differential equation of second order which is written as

$$\frac{d^2y}{dz^2} + \left(\frac{\gamma}{z} + \frac{\delta}{z-1} + \frac{\epsilon}{z-t}\right)\frac{dy}{dz} + \frac{\alpha\beta z - q}{z(z-1)(z-t)}y = 0, \tag{1.1}$$

with the relation

$$\gamma + \delta + \epsilon = \alpha + \beta + 1. \tag{1.2}$$

Here the parameters $\alpha, \beta, \gamma, \delta, \epsilon, q, t$ are constants. Heun's equation and the differential equations of its confluent type appear in several systems of physics including analysis of black holes, crystal transition ([**17**]), and fluid dynamics ([**1**]). From the viewpoint of mathematics, it is related with the Painlevé equations ([**17**]), Teichmüller spaces ([**8**]), and so on.

As is mentioned later, Heun's differential equation is Fuchsian, i.e. every singularity $\{0, 1, t, \infty\}$ of the equation on the Riemann sphere $\mathbb{C} \cup \{\infty\}$ is regular, and it is a canonical form of Fuchsian differential equations of second order with four singularities on the Riemann sphere. The canonical form of Fuchsian differential equations of second order with three singularities is the Gauss hypergeometric differential equation, which is given as

$$z(1-z)\frac{d^2y}{dz^2} + (\gamma - (\alpha + \beta + 1)z)\frac{dy}{dz} - \alpha\beta y = 0. \tag{1.3}$$

By dividing by $z(1-z)$, we have

$$\frac{d^2y}{dz^2} + \left(\frac{\gamma}{z} + \frac{\alpha + \beta - \gamma + 1}{z-1}\right)\frac{dy}{dz} + \frac{\alpha\beta}{z(z-1)}y = 0, \tag{1.4}$$

which resembles (1.1).

The Gauss hypergeometric differential equation has been studied very well for a long time, and global structures such as integral representations of solutions and explicit formulas on connection matrices which link local solutions around regular singularities were clarified. On the other hand, global structures for Heun's

---

This article originally appeared in Japanese in Sūgaku **60** (3) (2008), 272–294.
2000 *Mathematics Subject Classification.* Primary 33E10, 34M35; Secondary 34B30, 14H70.
The author was supported in part by JSPS Grant #19740089.

differential equation are much more difficult, although the appearance of Heun's differential equation is obtained just by adding a regular singularity to the Gauss hypergeometric differential equation. One of the reasons for the difficulty is that there is an accessory parameter q in Heun's differential equation (1.1). The accessory parameter q does not depend on local monodromy, and it may raise Heun's differential equation to be tough and fantastic.

In this paper, we will review fundamental facts on Heun's differential equation and the Gauss hypergeometric differential equation, and comment on what should be considered for further study. We will also explain recent results on Heun's differential equation by finite-gap integration.

2. Regular singularity and monodromy

Let us consider the following differential equation on the Riemann sphere $\mathbb{C} \cup \{\infty\}$:

$$\frac{d^2y}{dz^2} + p_1(z)\frac{dy}{dz} + p_2(z)y = 0, \tag{2.1}$$

where $p_1(z)$ and $p_2(z)$ are rational functions. We regard the neighborhood of the point at infinity $z = \infty$ as the neighborhood of the point $w = 0$ on the differential equation

$$\frac{d^2y}{dw^2} + \frac{2w - p_1(1/w)}{w^2}\frac{dy}{dw} + \frac{p_2(1/w)}{w^4}y = 0, \tag{2.2}$$

which is obtained by the change of variables $w = 1/z$.

Equation (2.1) is said to be singular about the point $z = a$ ($a \neq \infty$) if the function $p_1(z)$ or $p_2(z)$ is not holomorphic about $z = a$, and (2.1) is said to be singular about $z = \infty$ if the point $w = 0$ is singular on (2.2). If (2.1) is nonsingular (i.e., holomorphic) about $z = a$, then there exist solutions to the differential equation about $z = a$ whose dimension is two, and all solutions are holomorphic about $z = a$. Hence the holomorphy of solutions to a differential equation follows from the holomorphy of coefficients of the differential equation.

Here we recall a definition of regular singularity.

DEFINITION 2.1. The point $z = a$ is a regular singularity of the differential equation (2.1) if the point $z = a$ is singular, the degree of the pole of the function $p_1(z)$ about $z = a$ is no more than one and the degree of the pole of the function $p_2(z)$ about $z = a$ is no more than two, that is,

$$p_1(z) = \frac{(\text{holomorphic about } z = a)}{(z-a)}, \quad p_2(z) = \frac{(\text{holomorphic about } z = a)}{(z-a)^2}. \tag{2.3}$$

The point $z = \infty$ is a regular singularity of the differential equation (2.1) if the point $w = 0$ is a regular singularity of (2.2) which is obtained by the change of variables $w = 1/z$.

DEFINITION 2.2. A differential equation (2.1) is Fuchsian (on the Riemann sphere) if all singularities (on the Riemann sphere) of (2.1) are regular.

Now we observe how the definition of regular singularity reflects to solutions to the differential equation. Assume that (2.1) has a regular singularity about $z = a$

and write

$$p_1(z) = \frac{r_1}{z-a} + (\text{holomorphic about } z = a), \tag{2.4}$$
$$p_2(z) = \frac{r_2}{(z-a)^2} + \frac{(\text{holomorphic about } z = a)}{(z-a)}.$$

We call the quadratic $s^2 + (r_1 - 1)s + r_2 = 0$ determined by r_1 and r_2 in (2.4) the characteristic equation about $z = a$, and we call the roots α_1, α_2 of the quadratic the (characteristic) exponents. The characteristic equation has its origin from the equality

$$\left\{\frac{d^2}{dz^2} + \frac{r_1}{z-a}\frac{d}{dz} + \frac{r_2}{(z-a)^2}\right\}(z-a)^s = (s(s-1) + r_1 s + r_2)(z-a)^{s-2}. \tag{2.5}$$

We assume for simplicity that the exponents satisfy $\alpha_1 - \alpha_2 \notin \mathbb{Z}$. Then we can choose a basis of solutions to (2.1) as convergent power series written as

$$f_1(z) = (z-a)^{\alpha_1}(1 + c_1(z-a) + c_2(z-a)^2 + \dots), \tag{2.6}$$
$$f_2(z) = (z-a)^{\alpha_2}(1 + c_1'(z-a) + c_2'(z-a)^2 + \dots);$$

namely, we can choose a basis of solutions to (2.1) as the functions whose local behaviors about $z = a$ are approximated by $(z-a)^{\alpha_1}$ and $(z-a)^{\alpha_2}$. The coefficients $c_1, c_2, \dots, c_1', c_2', \dots$ are determined recursively by substituting (2.6) into (2.1).

We remark that there is another definition of the regular singularity such that every local solution of the differential equation about the singularity $z = a$ is bounded by the power function $C(z-a)^\rho$ (C, ρ: constants). By adapting this definition, the coefficients $p_1(z)$, $p_2(z)$ of the differential equation satisfy (2.3) (Fuchs's theorem).

We now adopt the above discussion to Heun's differential equation. Recall that Heun's differential equation is defined by

$$\frac{d^2y}{dz^2} + \left(\frac{\gamma}{z} + \frac{\delta}{z-1} + \frac{\epsilon}{z-t}\right)\frac{dy}{dz} + \frac{\alpha\beta z - q}{z(z-1)(z-t)}y = 0, \tag{2.7}$$

with the relation $\gamma + \delta + \epsilon = \alpha + \beta + 1$. Hence all the singularities on the Riemann sphere minus ∞ (i.e. the complex plane) are $z = 0, 1, t$, and these three points are regular because the degrees of the poles on the coefficients of dy/dz and y are one at most. We rewrite Heun's differential equation by the transformation $z = 1/w$. Then we have

$$\frac{d^2y}{dw^2} + \frac{2 - \gamma - \delta - \epsilon + (*)w + (*')w^2}{w(1-w)(1-wt)}\frac{dy}{dw} + \frac{\alpha\beta - qw}{w^2(1-w)(1-tw)}y = 0. \tag{2.8}$$

Since the degree of the poles about $w = 0$ on the coefficient of dy/dz (resp. y) is one (resp. two), the point $z = \infty$ is also a regular singularity. Hence all the singularities $\{0, 1, t, \infty\}$ of Heun's differential equation are regular; namely, Heun's differential equation is a Fuchsian differential equation with four singularities.

We now calculate the exponents of each singularity of Heun's differential equation. Since the characteristic equation about $z = 0$ is $s^2 + (\gamma - 1)s = 0$, the exponents about $z = 0$ are 0 and $1 - \gamma$. Similarly the exponents about $z = 1$ (resp. $z = t$) are 0 and $1 - \delta$ (resp. 0 and $1 - \epsilon$). It follows from (2.8) that the characteristic equation about $z = \infty$ (i.e. $w = 0$) is $s^2 + (2 - \gamma - \delta - \epsilon - 1)s + \alpha\beta = (s - \alpha)(s - \beta) = 0$, where we used the relation $\alpha + \beta + 1 = \gamma + \delta + \epsilon$. Hence the exponents about $z = \infty$ are α and β.

Thus the parameters $\alpha, \beta, \gamma, \delta, \epsilon$ are described by the exponents. On the other hand, the parameter q is independent of the exponents, and it is called an accessory parameter.

We will explain that Heun's differential equation is a canonical form of Fuchsian differential equation of second order with four singularities on the Riemann sphere. Let us pick up an arbitrary Fuchsian differential equation of second order with four singularities on the Riemann sphere. By a linear fractional transformation of the variable z, we can move three singularities to $0, 1, \infty$. We denote the other singularity by t. By multiplying the function $z^{-n_0}(z-1)^{-n_1}(z-t)^{-n_t}$ by the solution y where n_0 (resp. n_1, n_t) is one of the exponents about $z=0$ (resp. $z=1$, $z=t$), we may set one of the exponents about every point $z=0,1,t$ to be zero. On these transformations, the transformed equation is still Fuchsian, and it is written as

$$\frac{d^2y}{dz^2} + \frac{q_1(z)}{z(z-1)(z-t)}\frac{dy}{dz} + \frac{q_2(z)}{z(z-1)(z-t)}y = 0, \tag{2.9}$$

where $q_1(z)$ and $q_2(z)$ are polynomials. It follows from regularity of the singularity $z=\infty$ that the degree of $q_1(z)$ (resp. $q_2(z)$) is two (one) at most, and the differential equation is of the form of Heun's differential equation. Hence any Fuchsian differential equation of second order with four singularities on the Riemann sphere is reduced to Heun's differential equation, and we obtain that Heun's differential equation is a canonical form of Fuchsian differential equations of second order with four singularities on the Riemann sphere.

We now explain basic facts on monodromy. Let $y_1(z)$, $y_2(z)$ be a basis of solutions to a linear differential equation of second order about a holomorphic point $z=z_0$, C be a cycle (a closed path) with a base point $z=z_0$ which does not step over singularities of the differential equation, and $y_1^C(z)$, $y_2^C(z)$ be the functions which are analytically continued along the cycle C. Since the functions $y_1^C(z)$, $y_2^C(z)$ also satisfy the differential equation, they are expressed as linear combinations of $y_1(z)$ and $y_2(z)$, namely

$$(y_1^C(z)\ y_2^C(z)) = (y_1(z)\ y_2(z)) \begin{pmatrix} a_{11}^C & a_{12}^C \\ a_{21}^C & a_{22}^C \end{pmatrix}. \tag{2.10}$$

We call it the monodromy matrix of the differential equation with respect to the basis $y_1(z)$, $y_2(z)$ along the cycle C.

Let us consider the monodromy around a regular singularity. Let $z=a$ be a regular singularity of (2.1) whose exponents are α_1 and α_2, assume that $\alpha_1-\alpha_2 \notin \mathbb{Z}$ for simplicity, and let C be a cycle which turns the point $z=a$ counterclockwise and whose interior contains only the point $z=a$ as a singularity. We pick up a basis of solutions to (2.1) by the functions $f_1(z), f_2(z)$ in (2.6). Then the monodromy matrix along the cycle C is written as

$$(f_1^C(z)\ f_2^C(z)) = (f_1(z)\ f_2(z)) \begin{pmatrix} e^{2\alpha_1\pi\sqrt{-1}} & 0 \\ 0 & e^{2\alpha_2\pi\sqrt{-1}} \end{pmatrix}. \tag{2.11}$$

Therefore the local monodromy (the monodromy matrix around a regular singularity) is expressed in terms of exponents.

On the other hand, there are problems in studying the global structure of solutions of the differential equation, e.g. monodromy on the cycle, which turns multiple regular singularities, and the connection matrix, which links the respective bases of two regular singularities, which may be concerned with problems of other fields. For example they are related with the existence problem of solutions which are simultaneously holomorphic on several given singularities, or the problem to determine solutions which satisfy boundary conditions in quantum physics, i.e. a problem to find eigenfunctions on an appropriate functional space (e.g. L^2 space) and the eigenvalues. Those global structures are studied in detail for the case of three regular singularities, although they are very tough for the case of four regular singularities. In the next section, we will observe the case of three regular singularities, i.e. the case of a hypergeometric differential equation.

3. Global monodnomy of the hypergeometric differential equation

The Gauss hypergeometric differential equation is given by

$$z(1-z)\frac{d^2y}{dz^2} + (\gamma - (\alpha+\beta+1)z)\frac{dy}{dz} - \alpha\beta y = 0. \tag{3.1}$$

It has three singularities $\{0, 1, \infty\}$ on the Riemann sphere and they are all regular, i.e. the hypergeometric differential equation is Fuchsian, and it is known that the hypergeometric differential equation is a standard form of Fuchsian differential equations of second order with three singularities on the Riemann sphere. The exponents about $z = 0$ (resp. $z = 1$, $z = \infty$) are 0 and $1-\gamma$ (resp. 0 and $\gamma-\alpha-\beta$, α and β), and all parameters of the hypergeometric differential equation are determined by the exponents (the local monodromy); i.e., the hypergeometric differential equation is rigid, which is remarkably different from Heun's differential equation having an accessory parameter.

We now express local expansions of solutions to the hypergeometric differential equation about regular singularities for the case that the parameters α, β, γ are generic. Set

$$F(\alpha, \beta; \gamma; z) = \sum_{n=0}^{\infty} \frac{(\alpha)_n(\beta)_n}{n!(\gamma)_n} z^n, \quad (\lambda)_n = \prod_{i=0}^{n-1}(\lambda+i). \tag{3.2}$$

Then a basis of solutions about $z = 0$ is expressed as

$$\begin{aligned} F_1(z) &= F(\alpha, \beta; \gamma; z), \\ F_2(z) &= z^{1-\gamma}F(\alpha-\gamma+1, \beta-\gamma+1; 2-\gamma; z), \end{aligned} \tag{3.3}$$

and a basis of solutions about $z = 1$ is expressed as

$$\begin{aligned} F_3(z) &= F(\alpha, \beta; \alpha+\beta-\gamma+1; 1-z), \\ F_4(z) &= (1-z)^{\gamma-\alpha-\beta}F(\gamma-\beta, \gamma-\alpha; \gamma-\alpha-\beta+1; 1-z). \end{aligned} \tag{3.4}$$

We omit the case $z = \infty$ here. The connection matrix describes a global structure of solutions by linking the bases of solutions about two different singularities. For the case of the hypergeometric differential equation, the connection matrix which links the solutions about $z = 0$ to the solutions about $z = 1$ is expressed in terms of gamma functions, that is,

(3.5)

$$(F_1(z)\ F_2(z)) = (F_3(z)\ F_4(z)) \begin{pmatrix} \dfrac{\Gamma(\gamma)\Gamma(\gamma-\alpha-\beta)}{\Gamma(\gamma-\alpha)\Gamma(\gamma-\beta)} & \dfrac{\Gamma(2-\gamma)\Gamma(\gamma-\alpha-\beta)}{\Gamma(1-\alpha)\Gamma(1-\beta)} \\ \dfrac{\Gamma(\gamma)\Gamma(\alpha+\beta-\gamma)}{\Gamma(\alpha)\Gamma(\beta)} & \dfrac{\Gamma(2-\gamma)\Gamma(\alpha+\beta-\gamma)}{\Gamma(\alpha-\gamma+1)\Gamma(\beta-\gamma+1)} \end{pmatrix}.$$

It is known that the convergent radius of the local solution $F(\alpha, \beta; \gamma; z)$ is 1. On the other hand, it has an integral representation as follows:

$$F(\alpha, \beta; \gamma; z) = \frac{\Gamma(\gamma)}{\Gamma(\beta)\Gamma(\gamma-\beta)} \int_0^1 s^{\beta-1}(1-s)^{\gamma-\beta-1}(1-sz)^{-\alpha} ds. \tag{3.6}$$

The integrand is well defined as a multivalued function on $z \in \mathbb{C} \setminus \{0, 1\}$ if the real parts of the parameters α, β, γ satisfy some inequalities. Then the local solution about $z = 0$ on the left-hand side has a global expression as an integral on the right-hand side, and the integral representation is used to calculate the connection matrix. We recommend the textbook [**7**] for the monodromy matrix in (3.5) and the integral representation in (3.6).

We now consider the condition for existence of nonzero solutions to the hypergeometric differential equation which is holomorphic simultaneously on the points $z = 0$ and $z = 1$. On the manifold $(\mathbb{C} \cup \{\infty\}) \setminus \{0, 1, \infty\}$, the cycle turning two points $z = 0$, $z = 1$ simultaneously is homotopic to the cycle turning the point $z = \infty$ by the reverse orientation. If there exists a nonzero solution which is holomorphic simultaneously on the points $z = 0$ and $z = 1$, then there exists a nonzero solution which is invariant under the analytic continuation along the cycle turning two points $z = 0$, $z = 1$; namely, there exists a nonzero solution which is invariant under the analytic continuation along the cycle turning $z = \infty$. Hence one of the exponents α, β must be an integer. In fact, if $\alpha = -n$, $n \in \mathbb{Z}_{\geq 0}$ or $\beta = -n$, $n \in \mathbb{Z}_{\geq 0}$), then there exists a nonzero solution to the hypergeometric differential equation which is holomorphic simultaneously on the points $z = 0$ and $z = 1$, and the solution is expressed by the Jacobi polynomial $G_n(z)$. Here the Jacobi polynomials are orthogonal polynomials which include the Legendre polynomials and Chebyshev polynomials by specializing the parameters. In the theory of the hydrogen atom in quantum mechanics, the Legendre polynomials and the associated Legendre polynomials appear and the system of functions by those polynomials is treated as a basis of a certain Hilbert space. Thus we can understand the origin of appearance of them by monodromy of the hypergeometric differential equation. We now discuss the existence problem of holomorphic solutions simultaneously on the points $z = 0$ and $z = 1$ from the viewpoint of the connection matrix. If $\alpha \in \mathbb{Z}_{\leq 0}$ or $\beta \in \mathbb{Z}_{\leq 0}$, then the function $F_1(z)$ defined in (3.3) is a polynomial and the left lower element of the connection matrix in (3.5) is zero, and the function $F_1(z)$ is a scalar multiple of $F_3(z)$. If $\alpha \in \mathbb{Z}_{>0}$ or $\beta \in \mathbb{Z}_{>0}$, then the right upper element of the connection matrix is zero, and the function $F_2(z)$ is a scalar multiple of $F_4(z)$.

By the way, Heun's differential equation contains an accessory parameter which is independent of the exponents (the local monodromy), and the number of the singularities is four. Although the analysis of a Fuchsian equation with four singularities is much more difficult than the case with three singularities (the case of the hypergeometric differential equation), some results have been obtained. We will explain this in the rest of this paper.

4. Heun's differential equation in terms of elliptic functions

In this section, we rewrite Heun's differential equation in terms of elliptic functions. An elliptical expression was found by Darboux and Hermite in the nineteenth century. After the late 1980s, the relationship with the theory of finite-gap potential ([**27, 3, 18, 22**]) and with quantum mechanics, called the Inozemtsev model ([**5**]), is clarified by the expression in terms of elliptic functions.

Let $\wp(x)$ be the Weierstrass doubly periodic function defined by

$$\wp(x) = \frac{1}{x^2} + \sum_{(m,n)\in\mathbb{Z}\times\mathbb{Z}\setminus\{(0,0)\}} \left(\frac{1}{(x - 2m\omega_1 - 2n\omega_3)^2} - \frac{1}{(2m\omega_1 + 2n\omega_3)^2} \right). \tag{4.1}$$

Then the function $\wp(x)$ is doubly periodic with primitive periods $(2\omega_1, 2\omega_3)$, i.e. $\wp(x+2\omega_1) = \wp(x+2\omega_3) = \wp(x)$. It follows from double periodicity that the function $\wp(x)$ defines a map from the torus $\mathbb{C}/(2\omega_1\mathbb{Z} + 2\omega_3\mathbb{Z})$ to the Riemann sphere $\mathbb{C}\cup\{\infty\}$. By setting $\omega_0 = 0, \omega_2 = -\omega_1 - \omega_3$, the map is a double covering of the Riemann sphere with branching points $\{\omega_0, \omega_1, \omega_2, \omega_3\}$. Set

$$e_i = \wp(\omega_i) \quad (i=1,2,3), \quad t = \frac{e_3 - e_1}{e_2 - e_1}, \quad z = \frac{\wp(x) - e_1}{e_2 - e_1}. \tag{4.2}$$

The value $t = (e_3 - e_1)/(e_2 - e_1)$ depends on the ratio ω_3/ω_1 of the primitive periods, and conversely for a fixed value $t \in \mathbb{C}\setminus\{0,1\}$, there exists a ratio ω_3/ω_1 such that $t = (e_3 - e_1)/(e_2 - e_1)$ for the primitive periods $(2\omega_1, 2\omega_3)$ of the elliptic function, which is known in the theory of automorphic functions.

We rewrite Heun's differential equation by using the transformations defined by (4.2). By setting $\tilde{\Phi}(z) = z^{\frac{-l_1}{2}}(z-1)^{\frac{-l_2}{2}}(z-t)^{\frac{-l_3}{2}}$, $y\tilde{\Phi}(z) = f(x)$, Heun's differential equation (1.1) is rewritten as

$$\left(-\frac{d^2}{dx^2} + \sum_{i=0}^{3} l_i(l_i+1)\wp(x+\omega_i) - E \right) f(x) = 0, \tag{4.3}$$

where

(4.4)

$$\begin{aligned} &l_0 = \beta - \alpha - 1/2, \quad l_1 = -\gamma + 1/2, \quad l_2 = -\delta + 1/2, \quad l_3 = -\epsilon + 1/2, \\ &E = (e_2 - e_1)(-4q + (-(\alpha-\beta)^2 + 2\gamma^2 + 6\gamma\epsilon + 2\epsilon^2 - 4\gamma - 4\epsilon - \delta^2 + 2\delta + 1)/3 \\ &\qquad + (-(\alpha-\beta)^2 + 2\gamma^2 + 6\gamma\delta + 2\delta^2 - 4\gamma - 4\delta - \epsilon^2 + 2\epsilon + 1)t/3). \end{aligned}$$

We call it the elliptical representation of Heun's differential equation. The parameters $\alpha, \beta, \gamma, \delta, \epsilon$ concerned with the exponents (local monodromy) in (1.1) correspond to the parameters l_0, l_1, l_2, l_3 concerned with the coupling constants in (4.3), the accessory parameter q corresponds to the eigenvalue E, and the position t of the singularity corresponds to the ratio ω_3/ω_1 of the primitive periods of the elliptic function. The map from x to z defined by (4.2) is a double covering of the Riemann sphere from the torus, and the branching points $\{\omega_1, \omega_2, \omega_3, \omega_0\}$ which are also the poles of the potential are mapped to regular singularities $\{0, 1, t, \infty\}$ of Heun's differential equation (1.1) respectively. The cycle moving the point $x \to 2\omega_1$ in the x-plane corresponds to the cycle turning two points $0, \infty$ on the z-plane, and the cycle moving the point $x \to 2\omega_3$ corresponds to the cycle turning two points $0, 1$. Therefore the monodromy obtained by turning two singularities on Heun's differential equation (1.1) corresponds to the monodromy obtained by moving the point a period in the elliptical form. Note that we have other choices for setting up a correspondence between the two differential equations (1.1) and (4.3).

If three of l_0, l_1, l_2, l_3 are 0 (particularly $l_1 = l_2 = l_3 = 0$, i.e. $\gamma = \delta = \epsilon = 1/2$), then (4.3) (or (1.1)) is called Lamé's differential equation.

By the way, the Inozemtsev model ([**5**]) of type BC_N is a system of quantum mechanics with N particles whose Hamiltonian is given by

$$
(4.5) \qquad H = -\sum_{j=1}^{N} \frac{\partial^2}{\partial x_j^2} + 2l(l+1) \sum_{1\le j<k\le N} (\wp(x_j - x_k) + \wp(x_j + x_k)) + \sum_{j=1}^{N}\sum_{i=0}^{3} l_i(l_i+1)\wp(x_j + \omega_i),
$$

where the parameters l and l_i $(i = 0,1,2,3)$ are coupling constants. This model is a generalization of the elliptic Calogero-Moser-Sutherland system ([**13**]) of type BC_N. By restricting to the case $N = 1$, we recover the elliptical representation of Heun's differential equation (see (4.3)).

It is known that the Inozemtsev model of type BC_N is completely integrable. Namely, there exist N differential operators $P_1(= H), P_2, \dots, P_N$ such that they are algebraically independent and commute mutually. The expressions of commuting operators are obtained by Oshima ([**14**]) explicitly. He and his collaborators ([**12, 15**]) classified complete integrable systems which have the symmetry of the Weyl group of type B_N under some assumptions, and the systems are Inozemtsev models of type BC_N or the models obtained by degeneration of Inozemtsev models. In other words, an Inozemtsev model of type BC_N is a universal completely integrable system of type BC_N, and Heun's differential equation appears as the case of one particle. Despite a feeling of the word "completely integrable", spectral and eigenstates for the Inozemtsev model of type BC_N are not studied much. Complete integrability implies simply existence of sufficient numbers of commuting operators.

5. Local solutions for Heun's differential equation, Heun functions and Heun polynomials

In this section, we review solutions to Heun's differential equation, namely the Heun function and the Heun polynomial which were discussed in the book ([**16**]) edited by Ronveaux.

The Heun function is defined by a nonzero solution to Heun's differential equation which is holomorphic on a domain containing the points $z = 0$ and $z = 1$. Since one of the exponents about $z = 0$ (resp. $z = 1$) is 0, there exists a nonzero solution which is holomorphic about $z = 0$ (resp. $z = 1$), though the solution which is holomorphic about $z = 0$ may not coincide with the solution which is holomorphic about $z = 1$ in general. If the solution which is holomorphic about $z = 0$ is also holomorphic about $z = 1$, it is a Heun function. Although the definition of a Heun function in [**16**] is slightly different, we can regard the two definitions as equivalent by multiplying a function and replacing the singularities.

The Heun polynomial is defined by a nonzero solution to Heun's differential equation which is essentially expressed as a polynomial, i.e., a solution y expressed as

$$
(5.1) \qquad y = z^{\sigma_0}(z-1)^{\sigma_1}(z-t)^{\sigma_t}p(z),
$$

where $p(z)$ is a nonzero polynomial and σ_0 (resp. σ_1, σ_t) is one of the exponents about the regular singularity $z = 0$ (resp. $z = 1$, $z = t$). For the case $\sigma_0 = \sigma_1 = \sigma_t = 0$, the solution y is a polynomial and it is holomorphic on the domain containing three singularities $z = 0, 1, t$. If $\sigma_0, \sigma_1, \sigma_t$ do not satisfy $\sigma_0 = \sigma_1 = \sigma_t =$

0, then the polynomial part $p(z)$ of the Heun polynomial $y = z^{\sigma_0}(z-1)^{\sigma_1}(z-t)^{\sigma_t}p(z)$ is shown to be a solution to Heun's differential equation with different parameters. Hence the Heun polynomial is reduced to the one for the case $\sigma_0 = \sigma_1 = \sigma_t = 0$ with different parameters. The Lamé function and the Lamé polynomial are defined as the Heun function and the Heun polynomial in the case that Heun's differential equation is specialized to Lamé's differential equation. In some of the literature, the Heun polynomial (or Lamé polynomial) is called the Heun function (or Lamé function).

We now observe that Heun functions and Heun polynomials are related to local solutions to Heun's differential equation. Since the point $z = 0$ is a regular singularity of Heun's differential equation and the exponents are 0 and $1 - \gamma$, we can pick up a basis of local solutions as

$$f_1(z) = 1 + \sum_{i=1}^{\infty} c_i z^i, \quad f_2(z) = z^{1-\gamma}\left(1 + \sum_{i=1}^{\infty} c'_i z^i\right) \tag{5.2}$$

if γ is not an integer. Here we examine the coefficients of the expansion of $f_1(z)$. We multiply (1.1) by $z(z-1)(z-t)$ and substitute $y = f_1(z)$ to it. Then we have

$$\begin{aligned} z(z-1)(z-t)\sum_{i=2}^{\infty} i(i-1)c_i z^{i-2} + \{\gamma(z-1)(z-t) + \delta z(z-t) \\ + \epsilon z(z-1)\}\sum_{i=1}^{\infty} i c_i z^{i-1} + (\alpha\beta z - q)\left(1 + \sum_{i=1}^{\infty} c_i z^i\right) = 0. \end{aligned} \tag{5.3}$$

Set $c_0 = 1$. Then we obtain the recursive relations

$$\begin{aligned} & -q + t\gamma c_1 = 0, \\ & (i-1+\alpha)(i-1+\beta)c_{i-1} - [i\{(i-1+\gamma)(1+t) + t\delta + \epsilon\} + q]c_i \\ & \qquad + (i+1)(i+\gamma)tc_{i+1} = 0, \end{aligned} \tag{5.4}$$

for c_i $(i \geq 1)$ by observing the coefficients of z^0 and z^i. We denote the solution $f_1(z)$ by $Hl(t, q; \alpha, \beta, \gamma, \delta; z)$ (see [**16**]). Here Hl means "Heun local" and the parameter ϵ is omitted because it is recovered by the relation $\gamma + \delta + \epsilon = \alpha + \beta + 1$.

The radius of convergence of the power series $Hl(t, q; \alpha, \beta, \gamma, \delta; z)$ determined by (5.4) is 1 or $|t|$ if the series is not finite. We assume $|t| > 1$ for simplicity. If the radius of convergence is $|t|$, then the solution $Hl(t, q; \alpha, \beta, \gamma, \delta; z)$ is a Heun function, because it is holomorphic on the interior of the circle of convergence, which contains the points $z = 0$ and $z = 1$. Under the assumption $\gamma \notin \mathbb{Z}$, it is shown that the Heun function must be a scalar multiple of the solution $Hl(t, q; \alpha, \beta, \gamma, \delta; z)$ if the Heun function exists. Therefore the condition for the parameters that the Heun function exists is described as the radius of convergence for $Hl(t, q; \alpha, \beta, \gamma, \delta; z)$ being not 1. The radius of convergence is not determined by finitely many coefficients in principle. However we may determine the values of the accessory parameter q numerically on which the Heun function is realized by calculating finitely many coefficients. Note that the author does not know whether the convergence of the numerical calculation for the accessory parameter q has been justified rigorously as of the current writing. We have another method to find the condition for the existence of the Heun function; namely, we expand a solution to Heun's differential equation by an infinite sum of hypergeometric functions and observe the asymptotic of the coefficients of the expansion. Then the problem is replaced by the convergence of an infinite continued fraction ([**16**]). However we calculate it numerically by investigating finitely many terms for practice.

We now consider the case that the power series determined by (5.4) is indeed a finite series. Let $-\alpha = n \in \mathbb{Z}_{\geq 1}$ or $-\beta = n \in \mathbb{Z}_{\geq 1}$. It follows from (5.4) that c_i is a polynomial in the variable q of degree i. We fix a value of q as a solution to $c_{n+1} = 0$ which is an algebraic equation of degree $n+1$. Since the coefficient of c_n in (5.4) in the case $i = n+1$ is 0 and q is fixed to be $c_{n+1} = 0$, we have $c_{n+2} = 0$, and we also have $c_i = 0$ for $i > n+2$ by the recursive relation. Therefore, if $-\alpha = n \in \mathbb{Z}_{\geq 1}$ or $-\beta = n \in \mathbb{Z}_{\geq 1}$ and q satisfies the algebraic equation $c_{n+1} = 0$ of degree $n+1$, then we have a polynomial solution of degree z to Heun's differential equation. This is one of the constructions of the Heun polynomial. On the other hand, we may obtain the Heun polynomial by quasi-exact solvability.

Quasi-exact solvability is the situation that the Hamiltonian acts on a finite-dimensional subspace of the infinite-dimensional functional space (i.e. Hilbert space in theoretical physics) where the Hamiltonian acts naturally. In this case we can obtain eigenvalues and the eigenfunctions in principle by diagonalizing the matrix of finite size; thus we may understand a part of the eigenvalues and the eigenfunctions. Set

$$H^z = z(z-1)(z-t)\left\{\frac{d^2}{dz^2} + \left(\frac{\gamma}{z} + \frac{\delta}{z-1} + \frac{\epsilon}{z-t}\right)\frac{d}{dz}\right\} + \alpha\beta z. \tag{5.5}$$

Then Heun's differential equation is written as

$$H^z y = qy, \tag{5.6}$$

and the accessory parameter q plays the role of the eigenvalue. Since

(5.7)
$$H^z z^i = (i+\alpha)(i+\beta)z^{i+1} - i\{(i-1+\gamma)(1+t) + t\delta + \epsilon\}z^i + i(i-1+\gamma)tz^{i-1},$$

and the coefficient of z^{i+1} vanishes for the case $i = n$ if $-\alpha = n \in \mathbb{Z}_{\geq 1}$ or $-\beta = n \in \mathbb{Z}_{\geq 1}$, the operator H^z acts on the space of polynomials of degree no more than n. The eigenfunctions in this space are Heun polynomials, the eigenvalues represent the finitely many values for q, and the characteristic equation for the operator H^z coincides with the equation $c_{n+1} = 0$ for $Hl(t, q; \alpha, \beta, \gamma, \delta; z)$.

Quasi-exact solvability is extended to the Inozemtsev model of type BC_N ([**2, 19**]).

6. Finite-gap potential

We give definitions of the finite-gap potential and the algebro-geometric finite-gap potential, which are related to the monodromy of Heun's differential equation.

DEFINITION 6.1. Let $v(x)$ be a periodic, smooth, real function, H be the operator $-d^2/dx^2 + v(x)$, and $\sigma_b(H)$ be the set such that

$$E \in \sigma_b(H) \Leftrightarrow \text{every solution to } (H-E)f(x) = 0 \text{ is bounded on } x \in \mathbb{R}. \tag{6.1}$$

If the closure of the set $\sigma_b(H)$ can be written as $\overline{\sigma_b(H)} = [E_0, E_1] \cup [E_2, E_3] \cup \cdots \cup [E_{2g}, \infty)$, where $E_0 < E_1 < \cdots < E_{2g}$ for some $g \in \mathbb{Z}$, i.e. the number of bounded bands is finite, then $v(x)$ is called the finite-gap potential.

DEFINITION 6.2. Let $H = -d^2/dx^2 + v(x)$. If there exists an odd-order differential operator

$$A = \left(\frac{d}{dx}\right)^{2g+1} + \sum_{j=0}^{2g-1} b_j(x) \left(\frac{d}{dx}\right)^{2g-1-j} \tag{6.2}$$

which commutes with the operator H (i.e., $[A, H] = 0$), then $v(x)$ is called the algebro-geometric finite-gap potential.

Note that the equation $[A, H] = 0$ is equivalent to the function $v(x)$ being a solution to a stationary higher-order KdV equation ([**25**]). It was established in the 1970s by Novikov, etc. that, under the assumption that $v(x)$ is a periodic, smooth, real function, the potential $v(x)$ is finite-gap if and only if $v(x)$ is algebro-geometric finite-gap ([**25**]). This is a surprising theorem in the author's opinion, because the analytic property (the shape of the spectrum of the finite-gap potential) corresponds to the algebraic property (the existence of a commuting operator of algebro-geometric finite-gap potential).

For example, we consider the case $v(x) = 0$, i.e. the case $H = -d^2/dx^2$. Let $f(x)$ be an eigenfunction of the operator H with the eigenvalue E. Then we have $f''(x) + Ef(x) = 0$, and we solve it by dividing it into three cases. If $E < 0$, then the solutions are expressed as $f(x) = C_1 e^{\lambda x} + C_2 e^{-\lambda x}$ (C_1, C_2: constants) by setting $E = -\lambda^2$, and they are unbounded unless $C_1 = C_2 = 0$. If $E = 0$, then the solutions are expressed as $f(x) = C_1 + C_2 x$, and they are unbounded when $C_2 \neq 0$. If $E > 0$, then the solutions are expressed as $f(x) = C_1 \cos \lambda x + C_2 \sin \lambda x$ by setting $E = \lambda^2$, and every solution is bounded on $\mathbb{R}$. Thus we have $\sigma_b(H) = (0, \infty)$, $\overline{\sigma_b(H)} = [0, \infty)$, and the potential $v(x) = 0$ is finite-gap in the case $g = 0$. On the other hand, the operator $H = -d^2/dx^2$ is algebro-geometric finite-gap, because H commutes with the operator $A = d/dx$, which is of odd degree.

On explicit forms of nontrivial finite-gap potentials, Ince ([**4**]) established in 1940 that if $l_3 \in \mathbb{Z}_{\geq 0}$, $\omega_1 \in \mathbb{R} \setminus \{0\}$ and $\omega_3 \in \sqrt{-1}\mathbb{R} \setminus \{0\}$, then the potential of Lamé's operator $-d^2/dx^2 l_3(l_3+1)\wp(x+\omega_3)$ is finite-gap. After Ince, much progress in the study of finite-gap potentials was attained, e.g. the relationship with soliton theory and the Its-Matveev formula, which expresses finite-gap potentials in terms of theta functions of hyperelliptic functions, but other simple examples were not found for a long time.

Around 1990, Treibich and Verdier ([**27**]) established that the function $\sum_{i=0}^{3} l_i(l_i + 1)\wp(x + \omega_i)$ is finite-gap potential iff $l_0, l_1, l_2, l_3 \in \mathbb{Z}$, and it is now called the Treibich-Verdier potential. They invented the theory of elliptic solitons, which is based on the approach developed by Krichever, among others. The theory includes Jacobi varieties and tangential coverings and it seems to be difficult for most people including the author of the present article. Subsequently Gesztesy and Weikard ([**3**]), A. O. Smirnov([**18**]) and others have further studied this subject, but it seems that explicit formulas of monodromy were not studied very much.

7. Integral representation of solutions and formulas of monodromy in terms of the hyperelliptic integral

From now on, we investigate an elliptical representation of Heun's equation

$$\left(-\frac{d^2}{dx^2}+\sum_{i=0}^{3}l_i(l_i+1)\wp(x+\omega_i)-E\right)f(x)=0, \tag{7.1}$$

with the condition $l_0,l_1,l_2,l_3\in\mathbb{Z}$, which is related to the finite-gap potential. Since the equation is invariant under replacing l_i with $-l_i-1$ for each $i\in\{0,1,2,3\}$, we may restrict to the case $l_0,l_1,l_2,l_3\in\mathbb{Z}_{\geq 0}$. Note that the condition $l_0,l_1,l_2,l_3\in\mathbb{Z}$ corresponds to the condition $\gamma,\delta,\epsilon,\alpha-\beta\in\mathbb{Z}+1/2$ on the differential equation (1.1).

Observe that the product of two solutions to (7.1) satisfies the following differential equation of order three:

(7.2)
$$\left(\frac{d^3}{dx^3}-4\left(\sum_{i=0}^{3}l_i(l_i+1)\wp(x+\omega_i)-E\right)\frac{d}{dx}-2\sum_{i=0}^{3}l_i(l_i+1)\wp'(x+\omega_i)\right)h(x)=0.$$

Then (7.2) has a unique nonzero doubly-periodic solution $\Xi(x,E)$ for all E. More precisely, we have the following proposition:

PROPOSITION 7.1 ([**20**, Proposition 3.5]). *If $l_0,l_1,l_2,l_3\in\mathbb{Z}_{\geq 0}$, then (7.2) has a unique nonzero doubly-periodic solution $\Xi(x,E)$, which has the expansion*

$$\Xi(x,E)=c_0(E)+\sum_{i=0}^{3}\sum_{j=0}^{l_i-1}b_j^{(i)}(E)\wp(x+\omega_i)^{l_i-j}, \tag{7.3}$$

where the coefficients $c_0(E)$, $b_j^{(i)}(E)$ and $d^{(i')}(E)$ are polynomials in E, these polynomials do not share any common divisors, and the polynomial $c_0(E)$ is monic. We set $g=\deg_E c_0(E)$. Then the coefficients satisfy $\deg_E b_j^{(i)}(E)<g$ for all i, j.

A key point of the proof is that the local monodromy about each singularity $x=\omega_0,\omega_1,\omega_2,\omega_3$ is trivial. For details, see [**20**, Proposition 3.5]. Proposition 7.1 could be proved without using that the function $\sum_{i=0}^{3}l_i(l_i+1)\wp(x+\omega_i)$ is a finite-gap potential. The finite-gap property is shown by applying Proposition 7.1, which we will demonstrate in section 9.

We define a polynomial $Q(E)$ by

$$\begin{aligned}Q(E)=\;&\Xi(x,E)^2\left(E-\sum_{i=0}^{3}l_i(l_i+1)\wp(x+\omega_i)\right)\\&+\frac{1}{2}\Xi(x,E)\frac{d^2\Xi(x,E)}{dx^2}-\frac{1}{4}\left(\frac{d\Xi(x,E)}{dx}\right)^2,\end{aligned} \tag{7.4}$$

which plays important roles later. It is shown by differentiating the right-hand side of (7.4) and applying (7.2) that $Q(E)$ is independent of x, and $Q(E)$ is a monic polynomial in E of degree $2g+1$.

By using the function $\Xi(x,E)$ and the polynomial $Q(E)$, we may obtain an integral representation of solutions to (7.1).

THEOREM 7.2 ([**20**, Proposition 3.7]). *Let* $\Xi(x,E)$ *be the function defined in Proposition* 7.1 *and* $Q(E)$ *be the polynomial defined in* (7.4). *Then the function*

$$\Lambda(x,E)=\sqrt{\Xi(x,E)}\exp\int\frac{\sqrt{-Q(E)}}{\Xi(x,E)}dx, \tag{7.5}$$

is a solution to (7.1).

PROOF. By differentiating (7.5) and applying (7.4) we have

$$\frac{\Lambda'(x,E)}{\Lambda(x,E)}=\frac{1}{2}\frac{\Xi'(x,E)}{\Xi(x,E)}+\frac{\sqrt{-Q(E)}}{\Xi(x,E)}, \tag{7.6}$$

$$\begin{aligned}\frac{\Lambda''(x,E)}{\Lambda(x,E)}&=\frac{d}{dx}\left(\frac{\Lambda'(x,E)}{\Lambda(x,E)}\right)+\left(\frac{\Lambda'(x,E)}{\Lambda(x,E)}\right)^2 \\ &=\frac{1}{2}\frac{\Xi''(x,E)}{\Xi(x,E)}-\frac{1}{4}\left(\frac{\Xi'(x,E)}{\Xi(x,E)}\right)^2-\frac{Q(E)}{\Xi(x,E)^2}=\sum_{i=0}^{3}l_i(l_i+1)\wp(x+\omega_i)-E.\end{aligned} \tag{7.7}$$

Hence $\Lambda(x,E)$ satisfies (7.1). □

This expression of the integral was essentially written in the book by Whittaker and Watson ([**28**, §23.7]) for the case of Lamé's differential equation.

Here we observe examples. In the case $l_0=1$, $l_1=l_2=l_3=0$, the function $\Xi(x,E)$ and the polynomial $Q(E)$ are calculated as

$$\Xi(x,E)=\wp(x)+E,\quad Q(E)=(E+e_1)(E+e_2)(E+e_3). \tag{7.8}$$

Hence the function

$$\Lambda(x,E)=\sqrt{\wp(x)+E}\exp\int\frac{\sqrt{-(E+e_1)(E+e_2)(E+e_3)}}{\wp(x)+E}dx \tag{7.9}$$

is a solution to (7.1). In the case $l_0=2$, $l_1=l_2=l_3=0$, the function $\Xi(x,E)$ and the polynomial $Q(E)$ are calculated as

(7.10)

$$\Xi(x,E)=9\wp(x)^2+3E\wp(x)+E^2-\frac{9}{4}g_2,\quad Q(E)=(E^2-3g_2)\prod_{i=1}^{3}(E-3e_i),$$

where $g_2=-4(e_1e_2+e_2e_3+e_3e_1)$. By substituting these functions into (7.5), we obtain solutions to (7.1).

Since the differential equation (7.1) is periodic with respect to the primitive periods $(2\omega_1,2\omega_3)$, the functions $\Lambda(x+2\omega_1,E)$ and $\Lambda(x+2\omega_3,E)$ also satisfy (7.1). The problem of expressing the functions $\Lambda(x+2\omega_1,E)$ and $\Lambda(x+2\omega_3,E)$ in terms of the original solutions $\Lambda(x,E)$ and $\Lambda(-x,E)$ corresponds to the problem of determining the monodromy of solutions to Heun's differential equation (1.1) defined on the Riemann sphere with respect to the cycle enclosing two points $\{0,\infty\}$ (or $\{0,1\}$), which is a problem for studying a global structure of monodromy. The following theorem gives an explicit result on the problem.

THEOREM 7.3 ([**22**, Theorem 3.7]). *Assume* $l_0,l_1,l_2,l_3\in\mathbb{Z}_{\geq 0}$. *Let* E_0 *be the value such that* $Q(E_0)=0$. *Then there exists* $q_1,q_3\in\{0,1\}$ *such that* $\Lambda(x+$

$2\omega_k, E_0) = (-1)^{q_k}\Lambda(x, E_0)$ $(k = 1, 3)$, *and we have*

$$\Lambda(x+2\omega_k, E) = (-1)^{q_k}\Lambda(x, E)\exp\left(-\frac{1}{2}\int_{E_0}^{E}\frac{\int_{0+\varepsilon}^{2\omega_k+\varepsilon}\Xi(x,\tilde{E})dx}{\sqrt{-Q(\tilde{E})}}d\tilde{E}\right) \tag{7.11}$$

for all E and $k = 1, 3$. Here ε is a constant so determined as to avoid passing through the poles in the integration, and the left-hand side of (7.11) is independent from ε.

PROOF. It follows from $Q(E_0) = 0$ and (7.5) that $\Lambda(x, E_0)^2 = \Xi(x, E_0)$. Since $\Xi(x, E_0)$ is doubly-periodic, there exist $q_1, q_3 \in \{0, 1\}$ such that $\Lambda(x + 2\omega_k, E_0) = (-1)^{q_k}\Lambda(x, E_0)$ $(k = 1, 3)$. We have

$$\Lambda(x+2\omega_k, E) = \pm\Lambda(x, E)\exp\left(\int_{0+\varepsilon}^{2\omega_k+\varepsilon}\frac{\sqrt{-Q(E)}dx}{\Xi(x,E)}\right) \tag{7.12}$$

by (7.5). The sign $\pm$ is determined by the analytic continuation of the function $\sqrt{\Xi(x, E)}$. Set

$$R = R(x,E) = \frac{\Xi(x,E)}{\sqrt{-Q(E)}}, \quad v = \sum_{i=0}^{3} l_i(l_i+1)\wp(x+\omega_i), \quad R = \frac{\partial}{\partial x}R(x,E). \tag{7.13}$$

Then it follows from (7.4) that

$$R^2(v-E) - \frac{1}{2}RR'' + \frac{1}{4}(R')^2 = 1. \tag{7.14}$$

By differentiating (7.14) with respect to E and dividing by Q, we have

$$-\frac{\partial R''}{\partial E} = -\frac{R'}{R}\frac{\partial R'}{\partial E} + 2R + \frac{R''}{R}\frac{\partial R}{\partial E} - 4(v-E)\frac{\partial R}{\partial E}. \tag{7.15}$$

Hence

(7.16)

$$\begin{aligned}&\left(-\frac{\partial R'}{\partial E} + \frac{R'}{R}\frac{\partial R}{\partial E}\right)' \\ &= \left(-\frac{R'}{R}\frac{\partial R'}{\partial E} + 2R + \frac{R''}{R}\frac{\partial R}{\partial E} - 4(v-E)\frac{\partial R}{\partial E}\right) + \left(\frac{R''}{R} - \frac{(R')^2}{R^2}\right)\frac{\partial R}{\partial E} + \frac{R'}{R}\frac{\partial R'}{\partial E} \\ &= 2R + \frac{1}{R^2}\frac{\partial R}{\partial E}\left(2RR'' - 4R^2(v-E) - (R')^2\right) = 2R + 4\frac{\partial}{\partial E}\left(\frac{1}{R}\right).\end{aligned}$$

It follows from double-periodicity that

$$\begin{aligned}\int_{0+\varepsilon}^{2\omega_k+\varepsilon}&\left(-\frac{\partial R'}{\partial E} + \frac{R'}{R}\frac{\partial R}{\partial E}\right)' dx \\ &= \left(-\frac{\partial R'(2\omega_k+\varepsilon)}{\partial E} + \frac{R'(2\omega_k+\varepsilon)}{R(2\omega_k+\varepsilon)}\frac{\partial R(2\omega_k+\varepsilon)}{\partial E}\right) \\ &\quad - \left(-\frac{\partial R'(\varepsilon)}{\partial E} + \frac{R'(\varepsilon)}{R(\varepsilon)}\frac{\partial R(\varepsilon)}{\partial E}\right) = 0.\end{aligned} \tag{7.17}$$

Thus we have

$$\int_{0+\varepsilon}^{2\omega_k+\varepsilon} R dx + 2\int_{0+\varepsilon}^{2\omega_k+\varepsilon}\frac{\partial}{\partial E}\left(\frac{1}{R}\right)dx = 0. \tag{7.18}$$

By integrating with respect to E we see that

$$\exp\left(-\frac{1}{2}\int_{E_0}^{E}\frac{\int_{0+\varepsilon}^{2\omega_k+\varepsilon}\Xi(x,\tilde{E})dx}{\sqrt{-Q(\tilde{E})}}d\tilde{E}\right) \tag{7.19}$$

$$= \exp\left(\int_{0+\varepsilon}^{2\omega_k+\varepsilon}\frac{\sqrt{-Q(E)}}{\Xi(x,E)}dx\right)\Big/\exp\left(\int_{0+\varepsilon}^{2\omega_k+\varepsilon}\frac{\sqrt{-Q(E_0)}}{\Xi(x,E_0)}dx\right).$$

Therefore we have proved (7.11) in the theorem. □

We now rewrite the integral representation of monodromy in a simpler form. Since the function $\wp(x+\omega_i)^n$ is expressed as a linear combination of $(d/dx)^{2j}\wp(x+\omega_i)$ $(j=0,\dots,n)$, the function $\Xi(x,E)$ is expressed as

$$\Xi(x,E) = c(E) + \sum_{i=0}^{3}\sum_{j=0}^{l_i-1} a_j^{(i)}(E)\left(\frac{d}{dx}\right)^{2j}\wp(x+\omega_i). \tag{7.20}$$

Set

$$a(E) = \sum_{i=0}^{3} a_0^{(i)}(E). \tag{7.21}$$

Then it follows from Theorem 7.3 that

$$\Lambda(x+2\omega_k,E) = (-1)^{q_k}\Lambda(x,E)\exp\left(-\frac{1}{2}\int_{E_0}^{E}\frac{-2\eta_k a(\tilde{E})+2\omega_k c(\tilde{E})}{\sqrt{-Q(\tilde{E})}}d\tilde{E}\right), \tag{7.22}$$

for $k=1,3$, where $\eta_k=\zeta(\omega_k)$ $(k=1,3)$ and $\zeta(x)$ is the Weierstrass zeta function. The denominator of the integral is a square root of a polynomial of degree $2g+1$, and the numerator is a polynomial of degree g. Hence it follows from (7.22) that the global monodromy with respect to a shift of period is expressed as a hyperelliptic integral of the second kind with genus g.

Here we provide examples. For the case $l_0=1$, $l_1=l_2=l_3=0$, we set $E_0=-e_2$. Then $Q(E_0)=0$ and $\Lambda(x,E_0)=-\Lambda(x+2\omega_1,E_0)=-\Lambda(x+2\omega_3,E_0)$. Equation (7.22) in this case is written as

$$\Lambda(x+2\omega_k,E) = -\Lambda(x,E)\exp\left(-\frac{1}{2}\int_{-e_2}^{E}\frac{2\omega_k\tilde{E}-2\eta_k}{\sqrt{-(\tilde{E}+e_1)(\tilde{E}+e_2)(\tilde{E}+e_3)}}d\tilde{E}\right), \tag{7.23}$$

for $k=1,3$, and it is an expression of monodromy by an elliptic integral of the second kind.

For the case $l_0=2$, $l_1=l_2=l_3=0$, we set $E_0=\sqrt{3g_2}$. Then $Q(E_0)=0$ and $\Lambda(x,E_0)=\Lambda(x+2\omega_1,E_0)=\Lambda(x+2\omega_3,E_0)$. Equation (7.22) in this case is written as

$$\Lambda(x+2\omega_k,E) = \Lambda(x,E)\exp\left(-\frac{1}{2}\int_{\sqrt{3g_2}}^{E}\frac{\omega_k(2\tilde{E}^2-3g_2)-6\eta_k\tilde{E}}{\sqrt{-(\tilde{E}^2-3g_2)\prod_{i=1}^{3}(\tilde{E}-3e_i)}}d\tilde{E}\right), \tag{7.24}$$

for $k=1,3$, and it is an expression of monodromy by a hyperelliptic integral of the second kind with genus two.

8. Hermite-Krichever Ansatz and hyperelliptic-elliptic reduction formulas

Hermite and Halphen studied Lamé's differential equation by hypothesizing that solutions are expressed by elliptic Baker-Akhiezer functions (or Bloch functions), and Krichever ([**9**]) described quasi-periodic solutions to the Kadomtset-Petviashvili equation by setting solutions into the form of Baker-Akhiezer type functions.

In this section, we investigate solutions to the elliptical representation of Heun's differential equation (see (7.1)) by the Hermite-Krichever Ansatz. Set

$$\Phi_i(x,\alpha)=\frac{\sigma(x+\omega_i-\alpha)}{\sigma(x+\omega_i)}\exp(\zeta(\alpha)x)\qquad(i=0,1,2,3), \tag{8.1}$$

where $\zeta(x)$ (resp. $\sigma(x)$) is the Weierstrass zeta (resp. sigma) function. In our situation, the Hermite-Krichever Ansatz asserts that the differential equation has a solution that is expressed as a finite series in the derivatives of elliptic Baker-Akhiezer functions $\Phi_i(x,\alpha)$, multiplied by an exponential function. More precisely, we are going to find a nonzero solution to (7.1) of the form

$$f(x)=\exp(\kappa x)\left(\sum_{i=0}^{3}\sum_{j}\tilde{b}_j^{(i)}\left(\frac{d}{dx}\right)^j\Phi_i(x,\alpha)\right). \tag{8.2}$$

Then global monodromy (the behavior of solutions by the shift of periods $2\omega_1$ or $2\omega_3$) is described as

$$f(x+2\omega_k)=\exp(-2\eta_k\alpha+2\omega_k\zeta(\alpha)+2\kappa\omega_k)f(x), \tag{8.3}$$

for $k=1,3$. Hence, if we can determine explicit values of α and κ, then we may expect that the Hermite-Krichever Ansatz can be applied for investigating global structures of monodromy. The following theorem asserts that if $l_0,l_1,l_2,l_3\in\mathbb{Z}_{\geq 0}$, then there exists a nonzero solution to the elliptical representation of Heun's differential equation (7.1) which is expressed as a form of the Hermite-Krichever Ansatz.

THEOREM 8.1 ([**23**]). *Assume that $l_0,l_1,l_2,l_3\in\mathbb{Z}_{\geq 0}$. There exist polynomials $P_1(E),\dots,P_4(E)$ such that if $P_2(E)\neq 0$, then there exists a nonzero solution $f(x)$ to the elliptical representation of Heun's differential equation (7.1) which is expressed as*

$$f(x)=\exp(\kappa x)\left(\sum_{i=0}^{3}\sum_{j=0}^{l_i-1}\tilde{b}_j^{(i)}\left(\frac{d}{dx}\right)^j\Phi_i(x,\alpha)\right). \tag{8.4}$$

The values α and κ are expressed as

$$\wp(\alpha)=\frac{P_1(E)}{P_2(E)},\quad \kappa=\frac{P_3(E)}{P_4(E)}\sqrt{-Q(E)}, \tag{8.5}$$

and the periodicity of the solution $f(x)$ is expressed by (8.3). If $P_2(E)=0$, then there exists a nonzero solution written as

$$f(x)=\exp(\bar{\kappa}x)\left(\bar{c}+\sum_{i=0}^{3}\sum_{j=0}^{l_i-2}\bar{b}_j^{(i)}\left(\frac{d}{dx}\right)^j\wp(x+\omega_i)+\sum_{k=1}^{3}\bar{c}_k\frac{\wp'(x)}{\wp(x)-e_k}\right). \tag{8.6}$$

In the theorem, we can pick up the solution $f(x)$ in the form of the Hermite-Krichever Ansatz by the function $\Lambda(x,E)$ in (7.5), and it is a key point for the proof. In order to investigate monodromy, it is necessary to investigate the values α and κ, namely the polynomials $P_1(E),\dots,P_4(E)$ in (8.5). These polynomials are described by the Heun polynomial, the twisted Heun polynomial and the theta-twisted Heun polynomial, and the method for calculation was explained in [**23**]. Note that the twisted Heun polynomial and the theta-twisted Heun polynomial are generalizations of the twisted Lamé polynomial and the theta-twisted Lamé polynomial which were introduced by Maier ([**11**]).

If $l_0=1$, $l_1=l_2=l_3=0$, then (8.4) is written as

$$f(x)=\exp(\kappa x)\left(\tilde{b}_0^{(0)}\Phi_0(x,\alpha)\right), \tag{8.7}$$

and there exists a nontrivial solution (i.e. $\tilde{b}_0^{(0)}\neq 0$) of this form if α and κ satisfy

$$\wp(\alpha)=-E, \quad \kappa=0. \tag{8.8}$$

If $l_0=2$, $l_1=l_2=l_3=0$, then (8.4) is written as

$$f(x)=\exp(\kappa x)\left(\tilde{b}_0^{(0)}\Phi_0(x,\alpha)+\tilde{b}_1^{(0)}\left(\frac{d}{dx}\right)\Phi_0(x,\alpha)\right), \tag{8.9}$$

and the values α, κ are expressed as

(8.10)

$$\wp(\alpha)=e_1-\frac{(E-3e_1)(E+6e_1)^2}{9(E^2-3g_2)}, \quad \kappa=\frac{2}{3(E^2-3g_2)}\sqrt{-(E^2-3g_2)\prod_{i=1}^{3}(E-3e_i)}.$$

If $E\neq\pm\sqrt{3g_2}$, then there exists a nonzero solution in the form of (8.9). The ratio $\tilde{b}_1^{(0)}/\tilde{b}_0^{(0)}$ is also expressed as a function of E. In this case, the Heun polynomials are E^2-3g_2, $E-3e_i$ $(i=1,2,3)$, the twisted Heun polynomials are $E+6e_i$ $(i=1,2,3)$, and the theta-twisted Heun polynomial is 1.

Thus we have formulas of global monodromy of Heun's differential equation by the hyperelliptic integral (7.22) and by the Hermite-Krichever Ansatz (8.3). By comparing the two expressions, we can derive hyperelliptic-elliptic reduction formulas. Set

$$A=-2\alpha-\int_{E_0}^{E}\frac{a(\tilde{E})}{\sqrt{-Q(\tilde{E})}}d\tilde{E}, \tag{8.11}$$

$$B=2\zeta(\alpha)+2\kappa+\int_{E_0}^{E}\frac{c(\tilde{E})}{\sqrt{-Q(\tilde{E})}}d\tilde{E}, \tag{8.12}$$

where α, κ are the values determined by the Hermite-Krichever Ansatz and $Q(E)$, $a(E)$, $c(E)$ are the polynomials appearing in the hyperelliptic integral. By comparing (8.3) and (7.22) for each case $k=1$ and $k=3$, we have

$$\eta_1 A+\omega_1 B=\pi\sqrt{-1}(q_1+2n_1), \tag{8.13}$$

$$\eta_3 A+\omega_3 B=\pi\sqrt{-1}(q_3+2n_3), \tag{8.14}$$

for some integers n_1 and n_3. By Legendre's relation $\eta_1\omega_3 - \eta_3\omega_1 = \pi\sqrt{-1}/2$, it follows that

$$\alpha + \frac{1}{2}\int_{E_0}^{E} \frac{a(\tilde{E})}{\sqrt{-Q(\tilde{E})}} d\tilde{E} = -(q_1 + 2n_1)\omega_3 + (q_3 + 2n_3)\omega_1, \tag{8.15}$$

$$\zeta(\alpha) + \kappa + \frac{1}{2}\int_{E_0}^{E} \frac{c(\tilde{E})}{\sqrt{-Q(\tilde{E})}} d\tilde{E} = -(q_1 + 2n_1)\eta_3 + (q_3 + 2n_3)\eta_1. \tag{8.16}$$

We set $\xi = \wp(\alpha)$. It follows from the limit procedure $\alpha \to 0$ (mod $2\omega_1\mathbb{Z} \oplus 2\omega_3\mathbb{Z}$) as $E \to \infty$ and the relation $\int (1/\wp'(\alpha))d\xi = \int d\alpha$ that

$$\int_{\infty}^{\xi} \frac{d\tilde{\xi}}{\sqrt{4\tilde{\xi}^3 - g_2\tilde{\xi} - g_3}} = \alpha = -\frac{1}{2}\int_{\infty}^{E} \frac{a(\tilde{E})}{\sqrt{-Q(\tilde{E})}} d\tilde{E}. \tag{8.17}$$

Note that $Q(E)$ is a polynomial of degree $2g + 1$, while $a(E)$ is a polynomial of degree g. Hence (8.17) represents a formula which reduces a hyperelliptic integral of the first kind to an elliptic integral of the first kind. The transformation of variables is given by $\xi = P_1(E)/P_2(E)$ for polynomials $P_1(E)$ and $P_2(E)$ in (8.5). Let α_0 denote the value of α at $E = E_0$ ($Q(E_0) = 0$). Then α_0 is one of the half periods $\omega_0, \omega_1, \omega_2, \omega_3$. If $\alpha_0 = \omega_i$ ($i = 1, 2, 3$), then it follows from (8.16) that

$$\kappa = -\frac{1}{2}\int_{E_0}^{E} \frac{c(\tilde{E})}{\sqrt{-Q(\tilde{E})}} d\tilde{E} + \int_{e_i}^{\xi} \frac{\tilde{\xi}d\tilde{\xi}}{\sqrt{4\tilde{\xi}^3 - g_2\tilde{\xi} - g_3}} \tag{8.18}$$

(see [**23**]). Note that $Q(E)$ is a polynomial of degree $2g + 1$, $c(E)$ is a polynomial of degree $g+1$ and κ is expressed as $\kappa = \sqrt{-Q(E)}P_3(E)/P_4(E)$ (see (8.5)). Hence (8.18) represents a formula which reduces a hyperelliptic integral of the second kind to an elliptic integral of the second kind.

If $l_0 = 1, l_1 = l_2 = l_3 = 0$, then the monodromy formula by a hyperelliptic integral is the one by an elliptic integral. The change of variable is given by $\xi = -E$, and the transformation formula of the integrals is trivial.

If $l_0 = 2, l_1 = l_2 = l_3 = 0$, then we have the reduction formula of the hyperelliptic integral of genus two to the elliptic integral

(8.19)
$$-\frac{1}{2}\int_{\infty}^{E} \frac{3\tilde{E}}{\sqrt{-(\tilde{E}^2 - 3g_2)(\tilde{E}^3 - 9g_2\tilde{E}/4 - 27g_3/4))}} d\tilde{E} = \int_{\infty}^{\xi} \frac{d\tilde{\xi}}{\sqrt{4\tilde{\xi}^3 - g_2\tilde{\xi} - g_3}},$$

(8.20)
$$\frac{1}{2}\int_{3e_1}^{E} \frac{\tilde{E}^2 - 3g_2/2}{\sqrt{-(\tilde{E}^2 - 3g_2)(\tilde{E}^3 - 9g_2\tilde{E}/4 - 27g_3/4)}} d\tilde{E}$$
$$= -\frac{2}{3}\sqrt{\frac{-(E^3 - 9g_2E/4 - 27g_3/4)}{(E^2 - 3g_2)}} + \int_{e_1}^{\xi} \frac{\tilde{\xi}d\tilde{\xi}}{\sqrt{4\tilde{\xi}^3 - g_2\tilde{\xi} - g_3}},$$

by setting

$$\xi = e_1 - \frac{(E - 3e_1)(E + 6e_1)^2}{9(E^2 - 3g_2)}. \tag{8.21}$$

It follows from (8.19) and setting $\xi = -y/6$, $E = z$, $g_2 = a/3$, $g_3 = b/54$ that

$$\int \frac{zdz}{\sqrt{(z^2-a)(8z^3-6az-b)}} = \frac{1}{2\sqrt{3}} \int \frac{dy}{\sqrt{y^3-3ay+b}}, \tag{8.22}$$

by the change of variable $y = (2z^3-b)/(3(z^2-a))$, which reproduces Hermite's result of relations on elliptic integrals and hyperelliptic integrals of the 19th century. Moreover it follows from (8.20) that

(8.23)

$$\int \frac{(2z^2-a)dz}{\sqrt{(z^2-a)(8z^3-6az-b)}} = \frac{1}{3}\sqrt{\frac{8z^3-6az-b}{z^2-a}} + \frac{1}{2\sqrt{3}} \int \frac{ydy}{\sqrt{y^3-3ay+b}}.$$

9. Finite-gap potential and the polynomial $Q(E)$

In this section, we show that if $l_0, l_1, l_2, l_3 \in \mathbb{Z}_{\geq 0}$, then the function $\sum_{i=0}^{3} l_i(l_i+1)\wp(x+\omega_i)$ is an algebro-geometric finite-gap potential, and we investigate properties of the polynomial $Q(E)$.

Write the function $\Xi(x,E)$ defined in Proposition 7.1 as

$$\Xi(x,E) = \sum_{j=0}^{g} a_j(x) E^{g-j}. \tag{9.1}$$

THEOREM 9.1. *Assume that $l_0, l_1, l_2, l_3 \in \mathbb{Z}_{\geq 0}$. Write $v(x) = \sum_{i=0}^{3} l_i(l_i+1)\wp(x+\omega_i)$ and define the $(2g+1)$st-order differential operator A by*

$$A = \sum_{j=0}^{g} \left\{ a_j(x)\frac{d}{dx} - \frac{1}{2}\left(\frac{d}{dx}a_j(x)\right)\right\} \left(-\frac{d^2}{dx^2} + v(x)\right)^{g-j}, \tag{9.2}$$

where the $a_j(x)$ are determined in (9.1). Then the operator A commutes with the operator $H = -d^2/dx^2 + v(x)$. In other words, the function $v(x) = \sum_{i=0}^{3} l_i(l_i+1)\wp(x+\omega_i)$ is an algebro-geometric finite-gap potential.

PROOF. It follows from Proposition 7.1 that $a_0(x) = 1$ and A is a differential operator in the form of Definition 6.2. Since the function $\Xi(x,E)$ satisfies (7.2), the relation

$$a_j'''(x) - 4v(x)a_j'(x) - 2v'(x)a_j(x) + 4a_{j+1}'(x) = 0 \tag{9.3}$$

is obtained by observing the coefficient of E^{g-j} in (7.2) after substituting the function $\Xi(x,E)$ into it. Hence the commutativity of A and H is proved as

$$\begin{aligned} [A,H] &= \sum_{j=0}^{g} \left[a_j(x)\frac{d}{dx} - \frac{1}{2}a_j'(x), H\right] H^{g-j} \\ &= \sum_{j=0}^{g} \left(-2a_j'(x)\frac{d^2}{dx^2} + \frac{1}{2}a_j'''(x) - v'(x)a_j(x)\right) H^{g-j} \\ &= \sum_{j=0}^{g} \left(-2a_j'(x)(-H+v(x)) + \frac{1}{2}a_j'''(x) - v'(x)a_j(x)\right) H^{g-j} \\ &= \frac{1}{2}\sum_{j=0}^{g} \left(4a_{j+1}'(x) - 4v(x)a_j'(x) + a_j'''(x) - 2v'(x)a_j(x)\right) H^{g-j} = 0. \end{aligned} \tag{9.4}$$

□

For the differential operator $Q(H)$ obtained by substituting the operator $H = -d^2/dx^2 + \sum_{i=0}^{3} l_i(l_i+1)\wp(x+\omega_i)$ into the variable E of the polynomial $Q(E)$, we have the following relation.

PROPOSITION 9.2.

$$A^2 + Q(H) = 0. \tag{9.5}$$

PROOF. Let $f(x)$ be a solution to the second-order differential equation $(H - E)f(x) = 0$. From the definitions of A in (9.2) and (9.1), it follows that

$$\begin{aligned} Af(x) &= \sum_{j=0}^{g} \left\{ a_j(x)E^{g-j}\frac{d}{dx} - \frac{1}{2}\left(\frac{d}{dx}a_j(x)E^{g-j}\right)\right\} f(x) \\ &= \Xi(x,E)f'(x) - \frac{1}{2}\Xi'(x,E)f(x). \end{aligned} \tag{9.6}$$

From the commutativity of A and H we have $(H-E)Af(x) = 0$. Hence

$$\begin{aligned} A(Af(x)) &= \Xi(x,E)\frac{d}{dx}(Af(x)) - \frac{1}{2}\Xi'(x,E)Af(x) \\ &= \Xi(x,E)^2 f''(x) - \frac{1}{2}\Xi''(x,E)\Xi(x,E)f(x) + \frac{1}{4}\Xi'(x,E)^2 f(x). \\ &= \left(\Xi(x,E)^2(u(x)-E) - \frac{1}{2}\Xi''(x,E)\Xi(x,E) + \frac{1}{4}\Xi'(x,E)^2\right) f(x) \\ &= -Q(E)f(x) = -Q(H)f(x), \end{aligned} \tag{9.7}$$

where relation (7.4) was used. The operator $A^2 + Q(H)$ is a differential operator with degree at most $2g+1$. Since the equality $(A^2 + Q(H))f(x) = 0$ is valid for all eigenfunctions of H, it follows that $A^2 + Q(H) = 0$. □

Note that Theorem 9.1 and Proposition 9.2 had been established in ([**6, 25**]) for a smooth finite-gap potential. The hyperelliptic curve $\nu^2 = -Q(E)$ is called a spectral curve of the operator $H = -d^2/dx^2 + \sum_{i=0}^{3} l_i(l_i+1)\wp(x+\omega_i)$ related to the algebro-geometric finite-gap potential. The spectral curve $\nu^2 = -Q(E)$ appears not only in the relation $A^2 + Q(H) = 0$ but also as the form $\sqrt{-Q(E)}$ in the integral representation of solutions (see (7.5)), the monodromy formula represented by the hyperelliptic integral (see (7.22)) and the Hermite-Krichever Ansats.

We now investigate a relationship between the polynomial $Q(E)$ and the Heun polynomial. For this purpose, we quote the following proposition on the quasi-solvability of Heun's differential equation.

PROPOSITION 9.3 ([**21**, Proposition 5.1]). *Let α_i be a number such that $\alpha_i = -l_i$ or $\alpha_i = l_i+1$ for all $i \in \{0,1,2,3\}$, and set $d = -\sum_{i=0}^{3} \alpha_i/2$. Suppose that $d \in \mathbb{Z}_{\geq 0}$, and let $V_{\alpha_0,\alpha_1,\alpha_2,\alpha_3}$ be the $(d+1)$-dimensional space spanned by*

$$\left\{\widehat{\Phi}(\wp(x))\wp(x)^n\right\}_{n=0,\dots,d}, \tag{9.8}$$

where $\widehat{\Phi}(z) = (z-e_1)^{\alpha_1/2}(z-e_2)^{\alpha_2/2}(z-e_3)^{\alpha_3/2}$. Then the operator $H = -d^2/dx^2 + \sum_{i=0}^{3} l_i(l_i+1)\wp(x+\omega_i)$ preserves the space $V_{\alpha_0,\alpha_1,\alpha_2,\alpha_3}$.

Proposition 9.3 follows from the fact that the action of the operator $\widehat{H} = \widehat{\Phi}(z)^{-1} \circ H \circ \widehat{\Phi}(z)$ on monomials looks like (5.7), where $z = \wp(x)$. Note that the eigenfunctions of H on the space $V_{\alpha_0,\alpha_1,\alpha_2,\alpha_3}$ are Heun polynomials essentially.

We now describe the spaces determined by Proposition 9.3 for the case $l_0, l_1, l_2, l_3 \in \mathbb{Z}_{\geq 0}$. For simplicity we assume that $l_0 \geq l_1 \geq l_2 \geq l_3 \geq 0$. If $l_0 + l_1 + l_2 + l_3$ is even and $l_0 + l_3 \geq l_1 + l_2$, then

$$(9.9)\quad V = V_{-l_0,-l_1,-l_2,-l_3} \oplus V_{-l_0,-l_1,l_2+1,l_3+1} \oplus V_{-l_0,l_1+1,-l_2,l_3+1} \oplus V_{-l_0,l_1+1,l_2+1,-l_3}.$$

If $l_0 + l_1 + l_2 + l_3$ is even and $l_0 + l_3 < l_1 + l_2$, then

(9.10)
$$V = V_{-l_0,-l_1,-l_2,-l_3} \oplus V_{-l_0,-l_1,l_2+1,l_3+1} \oplus V_{-l_0,l_1+1,-l_2,l_3+1} \oplus V_{l_0+1,-l_1,-l_2,l_3+1}.$$

If $l_0 + l_1 + l_2 + l_3$ is odd and $l_0 \geq l_1 + l_2 + l_3 + 1$, then

(9.11)
$$V = V_{-l_0,-l_1,-l_2,l_3+1} \oplus V_{-l_0,-l_1,l_2+1,-l_3} \oplus V_{-l_0,l_1+1,-l_2,-l_3} \oplus V_{-l_0,l_1+1,l_2+1,l_3+1}.$$

If $l_0 + l_1 + l_2 + l_3$ is odd and $l_0 < l_1 + l_2 + l_3 + 1$, then

$$(9.12)\quad V = V_{-l_0,-l_1,-l_2,l_3+1} \oplus V_{-l_0,-l_1,l_2+1,-l_3} \oplus V_{-l_0,l_1+1,-l_2,-l_3} \oplus V_{l_0+1,-l_1,-l_2,-l_3}.$$

Since it follows from Proposition 9.3 that each component of V is invariant under the action of H, the space V is also invariant under the action of H. It is shown that the finite-dimensional invariant spaces of the operator H in the form of quasi-exact solvability are exhausted by the components of V (see [**20**]). On the space V and the polynomial $Q(E)$, we have

PROPOSITION 9.4 ([**24**]). *We assume that $l_0, l_1, l_2, l_3 \in \mathbb{Z}$ and $l_0 \geq l_1 \geq l_2 \geq l_3 \geq 0$. We define the space V by one of (9.9), (9.10), (9.11) or (9.12). Then the characteristic polynomial of $H = -d^2/dx^2 + \sum_{i=0}^{3} l_i(l_i+1)\wp(x+\omega_i)$ on the space V coincides with $Q(E) = 0$ defined by (7.4).*

We now define the nonnegative integer g by

(9.13)
$$g = \begin{cases} l_0, & l_0 + l_1 + l_2 + l_3 \text{ is even and } l_0 + l_3 \geq l_1 + l_2; \\ (l_0 + l_1 + l_2 - l_3)/2, & l_0 + l_1 + l_2 + l_3 \text{ is even and } l_0 + l_3 < l_1 + l_2; \\ l_0, & l_0 + l_1 + l_2 + l_3 \text{ is odd and } l_0 \geq l_1 + l_2 + l_3 + 1; \\ (l_0 + l_1 + l_2 + l_3 + 1)/2, & l_0 + l_1 + l_2 + l_3 \text{ is odd and } l_0 < l_1 + l_2 + l_3 + 1. \end{cases}$$

Then it follows from $\dim V = 2g + 1$ that the degree of $Q(E)$ is $2g + 1$. Hence the spectral curve $\nu^2 = -Q(E)$ is a hyperelliptic curve of genus g. The genus was also calculated in [**3, 26**], and it coincides with the value g which was used in the definition of $\Xi(x, E)$ (see (7.3)).

We can construct the odd-order commuting operator A another way. Namely, by considering the Darboux-Crum transformations for each component of V, deforming and composing them in a suitable way, we have another expression of A (see [**24**]).

Recall that Lamé's differential equation is a special case of Heun's differential equation. If $l_0 \in \mathbb{Z}_{>0}, l_1 = l_2 = l_3 = 0$, then the genus g of the spectral curve $\nu^2 = -Q(E)$ is l_0, and the space V is written as (9.9) (l_0: even) or (9.11) (l_0: odd) by setting $l_0 = g, l_1 = l_2 = l_3 = 0$. The odd-order commuting operator A admits the following expression, which was obtained in [**24**]:

$$(9.14)\qquad A = \frac{1}{2}\left(\frac{-1}{4}\right)^{g} \wp'\left(\frac{x}{2}\right)^{g+1} \cdot \left(\frac{2}{\wp'\left(\frac{x}{2}\right)}\frac{d}{dx}\right)^{2g+1} \cdot \wp'\left(\frac{x}{2}\right)^{g}.$$

If $l_0 = 1, l_1 = l_2 = l_3 = 0$, then we have $V = V_{-1,0,0,1} \oplus V_{-1,0,1,0} \oplus V_{-1,1,0,0} \oplus V_{-1,1,1,1}$ and $V_{-1,1,1,1} = \{0\}$. The characteristic polynomials of the operator $H = -d^2/dx^2 + 2\wp(x)$ on the spaces $V_{-1,0,0,1}$, $V_{-1,0,1,0}$, $V_{-1,1,0,0}$ are $E + e_3$, $E + e_2$, $E + e_1$, respectively, and the product of them is $Q(E)$. The odd-order commuting operator A is written as

$$A = -\left\{ \left(\frac{d}{dx}\right)^3 - 3\wp(x)\frac{d}{dx} - \frac{3}{2}\wp'(x) \right\}, \tag{9.15}$$

and another representation which follows from composing Darboux-Crum transformations is

$$\begin{aligned} A = &-\left(\frac{d}{dx} + \frac{1}{2}\frac{\wp'(x)}{\wp(x) - e_2}\right)\left(\frac{d}{dx} + \frac{1}{2}\frac{\wp'(x)}{\wp(x) - e_1} - \frac{1}{2}\frac{\wp'(x)}{\wp(x) - e_2}\right) \\ &\times \left(\frac{d}{dx} - \frac{1}{2}\frac{\wp'(x)}{\wp(x) - e_1}\right). \end{aligned} \tag{9.16}$$

If $l_0 = 2, l_1 = l_2 = l_3 = 0$, then we have $V = V_{-2,0,0,0} \oplus V_{-2,0,1,1} \oplus V_{-2,1,0,1} \oplus V_{-2,1,1,0}$. The characteristic polynomials of the operator $H = -d^2/dx^2 + 6\wp(x)$ on the spaces $V_{-2,0,0,0}$, $V_{-2,0,1,1}$, $V_{-2,1,0,1}$, $V_{-2,1,1,0}$ are $E^2 - 3g_2$, $E - 3e_1$, $E - 3e_2$, $E - 3e_3$ respectively, and the product of them is $Q(E)$. The odd-order commuting operator A is written as

$$A = \left(\frac{d}{dx}\right)^5 - 15\wp(x)\left(\frac{d}{dx}\right)^3 - \frac{45}{2}\wp'(x)\left(\frac{d}{dx}\right)^2 - 9\left(5\wp(x)^2 - \frac{3}{4}g_2\right)\frac{d}{dx}, \tag{9.17}$$

and another representation which follows from composing Darboux-Crum transformations is

$$\begin{aligned} A = &\left(\frac{d}{dx} + \frac{1}{2}\frac{\wp'(x)}{\wp(x) - e_1} + \frac{1}{2}\frac{\wp'(x)}{\wp(x) - e_2}\right)\left(\frac{d}{dx} + \frac{\wp'(x)}{\wp(x) - e_1} - \frac{1}{2}\frac{\wp'(x)}{\wp(x) - e_2}\right) \\ &\times \left(\frac{d}{dx} - \frac{\wp'(x)}{\wp(x) - e_1} + \frac{1}{2}\frac{\wp'(x)}{\wp(x) - e_2} + \frac{1}{2}\frac{\wp'(x)}{\wp(x) - e_3}\right) \\ &\times \left(\left(\frac{d}{dx}\right)^2 - \frac{1}{2}\left(\frac{\wp'(x)}{\wp(x) - e_1} + \frac{\wp'(x)}{\wp(x) - e_2} + \frac{\wp'(x)}{\wp(x) - e_3}\right)\frac{d}{dx}\right). \end{aligned} \tag{9.18}$$

10. Conclusion

In this article, results on solutions to Heun's differential equation and their monodromy are explained. Namely, the Heun function, the Heun polynomial and the elliptical representation of Heun's equation are introduced briefly and the integral representation of solutions and formulas on monodromy are explained. We now point out other approaches to Heun's differential equation. If the parameters $\alpha, \beta, \gamma, \delta, \epsilon, q, t$ satisfy some relations, then we may investigate solutions which are expressed by hypergeometric functions or algebraic functions by a suitable change of the variable z (see [**10**]). Another approach is to apply the method of perturbation by the limit $t \to \infty$, which changes Heun's differential equation into the hypergeometric differential equation (see [**21**]). Each method has its own available conditions for the parameters and its own results. We may choose methods in accordance with the subject of the study, and it would be interesting to compare results and their logic respectively for problems which allow several approaches.

Acknowledgements

The author would like to thank the referees for valuable comments.

References

1. R. V. Craster and V. H. Hoang, Applications of Fuchsian differential equations to free boundary problems, Roy. Soc. Lond. Proc. Ser. A Math. Phys. Eng. Sci., **454** (1998), 1241–1252. MR1631566 (99c:34005)
2. F. Finkel, D. Gomez-Ullate, A. Gonzalez-Lopez, M. A. Rodriguez and R. Zhdanov, New spin Calogero-Sutherland models related to BN-type Dunkl operators, Nuclear Phys. B, **613** (2001), 472–496. MR1859388 (2002h:81101)
3. F. Gesztesy and R. Weikard, Treibich-Verdier potentials and the stationary (m)KdV hierarchy, Math. Z, **219** (1995), 451–476. MR1339715 (96e:14030)
4. E. L. Ince, Further investigations into the periodic Lamé functions, Proc. Roy. Soc. Edinburgh, **60** (1940), 83–99. MR0002400 (2:46d)
5. V. I. Inozemtsev, Lax representation with spectral parameter on a torus for integrable particle systems, Lett. Math. Phys., **17** (1989), 11–17. MR990577 (90j:58058)
6. A. R. Its and V. B. Matveev, Schrödinger operators with the finite-band spectrum and the N-soliton solutions of the Korteweg-de Vries equation, Theoret. and Math. Phys., **23** (1975), 343–355. MR0479120 (57:18570)
7. K. Iwasaki, H. Kimura, S. Shimomura and M. Yoshida, From Gauss to Painlevé. A modern theory of special functions, Aspects of Mathematics, E16. Friedr. Vieweg & Sohn, Braunschweig, 1991. MR1118604 (92j:33001)
8. Y. Komori and T. Sugawa, Bers embedding of the Teichmüller space of a once-punctured torus. Conform. Geom. Dyn., **8** (2004), 115–142 MR2060380 (2005g:30055)
9. I. M. Krichever, Elliptic solutions of the Kadomcev-Petviašvili equations, and integrable systems of particles, Functional Anal. Appl., **14** (1980), 282–290. MR595728 (82e:58046)
10. R. S. Maier, On reducing the Heun equation to the hypergeometric equation, J. Differential Equations, **213** (2005), 171–203. MR2139342 (2006a:34245)
11. R. S. Maier, Lamé polynomials, hyperelliptic reductions and Lamé band structure, Preprint, math-ph/0309005, 2003. MR2377687 (2009i:37175)
12. H. Ochiai, T. Oshima and H. Sekiguchi, Commuting families of symmetric differential operators, Proc. Japan. Acad., **70** (1994), 62–66. MR1272672
13. M. A. Olshanetsky and A. M. Perelomov, Quantum integrable systems related to Lie algebras, Phys. Rep., **94** (1983), 313–404. MR708017 (84k:81007)
14. T. Oshima, Completely integrable systems with a symmetry in coordinates, Asian J. Math., **2** (1998), 935–955. MR1734134 (2000j:37095)
15. T. Oshima and H. Sekiguchi, Commuting families of differential operators invariant under the action of a Weyl group, J. Math. Sci. Univ. Tokyo, **2** (1995), 1–75. MR1348022 (96k:35006)
16. A. Ronveaux (ed.), Heun's differential equations, Oxford Science Publications, Oxford University Press, Oxford, 1995. MR1392976 (98a:33005)
17. S. Slavyanov and W. Lay, Special Functions., Oxford Science Publications, Oxford University Press, Oxford, 2000. MR1858237 (2004e:33003)
18. A. O. Smirnov, Elliptic solitons and Heun's equation, In: The Kowalevski property, CRM Proc. Lecture Notes, **32**, Amer. Math. Soc., Providence, RI, 2002, pp. 287–305. MR1916788 (2003f:34183)
19. K. Takemura, Quasi-exact solvability of Inozemtsev models, J. Phys. A **35** (2002), 8867–8881. MR1946978 (2004h:81264)
20. K. Takemura, The Heun equation and the Calogero-Moser-Sutherland system I: the Bethe Ansatz method, Comm. Math. Phys., **235** (2003), 467–494. MR1974511 (2004g:81055)
21. K. Takemura, The Heun equation and the Calogero-Moser-Sutherland system II: the perturbation and the algebraic solution, Electron. J. Differential Equations, **2004** (2004), no. 15 1–30. MR2036199 (2005d:81180)
22. K. Takemura, The Heun equation and the Calogero-Moser-Sutherland system III: the finite gap property and the monodromy, J. Nonlinear Math. Phys., **11** (2004), 21–46. MR2031210 (2005d:37124)

23. K. Takemura, The Heun equation and the Calogero-Moser-Sutherland system IV: the Hermite-Krichever Ansatz, Comm. Math. Phys., **258** (2005), 367–403. MR2171700 (2006k:33041)
24. K. Takemura, The Heun equation and the Calogero-Moser-Sutherland system V: generalized Darboux transformations, J. Nonlinear Math. Phys., **13** (2006), 584–611. MR2263090 (2007g:37061)
25. S. Tanaka and E. Date, KdV equation (Japanese), Kinokuniya Shoten, Tokyo, 1979.
26. A. Treibich, Hyperelliptic tangential covers, and finite-gap potentials, Russian Math. Surveys, **56** (2001), 1107–1151. MR1886721 (2003m:14049)
27. A. Treibich and J.-L. Verdier, Revêtements exceptionnels et sommes de 4 nombres triangulaires, Duke Math. J., **68** (1992), 217–236. MR1191559 (94f:14026)
28. E. T. Whittaker and G. N. Watson, A course of modern analysis. Fourth edition, Cambridge University Press, Cambridge, 1996.
MR1424469 (97k:01072)

Department of Mathematics, Chuo University, 1-13-27 Kasuga, Bunkyo-ku, Tokyo 112-8551, Japan

E-mail address: takemura@math.chuo-u.ac.jp

Translated by KOUICKI TAKEMURA

Amer. Math. Soc. Transl.
(2) Vol. **230**, 2010

Scattering Theory and Inverse Problems

Hiroshi Isozaki

ABSTRACT. This is an expository article on 2 topics, the quantum mechanical 3-body problem and the inverse boundary value problem. The main themes are the eigenfunction expansion theorem associated with the continuous spectrum of 3-body quantum systems and applications of hyperbolic geometry to the inverse boundary value problem.

1. Schrödinger operators

1.1. Energy operators for atoms and molecules. The Schrödinger operator is a differential operator of the form

$$H = -\Delta + V(x),$$

and is a fundamental tool in studying the quantum mechanical properties of small particles such as atoms or molecules. Here $x \in \mathbf{R}^n$ denotes the position of the particle, Δ is the Laplacian on $\mathbf{R}^n$, and $V(x)$ is a real function corresponding to the scalar electric field. Since H is selfadjoint in $L^2(\mathbf{R}^n)$, the spectrum of H is a closed subset in $\mathbf{R}$ and represents the energy of the particle.

The specrum of a system of finite particles has the following property. There exists some value $E_0 \in \mathbf{R}$ such that under E_0 there is only a discrete set of eigenvalues and above E_0 the particle can take arbitrary values of energy. The former is called the discrete spectrum and the latter the continuous spectrum.

1.2. Bound states and scattering states. The eigenvalue problem

$$(H - \lambda)\varphi(x, \lambda) = 0 \tag{1.1}$$

is important not only to compute the energy of the system but also to know its physical properties. If $\lambda < E_0$ is in the discrete spectrum, (1.1) has an eigenvector belonging to $L^2(\mathbf{R}^n)$. Normalized as $\int_{\mathbf{R}^n} |\varphi(x, \lambda)|^2 dx = 1$, for any domain $D \subset \mathbf{R}^n$, $\int_D |\varphi(x, \lambda)|^2 dx$ represents the probability that the particle exists in the region D. Therefore if λ is an eigenvalue, we may well think that the particle is localized in a finite region of $\mathbf{R}^n$. Hence we say that it is in a bound state.

When $\lambda > E_0$, in general, (1.1) has no eigenvector belonging to $L^2(\mathbf{R}^n)$. By the fundamental principle of quantum mechanics, there should exist a system of

---

This article originally appeared in Japanese in Sūgaku **59** (2) (2007), 113–130.
2000 *Mathematics Subject Classification.* Primary 35J10; Secondary 35P25, 47A40, 81U40.

functions on $\mathbf{R}^n$, which are solutions $\varphi(x,\lambda)$ of (1.1) not belonging to $L^2(\mathbf{R}^n)$ and, together with bound states, form a basis of $L^2(\mathbf{R}^n)$. The solution of the Schrödinger equation corresponding to the continuous spectrum will, as the time passes by, become asymptotically free and go away to infinity. In this case the particle is said to be in a scattering state.

1.3. Eigenfunction expansions. The eigenfunction expansion theory aims at constructing the complete system of the above eigenfunctions. This theorem for the multi-dimensional 2-body Schrödinger operator was first proved by Ikebe [**9**], which states as follows. If $V(x)$ decays sufficiently rapidly at infinity, there exist functions $\psi(x,\xi)$, $\xi \in \mathbf{R}^n$, satisfying $(H - |\xi|^2)\psi = 0$ and

$$\psi(x,\xi) \sim e^{ix\cdot\xi} + \frac{e^{i\sqrt{E}r}}{r^{(n-1)/2}}A(E,\theta,\omega), \quad r \to \infty$$

$(r = |x|, \theta = x/r, \omega = \xi/|\xi|, E = |\xi|^2)$. Moreover any $f \in L^2(\mathbf{R}^n)$ is expanded as follows:

$$\begin{aligned} f(x) &= (2\pi)^{-n/2}\int_{\mathbf{R}^n} \psi(x,\xi)\tilde{f}(\xi)d\xi + \sum_j (f,\psi_j)\psi_j(x), \\ \tilde{f}(\xi) &= (2\pi)^{-n/2}\int_{\mathbf{R}^n} \overline{\psi(x,\xi)}f(x)dx. \end{aligned} \tag{1.2}$$

Here (,) is the inner product of $L^2(\mathbf{R}^n)$, and ψ_j is the eigenvector associated with the discrete spectrum of H. If $V(x) = 0$, then $\psi(x,\xi) = e^{ix\cdot\xi}$ and (1.2) is the Fourier inversion formula. Therefore (1.2) is called the generalized Fourier transformation. $\psi(x,\xi)$ is called a **generalized eigenfunction** (henceforth called an eigenfunction for simplicity's sake), and $A(E,\theta,\omega)$ is called the **scattering amplitude**.

1.4. S-matrix. The process of the quantum mechanical scattering experiment is schematically as follows. One prepares a target particle and sends a beam of particles so that one particle of energy E from the ω-direction collides with the target per unit time. Then $|A(E,\theta,\omega)|^2$ numbers of the particles are scattered to the θ-direction. This is a physically observable quantity called the differential cross section. The integral operator

$$S(E)f(\theta) = f(\theta) - C(E)\int_{S^{n-1}} A(E,\theta,\omega)f(\omega)d\omega,$$

$C(E)$ being a suitable constant, is a unitary operator on $L^2(S^{n-1})$, called an **S-matrix**. What we can observe in the physical experiment is not the scattering amplitude, but the differential cross section. However, it is known that if $V(x)$ is sufficiently small we can construct the scattering amplitude from the differential cross section. Therefore (at least when $V(x)$ is small), one may think that S can be directly obtained from the physical experiment. The S-matrix is believed to be equivalent to the original physical system; namely, the Schrödinger operator $H = -\Delta + V(x)$ can be reconstructed from the S-matrix.

1.5. Fundamental problems for stationary Schrödinger operators. There are two main problems of quantum scattering theory. In the forward problem, we try to construct the above-mentioned complete system of eigenfunctions and the S-matrix from the given potential $V(x)$. The reverse process, reconstructing $V(x)$ from the given S-matrix, is called the inverse problem. They give a theoretical basis for determining quantum mechanical particles from the scattering experiment. The one-dimensional forward problem has a long history. The one-dimensional inverse problem was solved in the 1950s. The multi-dimensional forward problem for 2-particle systems was studied extensively in the 1960s and 1970s. We should point out the pioneering work of Kato for the foundation of the mathematical study of Schrödinger operators, and basic contributions by Ikebe, Kuroda ([**26**]) for 2-body problems. The temporal behavior of systems of more than 3 particles (the problem of asymptotic completeness) was studied deeply in the late 1980s. However, the analysis of eigenvalue problems (1.1) associated with the continuous spectrum for many particle systems remains open because of its difficulty. The multi-dimensional inverse scattering problem has not yet been solved completely even in the 2-particle case.

2. Eigenfunction expansion for 3-particle systems

2.1. 3-body Schrödinger operator. The work of Faddeev [**4**] is the landmark for the mathematical study of the 3-body problem, where the main theme was the proof of asymptotic completeness of wave operators by the eigenfunction expansion. This is the first brilliant success of the mathematical study of the many-body scattering theory; however, since in this work all the analysis is done in the Fourier transformed space, it is not easy to analyze the spatial asymptotics of eigenfunctions and properties of the S-matrix. Also it contains an implicit assumption on the 0-energy state for the 2-body subsystems.[1] Hence it is not obvious how to specify the class of potentials to which the results of the theorem can be applied.

In the late 1970s, Ginibre-Moulin [**7**], Howland [**8**], and Thomas [**37**] tried to study the 3-body problem in the x-space without passing to the Fourier transform. These works basically followed the idea of Faddeev [**4**], and the problems of asymptotic behavior of eigenfunctions, properties of the S-matrix and the implicit assumption on the 0-energy state for the 2-body subsystems remained untouched. Although the asymptotic completeness for N-particle systems was proved in the 1980s under a general assumption, the problem of constructing the complete system of eigenfunctions for the eigenvalue problem (1.1) of the 3-particle system and studying its spatial asymptotics remained unsolved.

In the work [**17**], we studied (1) the asymptotic expansion at infinity of the resolvent of the 3-particle Schrödinger operator, (2) the construction of the complete system of eigenfunctions and generalized Fourier transformation associated with the continuous spectrum for the 3-particle system, (3) the asymptotic expansion of eigenfunctions at infinity and the construction of the S-matrix. Its features are:

- characterization of the eigenspace for the eigenvalue problem associated with the continuous spectrum by the generalized Fourier transform,
- derivation of spatial asymptotics of eigenfunctions for all dierctions,
- clarification of the relation between the S-matrix and the asymptotics of eigenfunctions,

[1] The state ψ which satisfies the Schrödinger equation $(-\Delta + V(x))\psi(x) = 0$, and $\psi(x) \to 0$ as $|x| \to \infty$, but does not belong to $L^2(\mathbf{R}^n)$ is called the zero-resonance. The implicit assumption assumes that zero is neither the eigenvalue nor the resonance. However, this condition is related to the property of soluitons to the Schrödinger equation and cannot be directly examined from the potential.

- simple decay assumption of the potential avoiding the implicit 0-energy condition (nonexistence of zero-resonances).

Let us state its mathematical content. Assume that each particle moves in $\mathbf{R}^3$. Let $q^i \in \mathbf{R}^3$ $(i = 1, 2, 3)$ be the coordinate, and $m_i > 0$ the mass of i-th particle. Let $a = (ij)$ denote the pair i, j, and k the 3rd particle. The reduced masses m_a, n_a are defined by

$$\frac{1}{m_a} = \frac{1}{m_i} + \frac{1}{m_j}, \quad \frac{1}{n_a} = \frac{1}{m_i + m_j} + \frac{1}{m_k},$$

and the Jacobi coordinates are

$$x^a = \sqrt{2m_a}(q^i - q^j), \quad x_a = \sqrt{2n_a}\Big(q^k - \frac{m_i q^i + m_j q^j}{m_i + m_j}\Big).$$

Then the 3-body Schrödinger operator is written as

$$H_0 = -\Delta_{x_a} - \Delta_{x^a}, \quad H = H_0 + \sum_a V_a(x^a).$$

H_0 is the operator on $L^2(\mathbf{R}^6)$ corresponding to the kinetic energy for the 3-body system with center of mass removed, independent of the choice of the pair a. $\sum_a$ denotes the sum over 3 pairs $(ij), (jk), (ki)$. For different pairs a and b, x^b is written by a linear combination of x_a and x^a. Hence $\sum_a V_a(x^a)$ is also an operator on $L^2(\mathbf{R}^6)$. We put

$$r = (|x_a|^2 + |x^a|^2)^{1/2}, \quad r_a = |x_a|, \quad r^a = |x^a|.$$

r is independent of the choice of the pair a.

The system is in the scattering state if it has the energy belonging to the continuous spectrum. Let us explain the scattering experiment in a schematic way. Let A, B, C denote 3 particles, and suppose the 3rd partcle C collides to the pair A, B of the bound state. Then the following 5 results will appear:

$$(A,B) + C \Longrightarrow \begin{cases} (a) & (A,B) + C, \\ (b) & (A,B)^* + C_*, \\ (c) & (A,B)' + C, \\ (d) & (A,C) + B, \\ (e) & A + B + C. \end{cases}$$

In the case (a) the particles do not change their states, which is called the elastic scattering. The other cases are called inelastic scattering. In (b), the energy of the pair (A, B) increases or decreases (which is represented by the notation $(A, B)^*$), while the energy of C decreases or increases (denoted by C_*). In (c) the energy is unchanged; however, the pair (A, B) is transferred to a different eigenstate associated with the same eigenvalue (denoted by $(A, B)'$). The process (d) is a rearrangement of particles, and in (e) the bound state (A, B) disintegrates and 3 particles move freely without interactions. This is called the break-up process. Let us represent these states in terms of spherical wave solutions to the equation $(H - \lambda)u = 0$. Usually free particles are represented by a plane wave such as $e^{ix\cdot\xi}$. However, integrated as

$$\int_{S^{n-1}} e^{i\sqrt{\lambda}\omega\cdot x}\phi(\omega)d\omega,$$

the plane wave is asymptotically transformed to the spherical wave $r^{-(n-1)/2}e^{\pm i\sqrt{\lambda}r}$ as $r = |x| \to \infty$. Thus spherical waves also represent free particles. What is the spherical wave for the 3-body problem? In the scattering state (e), where 3 particles move freely without interaction, the spherical wave has the form

$$\frac{e^{\pm i\sqrt{\lambda}r}}{r^{5/2}}.$$

In the states $(a) \sim (d)$, where two particles $a = (i, j)$ form a bound state and k is free, it is

$$\frac{e^{\pm i\sqrt{\lambda-\lambda^{a,n}}r_a}}{r_a} \otimes \varphi^{a,n}(x^a).$$

Here $\varphi^{a,n}(x^a) \in L^2(\mathbf{R}^3)$ is an eigenvector for the 2-body Schrödinger operator

$$H^a = -\Delta_{x^a} + V_a(x^a)$$

associated with the eigenvalue $\lambda^{a,n}$. The paper [**17**] asserts that on the basis of physically natural assumptions, one can show that the eigenvalue problem of the 3-body Schrödinger equation

$$\Big(- \Delta_{x_a} - \Delta_{x^a} + \sum_a V_a(x^a) - \lambda \Big) u = 0$$

contains all of these solutions, which form a complete system.

2.2. Eigenfunction expansion. The following spaces $\mathcal{B}$, $\mathcal{B}^*$ introduced by Agmon-Hörmander [**1**] play a crucial role in the analysis of the 3-body problem. We remark that Murata had already discussed this space in his analysis of the continuous spectrum of differential operators (see e.g. [**28**]). The space $\mathcal{B}$ is the totality of f satisfying

$$\|f\|_{\mathcal{B}} = \sum_{j=0}^{\infty} 2^{j/2} \|f\|_{L^2(\Omega_j)} < \infty,$$

$$\Omega_j = \{x \in \mathbf{R}^n \,; r_{j-1} < |x| < r_j\}, \quad r_j = 2^j \ (j \geq 0), \quad r_{-1} = 0.$$

The dual space of $\mathcal{B}$ is identified with the set of all u satisfying

$$\|u\|_{\mathcal{B}^*} = \left(\sup_{R>1} \frac{1}{R} \int_{|x|<R} |u(x)|^2 dx \right)^{1/2} < \infty.$$

The following assumptions are imposed on the potential.

V_a is split into 2 parts : $V_a(x^a) = V_a^{(1)}(x^a) + V_a^{(2)}(x^a)$, where $V_a^{(1)}(y)$ is compactly supported, and $V_a^{(1)}(y) \in L^2(\mathbf{R}^3)$, $|\partial_y^\alpha V_a^{(2)}(y)| \leq C_\alpha (1 + |y|)^{-|\alpha| - \rho}$, $\rho > 5$, $\forall \alpha$.

With this assumption, the continuous spectrum of H is

$$\sigma_{cont}(H) = [\Sigma, \infty), \quad \Sigma = \inf_a \{\inf \sigma(H^a)\},$$

where $\sigma(H^a)$ is the spectrum of H^a.

The construction of the generalized Fourier transformation associated with H is summarized in the following theorem. Let N_a be the number of eigenvalues of H^a, and $\{\lambda^{a,n}\}_{n=1}^{N_a}$ the eigenvalues of H^a. It is known that the above assumption implies that $N_a < \infty$ and $\lambda^{a,n} < 0$. The set

$$\Lambda = \bigcup_a \{\lambda_{a,n} ; n = 1, \ldots, N_a\}$$

is called the set of thresholds of H, which are the values of the energy where properties of scattering states change.

Let us introduce the following L^2-space on the boundary at infinity:

$$\mathbf{h} = L^2(S^5) \oplus \left(\bigoplus_a \bigoplus_{n=1}^{N_a} L^2(S^2) \right).$$

In the following, for Banach spaces X, Y, $\mathbf{B}(X;Y)$ denotes the totality of bounded operators from X to Y. By the absolutely continuous subspace for H, we mean all u such that $(E(\lambda)u, u)$ is absolutely continuous with respect to the Lebesgue measure $d\lambda$, where $E(\lambda)$ is the spectral decomposition for H. In our case, it is the orthogonal complement of the linear span of eigenvectors for H.

THEOREM 2.1. *There exist bounded operators*

$$\mathcal{F}_0(\lambda) \in \mathbf{B}(\mathcal{B}; L^2(S^5)), \quad \mathcal{F}_{a,n}(\lambda) \in \mathbf{B}(\mathcal{B}; L^2(S^2)),$$

the former being defined for $\lambda > 0$, the latter for $\lambda > \lambda^{a,n}$ with $\lambda \notin \Lambda$, where $\lambda^{a,n}$ is an eigenvalue with eigenvector $\varphi^{a,n}$, having the following property.
(1) *For any $\varphi_0 \in L^2(S^5)$, $\varphi_{a,n} \in L^2(S^2)$,*

$$(H - \lambda)\mathcal{F}_0(\lambda)^*\varphi_0 = 0, \quad (H - \lambda)\mathcal{F}_{a,n}(\lambda)^*\varphi_{a,n} = 0.$$

(2) *For $f \in \mathcal{B}$, we put $(\mathcal{F}_0 f)(\lambda) = \mathcal{F}_0(\lambda)f$, $(\mathcal{F}_{a,n}f)(\lambda) = \mathcal{F}_{a,n}(\lambda)f$. Then $\mathcal{F} = (\mathcal{F}_0, \mathcal{F}_{a,1}, \ldots)$ is uniquely extended to a unitary operator*

$$\mathcal{F} : \mathcal{H}_{ac}(H) \to L^2((0,\infty); L^2(S^5)) \oplus \left(\bigoplus_a \bigoplus_{n=1}^{N_a} L^2((\lambda^{a,n}, \infty); L^2(S^2)) \right),$$

where $\mathcal{H}_{ac}(H)$ is the absolutely continuous subspace for H.
(3) *$\mathcal{F}$ diagonalizes H:*

$$(\mathcal{F}Hf)(\lambda) = \lambda(\mathcal{F}f)(\lambda).$$

(4) *For any $f \in \mathcal{H}_{ac}(H)$ the following inverson formula holds:*

$$f = \int_0^\infty \mathcal{F}_0(\lambda)^*(\mathcal{F}_0 f)(\lambda)d\lambda + \sum_a \sum_{n=1}^{N_a} \int_{\lambda^{a,n}}^\infty \mathcal{F}_{a,n}(\lambda)^*(\mathcal{F}_{a,n}f)(\lambda)d\lambda.$$

We define $\mathcal{F}_0(\lambda) = 0$ $(\lambda \leq 0)$, $\mathcal{F}_{a,n}(\lambda) = 0$ $(\lambda \leq \lambda^{a,n})$, and put $\mathcal{F}(\lambda) = (\mathcal{F}_0(\lambda), \mathcal{F}_{a,1}(\lambda), \ldots)$. Since $\mathcal{F}(\lambda) \in \mathbf{B}(\mathcal{B}; \mathbf{h})$, $\mathcal{F}(\lambda)^* \in \mathbf{B}(\mathbf{h}; \mathcal{B}^*)$. The following theorem shows that the eigenspace for the eigenvalue problem associated with the continuous spectrum for the 3-particle Schrödinger operator is described by this generalized Fourier transformation.

THEOREM 2.2. *If $\lambda \in \sigma_{cont}(H) \setminus \Lambda$, $\{u \in \mathcal{B}^* \mid (H - \lambda)u = 0\} = \mathcal{F}(\lambda)^*\mathbf{h}$.*

In the following, $f(x) \simeq g(x)$ means

$$\lim_{R\to\infty} \frac{1}{R} \int_{|x|<R} |f(x) - g(x)|^2 dx = 0.$$

THEOREM 2.3. *If $\lambda \in \sigma_{cont}(H) \setminus \Lambda$, and $u \in \mathcal{B}^*$ satisfies $(H - \lambda)u = 0$, there exists $\varphi^{(\pm)} = (\varphi_0^{(\pm)}, \varphi_{a,1}^{(\pm)}, \ldots) \in \mathbf{h}$ having the following asymptotic expansion:*

$$
\begin{aligned}
u \quad \simeq \quad & C(\lambda) r^{-5/2} e^{i\sqrt{\lambda}r} \varphi_0^{(+)}(\omega) + \overline{C(\lambda)} r^{-5/2} e^{-i\sqrt{\lambda}r} \varphi_0^{(-)}(\omega) \\
& + \sum_a \sum_{n=1}^{N_a} \Big[C_{a,n}(\lambda) r_a^{-1} e^{i\sqrt{\lambda - \lambda^{a,n}} r_a} \varphi_{a,n}^{(+)}(\omega_a) \otimes \varphi^{a,n}(x^a) \\
& + \overline{C_{a,n}(\lambda)} r_a^{-1} e^{-i\sqrt{\lambda - \lambda^{a,n}} r_a} \varphi_{a,n}^{(-)}(\omega_a) \otimes \varphi^{a,n}(x^a) \Big].
\end{aligned}
$$

Here $C(\lambda), C_{a,n}(\lambda)$ are constants, $\omega = x/r$, $\omega_a = x_a/r_a$. Furthermore

$$
\varphi^{(+)} = \hat{\mathbf{S}}(\lambda) \mathbf{J} \varphi^{(-)}.
$$

Here $\hat{\mathbf{S}}(\lambda) \in \mathbf{B}(\mathbf{h}; \mathbf{h})$ is the S-matrix, and $\mathbf{J}$ is the antipodal operator defined by

$$
\mathbf{J} : (\psi_0(\omega), \psi_{a,1}(\omega_a), \ldots) \to (\psi_0(-\omega), \psi_{a,1}(-\omega_a), \ldots).
$$

The above formula describes anisotropic asymptotic expansions, since the function $r_a^{-1} e^{i\sqrt{\lambda - \lambda^{a,n}} r_a} \varphi_{a,n}(\omega_a) \otimes \varphi^{a,n}(x^a)$ concentrates around the subspace $\{x^a = 0\}$. The properties of wave functions for the 3-particle Schrödinger operator are also studied by physicists (see e.g. [**31**], [**30**]). The above theorem makes it possible to understand these physical computations in a proper way in the structure of the 3-body problem.

The proof of the above theorem is based on the many results from the stationary theory of many-body Schrödinger operators. In particlular, the method of micro-localization, which localizes the solution of the Schrödinger equation to some regions of the phase space by using pseudo-differential operators or Fourier integral operators plays an important role. This was conspicuously the case for the two-body problem. However, the application of micro-local calculus for many-body Schrödinger operators was not developed directly. The first big achievement was the commutator method due to Mourre [**27**] which guarantees the limiting absorption principle for the resolvent. The next important development was the refinement of this commutator method which makes it possible to prove the asymptotic completeness of wave operators by estimating the unitary group e^{-itH} by commutators. The micro-local resolvent estimates of many-body Schrödinger operators were derived in two steps: first by introducing an algebra of commutators and then by translating the results obtained in this algebra in terms of an algebra of pseudo-differential operators, in e.g. [**6**]. The Sommerfeld type radiation condition, which guarantees the uniqueness of solutions to the Helmholtz equation and plays an imporatant role in the stationary theory for Schrödinger equations, was extended to the N-body problem in [**13**] by using the commutator method. The structure of the S-matrix, in particular that of the break-up process which describes the scattering process of collapsing particles is stated in [**12**]. One can see the detailed exposition of these results in the scattering theory for many-body problems in [**19**].

3. Faddeev theory for multi-dimensional inverse scattering

3.1. Direction dependent Green operators. Next we shall discuss the inverse scattering problem, and our main theme is the multi-dimensional inverse problem for two-body systems. For $n \geq 2$, the scattering amplitude explained in subsection 1.3 is written as follows:

$$
\begin{aligned}
A(E;\theta,\omega) &= \int_{\mathbf{R}^n} e^{-i\sqrt{E}(\theta-\omega)\cdot x}V(x)dx \\
&\quad - \int_{\mathbf{R}^n} e^{-i\sqrt{E}\theta\cdot x}V(x)R(E+i0)\left(V(y)e^{i\sqrt{E}\omega\cdot y}V(y)\right)dx,
\end{aligned}
\tag{3.1}
$$

where $R(z) = (H-z)^{-1}$ $(z \notin \mathbf{R})$, $R(E+i0) = \lim_{\epsilon\downarrow 0} R(E+i\epsilon)$. (This limit exists in an appropriate topology.) Letting $R_0(z) = (-\Delta - z)^{-1}$, one can rewrite it as $R(z) = R_0(z)(1+VR_0(z))^{-1}$. We know that $R(E+i0) \to 0$ when $E \to \infty$. Therefore for a given $\xi \in \mathbf{R}^n$, we choose $\theta, \omega \in S^{n-1}$ such that $\sqrt{E}(\theta-\omega) = \xi$, and let $E \to \infty$ to see that

$$
\lim_{E\to\infty} A(E;\theta,\omega) = \int_{\mathbf{R}^n} e^{-ix\cdot\xi}V(x)dx.
$$

Thus in the multi-dimensional case, one can reconstruct $V(x)$ from the high-energy limit of the scattering amplitude. This apparently gives the impression that the multi-dimensional inverse scattering is simpler than the 1-dimensional problem. However, this argument is not satisfactory in the following respects. First in the high energy region, the nonrelativistic Schrödinger equation does not describe the physics correctly. Second, in the 1-dimensional Gel'fand-Levitan-Marchenko theory, we have a necessary and sufficient condition which guarantees a given function to be the S-matrix associated with some potential. However, the above high energy limit does not give such a condition which characterizes the S-matrix. This is the most difficult issue of the multi-dimensional inverse scattering problem.

The crucial key for the 1-dimensional inverse scattering problem is a Volterra type integal equation called the Gel'fand-Levitan equation. To extend this idea to the multi-dimensional problem, Faddeev introduced the following Green operator:

$$
\widetilde{G}(\zeta)f(x) = (2\pi)^{-n/2}\int_{\mathbf{R}^n} \frac{e^{ix\cdot\xi}}{\xi^2 + 2\zeta\cdot\xi}\widehat{f}(\xi)d\xi,
\tag{3.2}
$$

where $\widehat{f}(\xi)$ is the Fourier transform of f, and $\zeta \in \mathbf{C}^n$ is a complex parameter satisfying $\zeta^2 = \sum_{i=1}^n \zeta_i^2 = E > 0$. This Green operator is a solution operator for the equation which is obtained by the gauge transformation $u = e^{i\zeta\cdot x}v$ in the equation $(-\Delta - E)u = f$. As is inferred by (3.1), the physical scattering amplitude is written by the Green operator $R(E+i0)$. With this in mind, let us introduce a new type of scattering amplitude by replacing the usual Green operator by Faddeev's Green operator, and call it the Faddeev scattering amplitude. Faddeev showed that there exists an integral equation between the physical scattering amplitude and the Faddeev scattering amplitude, and it is always solvable. Using this scattering amplitude, Faddeev derived a multi-dimensional analogue of the Gel'fand-Levitan equation and derived a procedure to compute the potential from the S-matrix ([**5**]). This deep, remarkable work explains very well the algebraic or operator-theoretical mechanism behind the multi-dimensional inverse problem. However, some parts of the arguments are formal, due to the lack of estimates of solutions to the Schrödinger equation, and we do not know whether we can accept it as it is. The work [**15**] is an exposition of Faddeev theory, and in [**18**] some proofs are given which justify a part of Faddeev's argument.

3.2. $\overline{\partial}$-theory. As is seen from (3.2), the Faddeev scattering amplitude contains a complex parameter ζ. Let it be denoted by $A_F(\sqrt{E};\theta,\omega,\zeta)$. Further, putting $\xi = \sqrt{E}(\theta - \omega)$, we denote $A_F(\sqrt{E};\theta,\omega,\zeta)$ by $T(\xi,\zeta)$. Nachman [**29**] and Khenkin-Novikov [**25**] proved that $T(\xi,\zeta)$ is, as a function of ζ, regarded as a function on the complex manifold

$$\mathcal{V}_\xi = \{\zeta \in \mathbf{C}^n \mid |\zeta| > C,\ \mathrm{Im}\,\zeta \neq 0,\ \xi^2 + 2\xi\cdot\zeta = 0\},$$

and satisfies the equation

$$\overline{\partial} T(\xi,\zeta) = \sum_{j=1}^n A_j(\xi,\zeta) d\overline{\zeta}_j,$$

$$A_j(\xi,\zeta) = -(2\pi)^{1-n/2}\int_{\mathbf{R}^n} T(\xi-\eta,\zeta+\eta)T(\eta,\zeta)\eta_j\delta(\eta^2+2\zeta\cdot\eta)d\eta.$$

Moreover, by using a Bochner-Martinelli type formula on $\mathcal{V}_\xi$, one can obtain the following integral expression for V:

$$\widehat{V}(\xi) = T(\xi,\zeta_0) + \int_{\partial\mathcal{V}_\xi} T(\xi,\zeta)K(\zeta,\zeta_0) + \int_{\mathcal{V}_\xi}\sum_{j=1}^n A_j(\xi,\zeta)d\overline{\zeta}_j\wedge K(\zeta,\zeta_0), \tag{3.3}$$

where $K(\zeta,\zeta_0)$ is a suitable differential form on $\mathcal{V}_\xi$. This is interesting in the following sense: There exists a transformation $V(x) \to T(\xi,\zeta)$ which assigns the function on the fibred space $\mathbf{R}^n \times \mathcal{V}_\xi$ to a function $V(x)$ on $\mathbf{R}^n$, and the inversion formla (3.3) holds. This is a nonlinear integral transformation.

3.3. Inverse boundary value problems. EIT (Electrical Impedance Tomography) is a technique used in engineering or medical problems. We consider a bounded electrical conductive body Ω in $\mathbf{R}^n$. Letting $\gamma(x)$ be the electric conductivity of Ω, and $u(x)$ the electric potential, we have the following equation:

$$\nabla\cdot(\gamma(x)\nabla u(x)) = 0 \quad \text{in} \quad \Omega. \tag{3.4}$$

Here $u(x)\big|_{\partial\Omega}$ represents the electric potential on the surface, $\gamma(x)\partial u/\partial n\big|_{\partial\Omega}$ is the electric current of the surface, n being the outer unit normal to $\partial\Omega$. The inverse boundary value problem for (3.4) is formulated as follows. Reconstruct $\gamma(x)$ from the set of boundary data

$$\left\{\left(u\Big|_{\partial\Omega}, \gamma\frac{\partial u}{\partial n}\Big|_{\partial\Omega}\right) \mid \nabla\cdot(\gamma\nabla u) = 0 \quad \text{in} \quad \Omega\right\}. \tag{3.5}$$

Given $u(x)\big|_{\partial\Omega}$, the Dirichlet problem (3.4) is uniquely solvable. The map Λ_γ which assigns $\gamma(x)\partial u/\partial n\big|_{\partial\Omega}$ to $u(x)\big|_{\partial\Omega}$ is called the Dirichlet-Neumann map. The procedure to obtain knowledge of $\gamma(x)$ from Λ_γ is the inverse boundary value problem. Of course, one can use the Neumann-Dirichlet map instead of the Dirichlet-Neumnan map. Letting $v = \sqrt{\gamma(x)}u$, one can reduce the problem to the Schrödinger equation

$$-\Delta v + qv = 0, \quad q = \frac{\Delta\sqrt{\gamma}}{\sqrt{\gamma}}.$$

The inverse boundary value problem and the inverse scattering problem for compactly supported potentials are equivalent. Namely, it is possible to reconstruct the scattering amplitude from the D-N map, and vice versa. Faddeev's Green function is also efficient in the inverse boundary value problems ([**36**], [**29**], [**25**]). To use a Green function such as (3.2) means to use $e^{ix\cdot\zeta}$, which is an exponentially growing solution to the Helmholtz equation. In fact, this function was used in the pioneering work of Calderón, who formulated the above inverse boundary value problem, and it was implemented as a powerful tool by Sylvester-Uhlmann [**36**].

3.4. Inverse scattering from a fixed energy. As is stated above, in the practical inverse scattering problem, it is better not to use the high-energy Born approximation. An excellent by-product of Faddeev's Green function is the application to the inverse scattering problem when the energy of the incident particle is fixed ([**29**], [**25**]). This method can also be applied to the inverse scattering problem for the wave equation in a stratified medium ([**14**]), and to that of the optical model in atomic physics ([**24**]).

The former deals with the inverse scattering problem for discontinuous perturbations of two media in $\mathbf{R}^3$ with an interface, and different constant speed of wave propagation in each medium. Such a case does happen in the atomosphere or ocean. Since the earth has a layer structure, this problem is also important as a first step for the inverse problem of seismic waves.

The latter deals with the inverse scattering problem for many-body systems in atomic physics. Since this is a multi-channel problem and it is impossible to solve the many-body Schrödinger equation explicitly, we need to introduce some approximate methods. The optical model aims at picking up the desired scattering channel by assuming a complex potential. It realizes actual experimental data, and is regarded as basic knowledge of atomic physics. Due to the spin of the nuclei, we need to consider the Schrödinger operator with spin-orbit interaction, which is a matrix type interaction and also will be energy dependent in the actual problem. This is why the theory of inverse scattering from a fixed energy is important. In [**24**], it is proved that one can reconstruct these interactions from a scattering amplitude of one fixed energy.

4. Inverse problems and hyperbolic spaces

4.1. Embedding in hyperbolic spaces. The work [**20**] points out that the hyperbolic space can play the role of a tool for solving the multi-dimensional inverse problems. For a solution u to the Schrödinger equation in $\mathbf{R}^n$,

$$(-\Delta + q(x))u = 0, \tag{4.1}$$

we put $v = x_n^{(n-2)/2}u$. Then v satisfies

$$\big(-x_n^2\Delta_x + (n-2)x_n\partial_n - \frac{n(n-2)}{4} + x_n^2 q(x)\big)v = 0, \tag{4.2}$$

where $\partial_n = \partial/\partial x_n$. We should note that

$$H_0 = -x_n^2\Delta + (n-2)x_n\partial_n$$

is the Laplace-Beltrami operator in $\mathbf{H}^n$. Therefore the Schrödinger equation (4.1) in the Euclidean space can be viewed as the Schrödinger equation (4.2) in the hyperbolic space. This gives rise to the following idea of solving the inverse boundary value problem. We consider the inverse boundary value problem for (4.1) on a bounded domain Ω in $\mathbf{R}^n$. Denote the points in $\mathbf{R}^n_+$ by (x,y), $x \in \mathbf{R}^{n-1}$, $y > 0$. (Step 1) By the change of dependent variable $v = y^{(n-2)/2}u$, consider the boundary value problem in $\mathbf{H}^n$:

$$(H_0 + \widetilde{q})v = 0 \quad \text{in} \quad \Omega, \quad v = f \quad \text{on} \quad \partial\Omega.$$

Here $\widetilde{q} = y^2 q - n(n-2)/4$.

(Step 2) Take a lattice Γ of rank $n-1$ in $\mathbf{R}^{n-1}$, and let it act on $\mathbf{H}^n$ as a translation. Let $\mathcal{M} = \Gamma\backslash\mathbf{H}^n$ be the quotient manifold. By taking Γ large enough, Ω can be assumed to be a one-coordinate patch of $\mathcal{M}$.
(Step 3) Take a parameter $\theta \in \mathbf{R}^{n-1}$ and let $u = e^{ix\cdot\theta}v$. Then it satisfies

$$(4.3)\qquad \begin{aligned} &(H_0(\theta) + \widetilde{q})u = 0 \quad \text{in} \quad \Omega, \quad u = e^{ix\cdot\theta}f \quad \text{on} \quad \partial\Omega, \\ &H_0(\theta) = -y^2(\partial_y^2 + (\partial_x + i\theta)^2) + (n-2)y\partial_y. \end{aligned}$$

(Step 4) Take $E > n(n-2)/4$ arbitrarily, and put $V = \widetilde{q}+E$ on Ω, $V = 0$ on Ω^c. The inverse boundary value problem and the inverse scattering problem are equivalent also in $\mathcal{M}$. Therefore one can construct the physical scattering amplitude $A(\theta)$ from the Dirichlet-Neumann map $\Lambda(\theta)$ for (4.3).
(Step 5) $A(\theta)$ is shown to be analytic with respect to θ. Hence letting θ tend to infinity along the imaginary axis, one can recover the Fourier transformation of V.

Let us note that $H_0(\theta)$ is an operator appearing in Floquet theory, which is the basic theory for periodic Schrödinger operators. The utility of considering quotient manifolds appears in Step 5. The Green operator of $H_0(\theta)$ on $\mathcal{M}$ is written by Bessel functions and has branch points with respect to θ. Since the lattice Γ is discrete, one can avoid these branch points in passing to the analytic continuation. These are the main contents of [**20**].

The last step of computing the Fourier transform of V works only for dimension $n \geq 3$. The inverse scattering problem for a fixed energy is unsolved even in the Euclidean space.

4.2. $\overline{\partial}$-theory. Due to the effect of the above-mentioned branch points, the Green operator of $H_0(\theta)$ in $\mathbf{H}^n$ satisfies a $\overline{\partial}$-equation, from which one can obtain a representation formula of the potential similar to (3.3) ([**22**]). For $\mathbf{H}^3$, the counterpart of the complex manifold $\mathcal{V}_\xi$ is the complex plane $\mathbf{C}$, and we only need the usual generalized Cauchy formula.

4.3. Hyperbolic isometry. One can make use of hyperbolic isometries to the inverse problem. Represent points $(x_1, x_2, x_3) \in \mathbf{H}^3 = \mathbf{R}^3_+$ by quaternions:

$$\mathbf{z} = x_1\mathbf{1} + x_2\mathbf{i} + x_3\mathbf{j} = \begin{pmatrix} x_1 + ix_3 & x_2 \\ -x_2 & x_1 - ix_3 \end{pmatrix}.$$

For $\beta = \begin{pmatrix} a & b \\ c & d \end{pmatrix} \in SL(2, \mathbf{C})$, the action

$$\mathbf{z} \to \beta \cdot \mathbf{z} = (a\mathbf{z} + b)(c\mathbf{z} + d)^{-1}$$

defines an isometry in $\mathbf{H}^3$, which maps geodesics to geodesics. The geodesics in $\mathbf{H}^3$ is a semicircle centered at $\{x_3 = 0\}$, or a straight line perpendicular to $\{x_3 = 0\}$. This plays an effective role in the inverse problem.

4.4. Identification of inclusions. In the inverse boundary value problems, sometimes it is important to identify the location of inclusions in the conductive body. Moreover in the practical case, we can place electrodes only on a part of the surface. This leads us to the following problem.

Suppose we know a normal electric conductivity $\gamma_0(x)$ of a bounded body Ω in $\mathbf{R}^n$ $(n = 2, 3)$. Assume that on a part Ω_1 of Ω, the conductivity is greater than the normal one: $\gamma_1(x) > \gamma_0(x)$. Can one detect the location of Ω_1 from the surface data whose supports are restricted on a part of the boundary H

There is an important contribution of Ikehata [**11**] to this problem. It is an algorithm for two-dimensions, which can detect the location of Ω_1 by using Mittag-Leffler's function in the theory of functions of one complex variable. The above hyperbolic isometry makes it possible to discuss this problem in three-dimensions. In [**10**] is obtained a result which can be applied to 2 and 3 dimensions. Let Λ, Λ_0 be the D-N maps associated with $\gamma(x), \gamma_0(x)$. Take x_0 outside the convex hull of Ω, and let $B(x_0,R) = \{|x-x_0| < R\}$. Let $(\ ,\ )_{\partial\Omega}$ be the inner product of $L^2(\partial\Omega)$.

THEOREM 4.1. *There exists a function $u_\tau(x)$ containing a large parameter $\tau > 0$ having the following properties.*
(1) *It solves the equation $\nabla\cdot(\gamma_0(x)\nabla u_\tau) = 0$.*
(2) *$u_\tau(x)$ grows exponentially with respect to $\tau > 0$ for $x \in B(x_0,R)$ and decays exponentially with respect to $\tau > 0$ for $x \notin \overline{B(x_0,R)}$.*
(3) *Let $f_\tau = u_\tau\big|_{\partial\Omega}$. If $\overline{B(x_0,R)}\cap\overline{\Omega_1} = \emptyset$, there exists $\delta > 0$ such that for sufficiently large $\tau > 0$, $0 \leq ((\Lambda-\Lambda_0)f_\tau, f_\tau)_{\partial\Omega} < e^{-\delta\tau}$ holds. If $B(x_0,R)\cap\Omega_1 \neq \emptyset$, there is a $\delta > 0$ such that for sufficiently large $\tau > 0$, $((\Lambda-\Lambda_0)f_\tau, f_\tau)_{\partial\Omega} > e^{\delta\tau}$ holds.*

Since f_τ decays exponentially outside $B(x_0,R)$, it can be regarded as 0 in the numerical computation. This means that one can decide whether $B(x_0,R)$ intersects with Ω_1 or not from the data $((\Lambda-\Lambda_0)f_\tau, f_\tau)_{\partial\Omega}$ obtained by the localized measurement. This theorem also holds for the case $\gamma_1(x) < \gamma_0(x)$ $(x \in \Omega_1)$ by a suitable modification.

The above function $u_\tau(x)$ is represented approximately as follows. Place Ω in the upper half-space of $\mathbf{R}^3$, and let $x_0 = 0$. Suppose $\gamma_0(x) = 1$. Then

$$u_\tau(x) \simeq \sqrt{\frac{y_3}{x_3}}e^{\tau(-y_1+i\tau y_3)},$$

$$y_1 = \frac{x_1^2+x_2^2+x_3^2-R^2}{(x_1+R)^2+x_2^2+x_3^2}, \tag{4.4}$$

$$y_3 = \frac{2x_3R}{(x_1+R)^2+x_2^2+x_3^2}.$$

This makes it possible to introduce the following algorithm for probing inclusions inside the body.

(1) Place Ω in the upper half-space $\mathbf{R}^3_+$.

(2) Insert the function (4.4) in the quadratic form $Q(\tau) = ((\Lambda-\Lambda_0)f_\tau, f_\tau)_{\partial\Omega}$.

(3) Draw the ball B_R of radius R centered at the origin.

(4) If $Q(\tau)$ grows up with τ, then B_R will intersect with Ω_1. If $Q(\tau)$ decays when τ grows up, then B_R will have no intersection with Ω_1.

In this way, we can obtain a sphere centered at x_0 which is tangent to the inclusion. By varying x_0, we get an envelope of spheres, which will approximate the shape of the inclusions. Figure 1 is a 2-dimensional numerical computation based on this algorithm.

The left-most column shows inclusions in the semidisc. The electric conductivity of the gray part (normal part or background part) is 1, and that of the dark part (inclusion) is 4. The boundary data has support only on the linear part of the semicircle (called the front part). We solved the boundary value problem (3.4) by the finite element method and computed the distance to the inclusion using the above algorithm. The gray part of the 2nd column is the region enclosed by the envelope of circles with center on the front part. This is the largest region in which

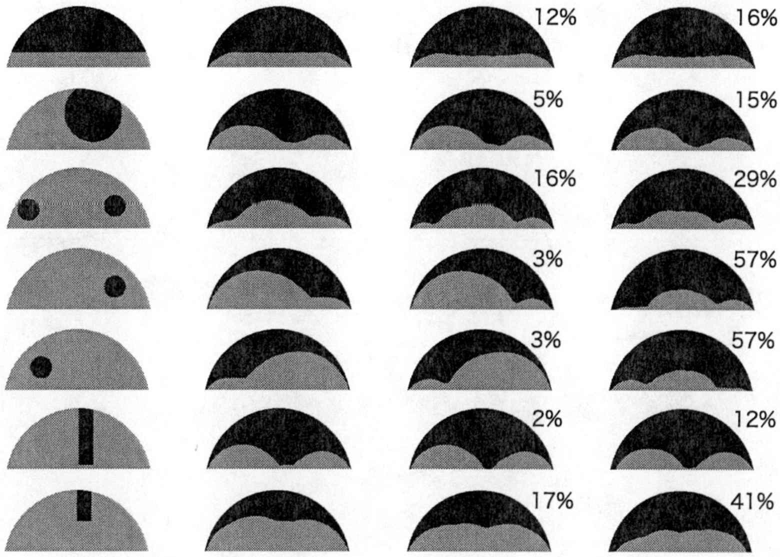

FIGURE 1

we can conclude geometrically the nonexistence of inclusions. The 3rd column is the numerical result without noise, and the right-most column is the one with 0.01 percent of noise. The digits on the shoulders of the 3rd and 4th columns are the errors compared to the 2nd column. The inclusions in the shallow part are easy to probe; however, it is not the case for the deep part. This is a physical feature of EIT.

The idea of Theorem 4.1 is to embed Ω in the upper half-space so that $x_0 = 0$. By the hyperbolic isometry explained in subsection 4.3, the hemisphere of radius R centered at the origin is mapped to the plane $\{x_2 = 0\}$ orthogonal to $\{x_3 = 0\}$. The Schrödinger equation (4.2) in the hyperbolic space is invariant by hyperbolic isometry. By the gauge transformation, we introduce a complex parameter θ and consider the equation (4.3). Put $\theta = (0, i\tau)$. We construct a Green function of $H_0(\theta)$ by using modified Bessel functions, and then solutions to (4.3) in terms of the Green function. By the inverse gauge transformation, we obtain solutions of the Schrödinger equation which are exponentially growing in one side of the plane $\{x_2 = 0\}$ and exponentially decaying in the opposite side. We then transform them back by the hyperbolic isometry.

4.5. Horosphere boundary value problem and Barber-Brown algorithm. Let us return to the problem of reconstruction of the conductivity $\gamma(x)$. When $n \geq 3$, in the late 1980s the uniqueness and the theoretical reconstruction procedure were shown ([**36**], [**29**], [**25**]). It transforms the inverse boundary value problem to the inverse scattering problem, and cannot be directly used in the numerical computation. (Note, however, that in the 2-dimensional inverse conductivity problem, Nachman's $\bar{\partial}$-method is partly numerically implemented [**34**].) On the other hand, in the early 1980s an algorithm for the numerical computation of $\gamma(x)$ had been proposed by the British engineers Barber and Brown. It gives a sufficiently good answer despite its low numerical cost and is the only one commercial method used so far. More interestingly, it has been pointed out by Santosa-Vogelius [**33**] that this problem is related to hyperbolic geometry.

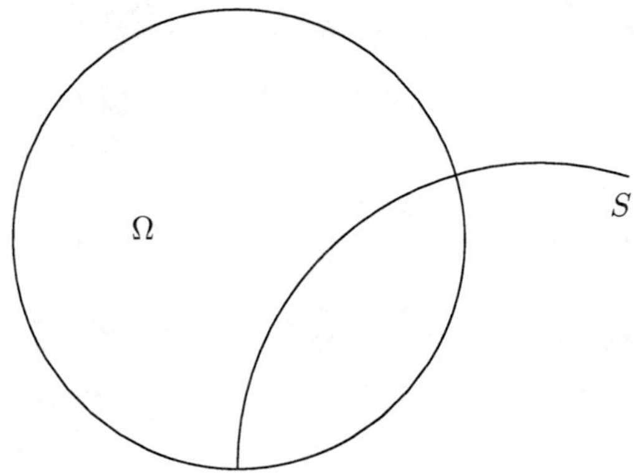

FIGURE 2

Let us recall their arguments briefly. (See Figure 2). They consider the linearized problem in the 2-dimensional disc $\Omega = \{|x| < 1\}$. Suppose $\gamma(x)$ is sufficiently close to a constant $\gamma_0 > 0$:

$$\gamma(x) = \gamma_0 + \epsilon\gamma_1(x) + \cdots .$$

Take a point $\omega = (\omega_1, \omega_2) \in \partial\Omega$ arbitrarily, and put $\omega^{\perp} = (-\omega_2, \omega_1)$. Let

$$u_0(x) = \frac{\omega^{\perp}\cdot x}{(\omega^{\perp}\cdot x)^2 + (1-\omega\cdot x)^2}$$

be a solution to the equation $\Delta u = 0$ with a singularity at $x = \omega$. They seek a solution to the equation $\nabla\cdot(\gamma(x)\nabla u) = 0$ having the asymptotic expansion

$$u(x) = u_0(x) + \epsilon u_1(x) + \cdots .$$

Barber-Brown's argorithm asserts that $\gamma_1(x)$ can be approximately written in the following way:

$$\gamma_1(x) \approx \int_{S^1} \varphi(x,\omega)\rho_1 d\omega,$$

where $\varphi(x,\omega)$ is a function computed from $u_1(x)$, and ρ_1 is a suitable density. Santosa and Vogelius noticed that $\varphi(x,\omega)$ can be written as follows:

$$\varphi(x,\omega) \approx K\left(\int_S \gamma_1(x)\rho_2 d\sigma\right),$$

where K is some convolution operator, and S is a circle orthogonal to $\partial\Omega$. Combining the above two formulas, we have

$$\gamma_1 \approx R^*KR\gamma_1, \quad Rf = \int_S f(x)\rho_2 d\sigma. \tag{4.5}$$

If we regard Ω as a Poincaré disc, then S is a geodesic, and (4.5) is very close to the Radon transform in the Poincaré disc. Berenstein-Tarabusi [**3**] analyzed Barber-Brown's algorithm and concluded that it is a sort of inversion formula of the Radon transform in the Poincaré disc.

These works suggest that there is a close connection between the inverse boundary value problem in the disc with hyperbolic geometry. However, further analysis, especially the 3-dimensional problem, and also the relation to the original inverse conductivity problem (3.4) have been left open.

The work [**23**] aims at justifying Barber-Brown's algorithm for the problem (3.4) by the method of hyperbolic geometry. Let us consider the Schrödinger equation $(-\Delta + q(x))u = 0$ in the 3-dimensional ball Ω. Take a sphere S orthogonal to $\partial\Omega$. Then the following theorem holds.

THEOREM 4.2. (1) *There exists a pseudo-differential operator $P(\tau)$ on $\partial\Omega$ containing a large parameter $\tau > 0$ such that the boundary value problem*

$$(-\Delta + q)u = 0 \quad \text{in} \quad \Omega, \qquad \frac{\partial u}{\partial n} = P(\tau)u + f \quad \text{on} \quad \partial\Omega$$

has a unique solution $u(\tau)$. Define the generalized Robin-Dirichlet map (GRD map) by

$$\mathcal{R}_q(\tau) : f \to u\big|_{\partial\Omega}.$$

(2) *There exist a boundary data $f(\tau)$ and a sequence $\{\tau_n\}$ such that*

$$\lim_{n\to\infty} \big((\mathcal{R}_q(\tau_n) - \mathcal{R}_0(\tau_n))f(\tau_n), f(-\tau_n)\big)_{\partial\Omega} = i\int_{S\cap\Omega} q(x)dS.$$

(3) *One can reconstruct $q(x)$ by using the inversion formula of the Radon transform in the disc model of $\mathbf{H}^3$.*

In the disc model of hyperbolic space, the Radon transform is written as

$$Rf(\xi) = \int_\xi f(x)dS(x),$$

where ξ is a sphere orthogonal to the boundary $\{|x| = 1\}$. The adjoint operator of R has the form

$$R^*\varphi(x) = \int_{x\in\xi} \varphi(\xi)d\mu(\xi) = \int_K \varphi(gk\cdot\eta)dk,$$

where $K = SO(3)$, $g \in SO(2,1)$, $g\cdot o = x$, o is the origin, and η is the plane passing through o. Then the inversion formula of the Radon transform is

$$f = -\frac{1}{2\pi}(1 + \Delta_g)R^*Rf,$$

Δ_g being the Laplace-Beltrami operator on $\mathbf{H}^3$. The operator $P(\tau)$, boundary data $f(\tau)$, and the sequence $\{\tau_n\}$ do not depend on the potential $q(x)$ and can be written explicitly.

To prove Theorem 4.2, we take Ω in the following way:

$$\Omega = \left\{(x_1 + R)^2 + x_2^2 + \left(x_3 - \frac{R}{\delta}\right)^2 < \left(\frac{R}{\delta}\right)^2\right\}.$$

This is a sphere in $\mathbf{R}^3_+$ tangent to $\{x_3 = 0\}$, namely the horosphere in $\mathbf{H}^3$. By a suitable choice of hyperbolic isometry in subsection 4.3, this sphere is mapped to the half-space $\{x_3 > \delta\}$. This reduces the problem to the Schrödinger equation $(H_0 + q)u = 0$ in $\{x_3 > \delta\}$. By the gauge transformation, we introduce a complex parameter θ. Using the Green function of $H_0(\theta)$ in $\{x_3 > \delta\}$, we can construct the solution u.

4.6. Problem of a boundary condition. In the mathematical formulation of the practical problem of EIT, we need to be careful about the boundary condition. In the real situation, the electrodes on the surface have a finite size, and one cannot make use of high-frequency waves unless one keeps some finite distance between these electrodes. The D-N map corresponds to the ideal case in which the infinitely small electrodes are put continuously on the surface. Therefore it does not match the real situation and has a discrepancy with the real data. As a more practical formulation, the gap model, the shunt model, and the complete electrode model are proposed. The last model has the best coincidence with the real data, and there, the Robin boundary condition is used in order to take into account the interaction between the electrode and the surface of the body ([**35**]). This seems to suggest to us to study the mathematical formulation of the inverse boundary value problem with more comparison with reality.

4.7. Conclusion. This inverse problem is said to be ill-posed. A small error in the observation results in a big variation of the physical quantity of the target. This is the reason why we need a rigorous mathematical theory. Even if we start from the real problem, it is not at all easy to do the theoretical study by taking into account the real situation. More than 20 years have passed from the pioneering works of Sylvester-Uhlmann, Nachman, and Khenkin-Novikov, who theoretically assured the reconstruction of electric conductivity in the inverse boundary value problem. The future direction may aim at getting good numerical results. In the above inclusion detection problem or the Barber-Brown algorithm, a 3-dimensional numerical computation will be the next goal. Interest in the multi-dimensional inverse problem has been increasing in recent years, and there are lots of important contributions other than the topics mentioned above. A remarkable advance is the boundary control method due to Belishev-Kurylev, which solved the inverse spectral problem on compact Riemannian manifolds [**2**]. The inverse scattering for an asymptotically hyperbolic metric contains interesting problems ([**32**]). The inverse scattering problem and the inverse boundary value problems should not be classified as simple applications of mathematics. They have deep roots in the classical stream of mathematics and physics.

References

[1] S. Agmon and L. Hörmander, Asymptotic properties of solutions of differential equations with simple characteristics, *J. d'Anal. Math.* 30 (1976), 1-38. MR0466902 (57:6776)

[2] M. Belishev and Y. Kurylev, To the reconstruction of a Riemannian manifold via its spectral data (BC-method), *Comm. in Partial Differential Equations* 17 (1992), 767-804. MR1177292 (94a:58199)

[3] C. A. Berenstein and E. C. Tarabusi, Integral geometry in hyperbolic spaces and electrical impedance tomography, *SIAM. J. Appl. Math.* 56 (1996), 755-764. MR1389752 (97d:53073)

[4] L. D. Faddeev, *Mathematical Aspects of the Three Body Problems in Quantum Scattering Theory*, Trudy Mat. Inst. Steklov **69** (1963), 122pp. MR01636985 (29:995)

[5] L. D. Faddeev, Inverse problems of quantum scattering theory, J. Sov. Math. 5 (1976), 334-396.

[6] C. Gérard, H. Isozaki, and E. Skibsted, N-body resolvent estimates, *J. Math. Soc. Japan* 48 (1996), 135-160. MR1361552 (96j:81131)

[7] J. Ginibre and M. Moulin, Hilbert space approach to the quantum mechanical three-body problem, *Ann. Inst. Henri Poincaré*, A 21 (1974), 97-145. MR0368656 (51:4897)

[8] J. Howland, Abstract stationary theory of multichannel scattering, *J. Functional Anal.* 22 (1976), 252-282. MR0461175 (57:1160)

[9] T. Ikebe, Eigenfunction expansions associated with the Schrödinger operators and their applications to scattering theory, *Arch. Rational Mech. Anal.* 5 (1960), 1-34. MR0128355 (23:B1398)

[10] T. Ide, H. Isozaki, S. Nakata, S. Siltanen and G. Uhlmann, Probing for electrical inclusions with complex spherical waves, Comm. Pure and Appl. Math. Vol LX, 1415-1442 (2007) . MR2342953 (2008j:35186)

[11] M. Ikehata, Mittag-Leffler's function and extracting from Cauchy data, *Contemporary Mathematics (Inverse Problems and Spectral Theory, ed. H. Isozaki)* 348 (2004), 41-52. MR2066480 (2005e:35245)

[12] H. Isozaki, Structures of S-matrices for three-body Schrödinger operators, *Commun. Math. Phys.* 146 (1992), 241-258. MR1165182 (93h:81146)

[13] H. Isozaki, A generalization of the radiation condition of Sommerfeld for N-body Schrödinger operators, *Duke Math. J.* 74 (1994), 557-584. MR1272984 (95d:81148)

[14] H. Isozaki, Inverse scattering theory for wave equations in stratified media, *J. Diff. Eq.* 138 (1997), 19-54. MR1458455 (98f:35156)

[15] H. Isozaki, Inverse problems of scattering theory, Scientific technique collaboration research center, Ryuukoku University (1997), in Japanese.

[16] H. Isozaki, Inverse problem of quantum scattering theory, Suugaku 50 (1998), 163-180, in Japanese. MR1648436 (99k:81335)

[17] H. Isozaki, Asymptotic properties of solutions to 3-particle Schrödinger equations, *Commun. Math. Phys.* 222 (2001), 371-413. MR1859604 (2002h:81285)

[18] H. Isozaki, Inverse spectral theory, *Topics in the Theory of Schrödinger Operators*, eds. H. Araki and H. Ezawa, World Scientific (2003), 93-143. MR2117108

[19] H. Isozaki, *Many-body Schrödinger Equations*, Springer, Tokyo, 2004 (in Japanese).

[20] H. Isozaki, Inverse spectral problems on hyperbolic manifolds and their applications to inverse boundary value problems in Euclidean space, *Amer. J. of Math.* 126 (2004), 1261-1313. MR2102396 (2005m:35311)

[21] H. Isozaki, Inverse problems and hyperbolic manifolds, Contemporary Mathematics 348, *Inverse Problems and Spectral Theory*, ed. H. Isozaki, Amer. Math. Soc., Providence, Rhode Island, 2004, 181-197. MR2066489 (2006b:35339)

[22] H. Isozaki, The $\overline{\partial}$-theory for inverse problems associated with Schrödinger operators on hyperbolic spaces, *Publ. RIMS Kyoto Univ.* Vol 43, No 1 (2007), 201–240. MR2319543 (2009d:35339)

[23] H. Isozaki, Inverse boundary value problems in the horosphere — A link between hyperbolic geometry and electrical impedance tomography, *Inverse Problems and Imaging* 1 (2007), 55-82. MR2262749 (2008j:35188)

[24] H. Isozaki, H. Nakazawa and G. Uhlmann, Inverse scattering problem in nuclear physics, *J. Math. Phys.* 45 (2004), 2613-2632. MR2067577 (2005c:81251)

[25] G. M. Khenkin and R. G. Novikov, The $\overline{\partial}$-equation in the multi-dimensional inverse scattering problem, *Russ. Math. Survey* 42 (1987), 109-180. MR896879 (89d:35043)

[26] S. T. Kuroda, Scattering theory for differential operators, I, II, *J. Math. Soc. Japan* (1973), 75-104, 222-234. MR0326435 (48:4779); MR0326436 (48:4780)

[27] E. Mourre, Absence of singular continuous spectrum for certain self-adjoint operators, *Commun. Math. Phys.* 78 (1981), 391-408. MR603501 (82c:47030)

[28] M. Murata, Lower bounds of growth order of solutions of partial differential equations at infinity, Suugaku, 32 (1980), 1-16, in Japanese.

[29] A. Nachman, Reconstructions from boundary measurements, *Ann. of Math.* (2) 128 (1988), 531-576. MR970610 (90i:35283)

[30] R. G. Newton, The asymptotic form of three-particle wave functions and the cross sections, *Ann. Phys.* 74 (1972), 324-351. MR0321445 (47:9978)

[31] J. Nuttal, Asymptotic form of the three-particle scattering wave functions for free incident particles, *J. Math. Phys.* 12 (1971), 1891-1899.

[32] A. Sa Barreto, Radiation fields, scattering, and inverse scattering on asymptotically hyperbolic manifolds, *Duke Math. J.* 129 (2005), 407-480. MR2169870 (2006g:58058)

[33] F. Santosa and M. Vogelius, A back projection algorithm for electrical impedance tomography, *SIAM. J. Appl. Math.* 50 (1990), 216-243. MR1036240 (91c:35159)

[34] S. Siltanen, J. Mueller and D. Isaacson, An implementation of the reconstruction algorithm of A. Nachman for the 2-D inverse conductivity problem, *Inverse Problems* 16 (2000), 681-699. MR1766226 (2001g:35269)

[35] E. Somersalo, M. Cheney and D. Issacson, Existence and uniqueness for electrode models electric current computed tomography, *SIAM. J. Appl. Math.* 52 (1992), 1023-1040. MR1174044 (93f:35052)

[36] J. Sylvester and G. Uhlmann, A global uniqueness theorem for an inverse boundary value problem, *Ann. of Math.* (2) 125 (1987), 153-169. MR873380 (88b:35205)

[37] L. Thomas, Asymptotic completeness in two and three particle quantum scattering theory, *Ann. Phys.* 90 (1975), 127-165. MR0424082 (54:12050)

Institute of Mathematics, University of Tsukuba, Tsukuba, Ibaraki 305-8521 Japan

Translated by HIROSHI ISOZAKI

Amer. Math. Soc. Transl.
(2) Vol. **230**, 2010

Nondoubling Measure and Harmonic Analysis

Yasuo Komori

ABSTRACT. We survey recent developments in harmonic analysis on nondoubling measure spaces. In particular, we explain theories by Nazarov, Treil and Volberg, and Tolsa.

1. Introduction

Harmonic analysis on nondoubling measure spaces has been in a state of rapid development since the latter half of the 1990s. This field is a mixture of complex analysis (the theory of analytic capacity), geometric measure theory (the theory of Hausdorff measure) and harmonic analysis (the theory of singular integrals). This article is not intended to be exhaustive and is organized in the following way. In Section 2, we survey the so-called Painlevé problem, which is a 100-year-old problem in complex analysis, and we explain why we need to study harmonic analysis on nondoubling measure spaces for the purpose of solving this problem. Next in Section 3, we recall the theories of singular integral operators and Hardy spaces on Euclidean spaces which had already been obtained before the 1990s. In particular, we explain the Calderón–Zygmund theory, the theory of David and Journé, and the Fefferman–Stein theorem. In Section 4, we discuss how to develop these theories on nondoubling measure spaces; that is, we explain theories by Nazarov, Treil and Volberg, and Tolsa.

DEFINITION 1. Let (X, d, μ) be a topological space endowed with a distance d and a measure μ. We say that μ is **doubling** if there exists a positive constant C_0 such that

$$\mu(B(x,2r)) \leq C_0\mu(B(x,r)) \quad \text{for } x \in \operatorname{supp}(\mu), \tag{1}$$

where $B(x,r) = \{y \in X \; : \; d(x,y) \leq r\}$.

Of course the Lebesgue measure is doubling.

2. The Painlevé problem

In this section we survey a history of the Painlevé problem. The article [**83**] is a nice expository book about this problem (see also [**21**], [**104**], [**112**] and [**113**]).

This article originally appeared in Japanese in Sūgaku **61** (4) (2009), 376–394.
2000 *Mathematics Subject Classification.* Primary 42B20, 42B25, 30C85.

Let $E \subset \mathbb{C}$ be a compact set. We say that E is **removable** for bounded analytic functions if, for any open set $U \supset E$, any bounded analytic function f on $U \setminus E$ has an analytic extension to the whole U. The **Painlevé problem** can be stated as follows:

Find a geometric/metric characterization of such removable sets.

We give two simple examples: a singleton is removable and a circle is not removable.

In 1947, Ahlfors introduced the notion of analytic capacity of a set and gave the following answer.

DEFINITION 2. The analytic capacity of a compact set $E \subset \mathbb{C}$ is

$$\gamma(E) = \sup\{|f'(\infty)| \ : \ f \text{ is bounded and analytic on } \mathbb{C} \setminus E, \|f\|_\infty \leq 1\},$$

where $f'(\infty) = \lim_{z\to\infty} z(f(z) - f(\infty))$.

THEOREM 1 (Ahlfors [**1**]). *A compact set E is removable if and only if* $\gamma(E) = 0$.

By Ahlfors' result, Painlevé's problem turns out to be equivalent to describing compact sets with positive analytic capacity in geometric/metric terms.

In the 1950s and 1960s, Vituškin [**115**], [**116**] showed that analytic capacity plays an important role in problems of uniform rational approximation on compact sets of the complex plane (see also [**28**] and [**110**]). Related to these problems he raised the following question.

PROBLEM 1 (Semiadditivity). Does there exist a C_0 such that $\gamma(E \cup F) \leq C_0(\gamma(E) + \gamma(F))$?

This problem will be solved in Section 2.5.

2.1. Analytic capacity. First we state some elementary properties pertaining to analytic capacity. We should keep in mind that, in a sense, analytic capacity measures the size of a set as a nonremovable singularity for bounded analytic functions. The following properties are easy. If $E \subset F$, then $\gamma(E) \leq \gamma(F)$, $\gamma(z + E) = \gamma(E)$ and $\gamma(zE) = |z|\gamma(E)$. Furthermore if E is connected, then

$$\mathrm{diam}(E)/4 \leq \gamma(E) \leq \mathrm{diam}(E).$$

The first inequality is a consequence of Koebe's 1/4 theorem (see [**38**]). This implies that any removable compact set must be totally disconnected.

The relationship between Hausdorff dimension and analytic capacity is the following:

(i) If $\dim_H(E) > 1$, then $\gamma(E) > 0$.
(ii) If $\dim_H(E) < 1$, then $\gamma(E) = 0$.

Precisely speaking, if $\mathcal{H}^1(E) = 0$, then $\gamma(E) = 0$, where $\mathcal{H}^1$ is the one-dimensional Hausdorff measure. Therefore the dimension 1 is the critical index and a natural question arises.

Is E removable if and only if $\mathcal{H}^1(E) = 0$?

Unfortunately the answer is no. Vituškin [**114**] showed that there are sets such that $\mathcal{H}^1(E) > 0$ and $\gamma(E) = 0$. Garnett [**37**] gave a simple example (see also Murai [**76**], p. 79). A typical example of such a set is the so-called corner quarters Cantor set (see the figure in [**83**]). Therefore we need to analyze such singular sets. For this purpose we introduce some terminology.

DEFINITION 3. A set E is said to be **1-rectifiable** if it is $\mathcal{H}^1$-almost all contained in a countable union of rectifiable curves. It is called **purely 1-unrectifiable** if, for every rectifiable curve Γ, $\mathcal{H}^1(E \cap \Gamma) = 0$.

Since Vituškin's example and the corner quarters Cantor set are purely 1-unrectifiable, Vituškin conjectured the following.

THE VITUŠKIN CONJECTURE . *If $E \subset \mathbb{C}$ is a compact set with $\mathcal{H}^1(E) < \infty$, then $\gamma(E) = 0$ if and only if E is purely 1-unrectifiable.*

If this is true, we obtain a geometric solution of the Painlevé problem in the case that $\mathcal{H}^1(E) < \infty$. For a while we focus on the Vituškin conjecture.

2.2. The Cauchy transform. From the definition of analytic capacity, we find that it is important to construct bounded and analytic functions on $\mathbb{C} \setminus E$. Therefore we need to consider the Cauchy transform and this leads us to the study of harmonic analysis.

DEFINITION 4. Given a finite complex Radon measure μ, the Cauchy transform of μ is defined by

$$\mathcal{C}\mu(z) = \int \frac{d\mu(\xi)}{\xi - z}, \qquad z \in \mathbb{C} \setminus E.$$

Then $\mathcal{C}\mu$ is analytic out of E. Since $(\mathcal{C}\mu)'(\infty) = -\int d\mu$ we define another analytic capacity γ_+.

DEFINITION 5.

$$\gamma_+(E) = \sup\{\|\mu\| \ : \ \mu \text{ is a positive Radon measure, } \operatorname{supp}(\mu) \subset E, \ \|\mathcal{C}\mu\|_\infty \leq 1\}.$$

It is trivial that $\gamma_+(E) \leq \gamma(E)$. The next problem is important.

PROBLEM 2. Does there exist a C_0 such that $\gamma(E) \leq C_0\gamma_+(E)$?

This problem will be solved in Section 2.5.

We introduce some terminology about the Cauchy transform. Let μ be a positive Radon measure. For a μ-measurable function f, $\mathcal{C}_\mu f(z) = \mathcal{C}(f d\mu)(z)$. $\mathcal{C}\mu(z)$ is not defined for $z \in E$, but for $\varepsilon > 0$,

$$\mathcal{C}_\varepsilon\mu(z) = \int_{|\xi - z| > \varepsilon} \frac{d\mu(\xi)}{\xi - z} \tag{2}$$

is defined for all $z \in \mathbb{C}$. We say that '$\mathcal{C}_\mu$ is bounded on $L^2(\mu)$' if $\|\mathcal{C}_\varepsilon(f d\mu)\|_{L^2(\mu)} \leq C_0\|f\|_{L^2(\mu)}$ uniformly in ε. Concerning pointwise convergence of $\mathcal{C}_\varepsilon\mu(z)$, see [**67**], [**70**], [**91**], [**93**] and [**109**].

2.3. Menger curvature. Menger curvature was introduced by Menger [**51**] in a purely geometric problem in the 1930s. In 1995, Melnikov [**73**] rediscovered it and used it to study the semi-additivity of analytic capacity.

DEFINITION 6. Menger curvature $c(x, y, z)$ for three noncollinear points x, y and z is defined by

$$c(x, y, z) = \frac{1}{R(x, y, z)},$$

where $R(x, y, z)$ denotes the radius of the circle passing through x, y and z.

The curvature of the measure μ is defined as follows.

DEFINITION 7.

$$c^2(\mu) = \iiint c(x,y,z)^2 d\mu(x)d\mu(y)d\mu(z).$$

The notion of curvature is connected to the Cauchy transform by the following result.

PROPOSITION 1 (Melnikov and Verdera [**74**]).

$$\|\mathcal{C}\mu\|^2_{L^2(\mu)} = \frac{1}{6}c^2(\mu).$$

REMARK. Precisely speaking, we need to consider $C_\varepsilon \mu$ as in (2), but for simplicity we give a rough sketch of the proof (see [**83**], p. 64).

PROOF.

$$\begin{aligned}\|\mathcal{C}\mu\|^2_{L^2(\mu)} &= \int\left|\int \frac{1}{y-x}d\mu(y)\right|^2 d\mu(x) = \iiint \frac{1}{(y-x)\overline{(z-x)}}d\mu(x)d\mu(y)d\mu(z)\\ &= \frac{1}{6}\iiint \sum_{\sigma\in S_3} \frac{1}{(z_{\sigma(2)}-z_{\sigma(1)})\overline{(z_{\sigma(3)}-z_{\sigma(1)})}}d\mu(z_1)d\mu(z_2)d\mu(z_3)\\ &= \frac{1}{6}c^2(\mu).\end{aligned}$$

In the proof we use the following magic formula.

LEMMA 1.

$$c(z_1,z_2,z_3)^2 = \left(\frac{4S(z_1,z_2,z_3)}{|z_1-z_2||z_2-z_3||z_3-z_1|}\right)^2 = \sum_{\sigma\in S_3}\frac{1}{(z_{\sigma(2)}-z_{\sigma(1)})\overline{(z_{\sigma(3)}-z_{\sigma(1)})}},$$

where $S(z_1,z_2,z_3)$ is the area of the triangle with vertices z_1, z_2 and z_3, and S_3 is the permutation group of three elements.

The following theorem shows the relation between the notion of curvature and rectifiability.

THEOREM 2 (Léger [**56**]). *Let E be compact and $\mathcal{H}^1(E) < \infty$. If $c^2(\mathcal{H}^1|_E) < \infty$, then E is 1-rectifiable.*

This theorem was proved by using geometric techniques developed by Jones [**49**] (see also [**69**]), when he solved the geometric Traveling Salesman Problem: Under which conditions is a compact set E contained in a rectifiable curve?

By Proposition 1 and Theorem 2 we know the following. When $\mathcal{H}^1(E) < \infty$, if the Cauchy transform is bounded on $L^2(\mathcal{H}^1|_E)$, then E is 1-rectifiable (see [**69**]).

Now we have found the importance of the L^2 boundedness of the Cauchy transform.

2.4. Solution of the Vituškin conjecture. For some historical reasons we split the result into two parts.

THEOREM 3 (Calderón [**4**] (1977)). *Let E be a subset of some rectifiable curve Γ. Then $\gamma(E) = 0$ if and only if $\mathcal{H}^1(E) = 0$.*

REMARK. It has already been proved that if $\mathcal{H}^1(E) = 0$, then $\gamma(E) = 0$ (see Section 1.1).

THEOREM 4 (David [**22**] (1998)). *Let E be compact and $\mathcal{H}^1(E) < \infty$. If $\gamma(E) > 0$, then there exists a rectifiable curve Γ such that $\mathcal{H}^1(E \cap \Gamma) > 0$.*

REMARK. Mattila, Melnikov and Verdera (1996) [**69**] proved the theorem by assuming that E is an Ahlfors 1-regular set (see also [**25**] and [**102**]).

DEFINITION 8. A set E is said to be Ahlfors 1-regular if

$$C_1 r \leq \mathcal{H}^1(E \cap B(z,r)) \leq C_2 r \quad \text{for all} \quad z \in \mathbb{C} \quad \text{and} \quad r > 0. \tag{3}$$

Sketch of proof of Theorem 3. We shall prove that if $\mathcal{H}^1(E) > 0$, then $\gamma(E) > 0$. By the next lemma, it is sufficient to show the theorem in the special case where Γ is a Lipschitz graph.

LEMMA 2. *Let E be a subset of a rectifiable curve such that $\mathcal{H}^1(E) > 0$. Then there exists a Lipschitz graph Γ such that $\mathcal{H}^1(E \cap \Gamma) > 0$.*

To prove this lemma we use 'Rising Sun Lemma' due to F. Riesz (see [**83**], p. 85).

When Γ is a Lipschitz graph we write the Cauchy transform as follows:

$$\mathcal{C}_\Gamma f(z) = \int_\Gamma \frac{f(w)}{w - z} d\mathcal{H}^1(w). \tag{4}$$

THEOREM 5 (Calderón [**4**]). *If the Lipschitz constant of Γ is sufficiently small, then $\mathcal{C}_\Gamma$ is bounded on $L^2(\mathcal{H}^1|_\Gamma)$.*

This restriction about the Lipschitz constant is not serious when we solve the Vituškin conjecture, because the Lipschitz constant of Γ can be chosen as small as desired in the proof of Lemma 2. It had been an important problem in real analysis to remove this restriction. We shall consider this problem again in the next section (see [**7**], [**14**], [**15**], [**19**], [**20**], [**23**], [**24**], [**72**] and [**76**]).

To accomplish the proof of Theorem 3, we need another magic trick which relates the Cauchy transform to analytic capacity. Since it is a little complicated to state precisely, we give a rough expression (see [**7**], Chapter 7, [**83**], p. 82 and see also Section 3.1).

PROPOSITION 2 ([**29**], [**108**]). *If $\mathcal{C}_\Gamma$ is L^2 bounded, then there exists a nonconstant bounded analytic function on $\mathbb{C} \setminus \Gamma$, that is, $\gamma(\Gamma) > 0$.*

□

Sketch of proof of Theorem 4. In the proofs of Theorem 3 and a partial result for Theorem 4 [**69**], the Hausdorff measure restricted on a Lipschitz graph or on an Ahlfors 1-regular set (3) is doubling. The difficulty of the proof of Theorem 4 is that we need to consider nondoubling measures on very irregular sets. First we can find a function bounded and analytic on $\mathbb{C} \setminus E$ such that $f'(\infty) = \gamma(E)$. We call such a function an Ahlfors function. Since $\mathcal{H}^1(E) < \infty$, we can find a bounded function g such that $f(z) = \mathcal{C}(gd\mathcal{H}^1|_E)(z)$, $z \in \mathbb{C} \setminus E$. Next, and this part is the most difficult, there exists a set $F \subset E$ such that $\mathcal{H}^1(F) > 0$ and the Cauchy transform is bounded on $L^2(\mathcal{H}^1|_F)$. (A similar result had been obtained by Christ [**8**] (1990)). By Proposition 1 we have $c^2(\mathcal{H}^1|_F) < \infty$. By Theorem 2 we obtain that F is 1-rectifiable. □

Now the Vituškin conjecture has been solved. As we have said several times, the L^2 boundedness of the Cauchy transform is important; furthermore the boundedness on nondoubling measure spaces is essential. We shall explain this in full detail in Sections 3 and 4.

2.5. Solution of the Painlevé problem. In this subsection we explain that the Painlevé problem (the case $\mathcal{H}^1(E) = \infty$), as well as Problems 1 and 2 are all solved. We define some notation.

(5) $\Sigma(E) =$

$\{\mu : \text{a positive Radon measure, } \operatorname{supp}(\mu) \subset E,\ \mu(B(z,r)) \le r,\ z \in \mathbb{C}, r > 0\}.$

Tolsa [**92**] proved the following.

THEOREM 6 (Tolsa).

$$\gamma_+(E) \approx \sup\{\mu(E) \ :\ \mu \in \Sigma(E), c^2(\mu) \le \mu(E)\} \tag{6}$$

$$\approx \sup\{\mu(E) \ :\ \mu \in \Sigma(E), \|\mathcal{C}\mu\|_{L^2(\mu)\to L^2(\mu)} \le 1\}. \tag{7}$$

By (7), we obtain the following.

COROLLARY 1 (Countable semiadditivity).

$$\gamma_+\Big(\bigcup_{i=1}^{\infty} E_i\Big) \le C_0 \sum_{i=1}^{\infty} \gamma_+(E_i).$$

Tolsa [**100**] solved Problem 2.

THEOREM 7 (Tolsa (2003)).

$$\gamma(E) \approx \gamma_+(E).$$

REMARK. Jones (see [**83**], Chapter 3) and Mateu, Tolsa and Verdera [**64**] had proved the equivalence of two capacities in special cases (see also [**65**] and [**98**]).

Combining with Corollary 1, we can solve Problem 1. Furthermore we obtain better result (see also [**101**]).

COROLLARY 2 (Countable semiadditivity).

$$\gamma\Big(\bigcup_{i=1}^{\infty} E_i\Big) \le C_0 \sum_{i=1}^{\infty} \gamma(E_i).$$

By Theorem 6 (6) and Theorem 7, We can solve the Painlevé problem.

COROLLARY 3 (Solution of Painlevé's problem). *Let E be a compact set. Then $\gamma(E) > 0$ if and only if there exists a nonzero Radon measure μ such that $\mu \in \Sigma(E)$ and $c^2(\mu) < \infty$.*

Furthermore Tolsa [**103**] showed that the property $\gamma(E) > 0$ is invariant under bi-Lipschitz maps (see also [**40**]). We say that f is a bi-Lipschitz map if

$$C_1|x-y| \le |f(x) - f(y)| \le C_2|x-y|.$$

For example, Hausdorff dimension is invariant under bi-Lipschitz maps.

We conclude this section by stating one open problem.

PROBLEM. Under what conditions do we have $\gamma(E \cup F) \le \gamma(E) + \gamma(F)$?

The following result by Suita [**90**], together with Corollary 2, suggests this may be true: If E and F are compact, connected and $E \cap F = \emptyset$, then $\gamma(E \cup F) \le \gamma(E) + \gamma(F)$.

However, even when $\gamma(F) = 0$, the identity $\gamma(E \cup F) = \gamma(E)$ remains unproved.

We can consider the results in this section on $\mathbb{R}^d$ by introducing the notion of Lipschitz harmonic capacity ([**71**]). Volberg [**117**] proved a generalization of Theorem 7. The main difficulty arises from the fact that in higher-dimensional cases we do not have any good substitute for the notion of Menger curvature (see [**39**], [**62**], [**63**], [**66**] and [**84**]). About removability on $\mathbb{R}^d$, see [**26**], [**27**] and [**117**].

3. Harmonic analysis on Euclidean spaces

In this section we go over theories in real analysis which had been obtained before the 1990s; in particular, we focus on the L^2 boundedness of singular integral operators (see [**41**] and [**106**]).

3.1. Classical singular integral operators.

DEFINITION 9. We say that T is a classical singular integral operator with kernel $K(x)$ if

$$Tf(x) = \text{p.v.} \int_{\mathbb{R}^d} K(x-y)f(y)dy, \quad \text{where} \quad f \in \mathcal{S}(\mathbb{R}^d),$$

and if K satisfies the following:

(i) $$|K(x)| \le \frac{C_0}{|x|^d},$$

(ii) $$|\nabla K(x)| \le \frac{C_0}{|x|^{d+1}},$$

(iii) $$\int_{\varepsilon<|x|<N} K(x)dx = 0, \qquad \text{for all } 0<\varepsilon<N<\infty.$$

EXAMPLE (Hilbert transform).

$$Hf(x) = \text{p.v.} \int_{\mathbb{R}^1} \frac{1}{x-y} f(y)dy.$$

PROPOSITION 3 (Calderón and Zygmund [**5**]). *A classical singular integral operator is bounded on $L^p(\mathbb{R}^d)$ where $1<p<\infty$.*

PROOF. By the conditions (i)–(iii) we have $\hat{K} \in L^\infty$; therefore we obtain the L^2 boundedness of T. Once we know the L^2 boundedness we obtain the weak $(1,1)$ boundedness of T by the Calderón–Zygmund theory. By using the interpolation theorem and a duality argument we obtain the L^p boundedness of T. □

3.2. Generalized singular integral operators.

DEFINITION 10. We say that T is a generalized singular integral operator with kernel $K(x,y)$ if

$$Tf(x) = \text{p.v.} \int_{\mathbb{R}^d} K(x,y)f(y)dy, \quad \text{where} \quad f \in \mathcal{S}(\mathbb{R}^n),$$

and if K satisfies the following:

(i′) $$|K(x,y)| \le \frac{C_0}{|x-y|^d},$$

(ii′) $$|\nabla_x K(x,y)| + |\nabla_y K(x,y)| \le \frac{C_0}{|x-y|^{d+1}}.$$

REMARK. Different from the classical cases, the limit p.v. $\int_{\mathbb{R}^d} K(x,y)f(y)dy$ may not exist in the usual sense, but for simplicity we use this definition. For the precise definition, see [**41**], Chapter 8 and [**75**], Chapter 7.

The Cauchy transform (4) is a generalized singular integral operator. We redefine it as follows.

DEFINITION 11.

$$\widetilde{C}_A f(x) = \text{p.v.} \int_{\mathbb{R}^1} \frac{f(y)(1+iA'(y))}{x-y+i(A(x)-A(y))} dy, \quad \text{where} \quad A' \in L^\infty(\mathbb{R}^1). \tag{8}$$

Since $A' \in L^\infty(\mathbb{R}^1)$, the factor $1+iA'(y)$ may be omitted. Therefore we discuss the following operator.

DEFINITION 12 (Cauchy integrals on Lipschitz curves).

$$C_A f(x) = \text{p.v.} \int_{\mathbb{R}^1} \frac{f(y)}{x-y+i(A(x)-A(y))} dy, \quad \text{where} \quad A' \in L^\infty(\mathbb{R}^1). \tag{9}$$

We write $x-y+i(A(x)-A(y)) = (x-y)\big(1+i(A(x)-A(y))/(x-y)\big)$, and expanding in a geometric series formally (see Section 3.4), we obtain the following formula:

$$C_A f(x) \sim \sum_{k=0}^{\infty} (-i)^k \int_{\mathbb{R}^1} \frac{(A(x)-A(y))^k}{(x-y)^{k+1}} f(y)dy. \tag{10}$$

We define the next operators.

DEFINITION 13 (kth-order Calderón commutator).

$$T_A^k f(x) = \text{p.v.} \int_{\mathbb{R}^1} \frac{(A(x)-A(y))^k}{(x-y)^{k+1}} f(y)dy. \tag{11}$$

REMARK. T_A^0 is the Hilbert transform, and we write $T_A^1 = T_A$ for short.

The name 'Calderón's commutator' results from the following. In his study of the symbolic calculus of pseudodifferential operators, Calderón [**3**] proposed the next problem (see [**10**]). We explain the one-dimensional case.

PROBLEM. Let M_A be the multiplication operator by A, and let H be the Hilbert transform. Is the commutator operator $[H, M_A] = H \cdot M_A - M_A \cdot H$ smoothing of order 1 when $A' \in L^\infty$?

Since the following formula holds:

$$\frac{d}{dx}[H, M_A]f(x) = T_A f(x) - A'(x)Hf(x),$$

the problem is reduced to showing the L^2 boundedness of T_A (see [**7**], p. 9).

Next we define Calderón–Zygmund operators.

DEFINITION 14. We say that a generalized singular integral operator T is a Calderón–Zygmund operator if T satisfies the following:

$$T \text{ is bounded on } L^2(\mathbb{R}^d). \tag{iii$'$}$$

The following proposition about Calderón–Zygmund operators is easy.

PROPOSITION 4. *Calderón–Zygmund operators are bounded on* $L^p(\mathbb{R}^d)$, *where* $1 < p < \infty$.

Since we assume the L^2 boundedness of T in (iii′), we obtain the L^p boundedness of T by the Calderón–Zygmund theory. Therefore the following problem has arisen.

PROBLEM. Find the conditions under which generalized singular integral operators are bounded on L^2.

This is one of the most important problems in the theory of singular integral operators in the latter half of the 20th century.

The difficulty of this problem comes from the following. Different from the classical cases, we cannot use the method of Fourier transforms (see the proof of Proposition 3).

For simplicity we make one more assumption on T:

(iv) $$K(x,y) = -K(y,x) \qquad \text{(antisymmetric condition)}.$$

REMARK. This assumption is not essential and is only for simplicity. Of course, the Cauchy integrals and Calderón's commutators satisfy this condition.

3.3. The $T1$ theorem. Before we state the celebrated $T1$ theorem by David and Journé, we make some preparations.

PROPOSITION 5 (Stein [**89**]). *If T is a Calderón–Zygmund operator, then*

$$\|Tf\|_{BMO} \le C_0 \|f\|_{L^\infty}, \quad \textit{where} \quad f \in L^2 \cap L^\infty.$$

REMARK. T is not bounded on L^∞.

BMO is the function space defined by John and Nirenberg [**47**].

DEFINITION 15 (BMO (bounded mean oscillation)). Let Q be a d-dimensional closed ball and denote the Lebesgue measure of Q by $|Q|$. We write $m_Q(f) = |Q|^{-1}\int_Q f(x)dx$.

$$BMO(\mathbb{R}^n) = \left\{ f \in L^1_{loc}(\mathbb{R}^d) \ : \ \|f\|_{BMO} = \sup_Q \frac{1}{|Q|}\int_Q |f(x) - m_Q(f)|dx < \infty \right\}.$$

REMARK. $L^\infty \subsetneq BMO$ and $\log|x| \in BMO(\mathbb{R}^d)$.

David and Journé [**23**] proved the following.

THEOREM 8 ($T1$ theorem, David and Journé (1984)). *Let T be a generalized singular integral operator satisfying (i′), (ii′) and (iv). Then T is bounded on L^2 if and only if*

$$T1 \in BMO(\mathbb{R}^d).$$

REMARK. About the meaning of $T1$, see [**75**], p. 49. The necessity has already been proved in Proposition 5.

By using this theorem we can prove the L^2 boundedness of Calderón's commutators.

COROLLARY 4 ([**23**]). *If $A' \in L^\infty(\mathbb{R}^1)$, then T_A^k is bounded on $L^2(\mathbb{R}^1)$; that is, T_A^k is a Calderón–Zygmund operator.*

PROOF. We prove the case $k = 1$. To calculate $T_A(1)$, at least formally, integration by parts gives

$$T_A(1) = \int \frac{A(x) - A(y)}{(x-y)^2} dy = \int \frac{A'(y)}{x-y} dy = H(A')(x).$$

By Proposition 5 the Hilbert transform is bounded from L^∞ to BMO; therefore we obtain $T_A(1) \in BMO$. □

Calderón's problem has a long history (see **[3]**, **[15]** and **[106]**, p. 416).

3.4. The Tb theorem. In this subsection we discuss the L^2 boundedness of the Cauchy integral C_A; that is, we remember (10) in Section 3.2.
1) We need $\|A'\|_{L^\infty} < 1$.
2) We need precise estimates of the operator norms $\|T_A^k\|_{L^2 \to L^2}$.

There is another method. First, note that the $T1$ theorem is not very useful in this case. It is not easy to show that $C_A(1) = \int (x - y + i(A(x) - A(y)))^{-1} dy$ is a BMO function. Compare this with the proof of Corollary 4. However we have formally

$$\widetilde{C}_A(1) = \int \frac{(1 + iA'(y))}{x - y + i(A(x) - A(y))} dy = 0.$$

This motivates the Tb theorem.

REMARK. One may think that one can apply the $T1$ theorem for the kernel $K(x, y) = (1 + iA'(y))/(x - y + i(A(x) - A(y)))$, but this kernel does not satisfy (ii′).

DEFINITION 16. We say that a bounded complex function b is accretive if there exists a $c > 0$ such that

$$\text{Re } b(x) \geq c.$$

REMARK. We keep the function $b(x) = 1 + iA'(x)$ in mind.

THEOREM 9 (Tb theorem, David, Journé and Semmes **[24]**). *We assume that T satisfies the conditions (i′), (ii′) and (iv). If there exists a bounded accretive function b such that*

$$Tb \in BMO(\mathbb{R}^d), \text{ then } T \text{ is bounded on } L^2(\mathbb{R}^d).$$

As a corollary of this theorem we can show the L^2 boundedness of the Cauchy integrals.

COROLLARY 5 (McIntosh and Meyer **[72]**). *If $A' \in L^\infty(\mathbb{R}^1)$, then C_A is bounded on $L^2(\mathbb{R}^1)$.*

The L^2 boundedness of the Cauchy integrals on Lipschitz curves has a lot of applications to partial differential equations on nonsmooth domains (see Kenig **[52]**).

There are other results for the problems 1 and 2. David **[19]** used a bootstrap argument to remove the restriction in 1. We have the following result for problem 2 (see **[3]**, **[9]**, **[15]**, **[23]** and **[76]**).

THEOREM 10 (Coifman, McIntosh and Meyer **[14]**).

$$\|T_A^k\|_{L^2 \to L^2} \leq C_0 k^4 \|A'\|_{L^\infty}^k.$$

We find other proofs of L^2 boundedness of the Cauchy integrals in **[4]**, **[13]**, **[19]**, **[48]** and **[76]**. Murai **[76]** obtained the optimum estimate.

3.5. Hardy spaces. Singular integral operators are not bounded on L^1. The Hardy space H^1 is a substitute for this. In this section we explain some elementary properties of the Hardy space H^1. We do not discuss H^p.

DEFINITION 17. Let $P_y(x) = y\cdot(|x|^2+y^2)^{-(d+1)/2}$ and define $f^*(x) = \sup_{y>0} |P_y * f(x)|$. Define

$$H^1(\mathbb{R}^d) = \{f \in L^1(\mathbb{R}^d) \ : \ \|f\|_{H^1} = \|f^*\|_{L^1} < \infty\}.$$

The following results are well known.

THEOREM 11 ([**31**]). *Classical singular integral operators are bounded on $H^1(\mathbb{R}^d)$.*

THEOREM 12 ([**31**]). *The dual space of $H^1(\mathbb{R}^d)$ is $BMO(\mathbb{R}^d)$.*

DEFINITION 18. We say that a function $a(x)$ is an H^1–atom if there exists a ball $B(x_0, r)$ such that

$$\operatorname{supp}(a) \subset B(x_0, r), \ \|a\|_\infty \le r^{-d} \text{ and } \int a(x)dx = 0.$$

THEOREM 13 (Atomic decomposition [**11**], [**55**]). *For any function $f \in H^1(\mathbb{R}^d)$, there exist constants c_j and H^1–atoms $a_j(x)$ such that*

$$f(x) = \sum_{j=1}^{\infty} c_j a_j(x) \quad \textit{and} \quad \sum_{j=1}^{\infty} |c_j| \le C_0 \|f\|_{H^1}.$$

3.6. Harmonic analysis on doubling measure spaces. In Section 3 we have considered only the Lebesgue measure on $\mathbb{R}^d$ for simplicity. Almost all the results above hold on (X, d, μ), where the measure μ is doubling (1). Coifman and Weiss [**17**] (1971), [**18**] called these spaces **spaces of homogeneous type** and showed that it is possible to make a unified approach to many function spaces. Thirteen examples are listed in [**18**], p. 588. We give the results corresponding to the ones stated above:

The Calderón–Zygmund theory holds (Coifman and Weiss [**17**]).

The $T1$ theorem and the Tb theorem hold (Christ [**7**], [**8**]).

The theory of Hardy spaces (atomic decomposition) and the theory of BMO hold (Coifman and Weiss [**18**], Macías and Segovia [**58**], [**59**] and Uchiyama [**107**]).

4. Harmonic analysis on nondoubling measure spaces

We summarize what we have known in the previous sections as follows. To solve the Painlevé problem we need to show the L^2 boundedness of singular integral operators (the Cauchy transforms) on irregular measure spaces. As results of real analysis we know the theories ($T1$ theorem and Tb theorem) which prove the L^2 boundedness of singular integral operators on spaces of homogeneous type. Therefore we must construct a theory of singular integral operators on nondoubling measure spaces.

In the following, we assume that μ is a positive Radon measure on R^d.

4.1. The Calderón–Zygmund theory. The theories on spaces of homogeneous type were successful, because the Hardy–Littlewood maximal operator is weak $(1,1)$ bounded.

DEFINITION 19 (Hardy–Littlewood maximal operator).

$$M_\mu f(x) = \sup_{Q \ni x} \frac{1}{\mu(Q)} \int_Q |f(y)| d\mu(y),$$

where the supremum is taken over all closed balls containing x.

PROPOSITION 6. *When μ is doubling,*

$$\mu(\{x \in \mathbb{R}^d : M_\mu f(x) > \lambda\}) \leq \frac{C_0}{\lambda} \int_{\mathbb{R}^d} |f(x)| d\mu(x). \tag{12}$$

To prove this we use the following very simple property:

$$\frac{1}{\mu(B(x,r))} \int_{B(x,r)} |f(y)| d\mu(y) \leq \frac{C_0}{\mu(B(x,2r))} \int_{B(x,2r)} |f(y)| d\mu(y). \tag{13}$$

By this property we can obtain appropriate estimates on the annuli $\{x \in \mathbb{R}^d : 2^k < |x - x_0| \leq 2^{k+1}\}_{k=-\infty}^{\infty}$.

The most important point of the Calderón–Zygmund theory is the following fact. If T is bounded on L^2, then T is weak $(1,1)$ bounded (see the proof of Proposition 3). In fact when we proved the Vituškin conjecture, in the proof of Proposition 2 we used the following fact: if C_Γ is bounded on L^2, then C_Γ is weak $(1,1)$ bounded.

In the case of nondoubling measure we do not have (13). Therefore (12) does not hold in general (see [**50**], p. 10, and see also [**2**]). As a substitute for this we obtain the next proposition.

DEFINITION 20 (Modified Hardy–Littlewood maximal operator).

$$\widetilde{M}_\mu f(x) = \sup_{Q \ni x} \frac{1}{\mu(3Q)} \int_Q |f(y)| d\mu(y),$$

where $3Q$ denotes the ball which has the same center as Q and the radius is three times that of Q.

PROPOSITION 7.

$$\mu(\{x \in \mathbb{R}^d : \widetilde{M}_\mu f(x) > \lambda\}) \leq \frac{C_0}{\lambda} \int_{\mathbb{R}^d} |f(x)| d\mu(x).$$

This proposition is easily proved by using Besicovitch's covering lemma (see [**68**]). The constant 3 is not essential. The ρ times ($\rho > 1$) will do (see [**85**]). When $d = 1$, (12) is true even if μ is a nondoubling measure [**88**].

As mentioned above, it is important to expand a ball a little when we study nondoubling measures.

The following result by Nazarov, Treil and Volberg and Tolsa is the first step to success.

DEFINITION 21. We say that μ has growth of degree n if

$$\mu(B(x,r)) \leq C_0 r^n. \tag{14}$$

REMARK. Please remember the condition (5) in Section 2.5. If $n > d$, then the measure which has growth of degree n is zero.

DEFINITION 22 (n-dimensional singular integral operators). We say that T_μ is an n-dimensional singular integral operator with kernel K if

$$T_\mu f(x) = \int_{\mathbb{R}^d} K(x,y)f(y)d\mu(y), \qquad x \in \mathbb{R}^d \setminus \operatorname{supp}(f)$$

and if K satisfies the following:

$$|K(x,y)| \leq \frac{C_0}{|x-y|^n},$$
$$|\nabla_x K(x,y)| + |\nabla_y K(x,y)| \leq \frac{C_0}{|x-y|^{n+1}}.$$

$T_{\mu,\varepsilon}$ and the terminology 'T_μ is bounded on $L^2(\mu)$' are defined the same as (2) in Section 2.2.

When T is a classical singular integral, $n = d$. Now we keep the following case in mind: $\mathbb{C} = \mathbb{R}^2$, that is, $d = 2$ and $n = 1$.

THEOREM 14 ([**78**], [**92**], [**96**]). *We assume that μ satisfies (14) and that T_μ is an n-dimensional singular integral operator.*

If T_μ is bounded on $L^2(\mu)$, then T_μ is weak $(1,1)$ bounded.

REMARK. The condition (14) is natural, because when T_μ has a nontrivial kernel such that $|K(x,y)| \geq |x-y|^{-n}$, if T is bounded on $L^2(\mu)$, then μ must satisfy (14) (see [**20**], p. 56).

In [**78**], Nazarov, Treil and Volberg proved this theorem by showing a variant of Cotlar's inequality ([**75**], p. 34). In [**96**], Tolsa proved a variant of the Calderón–Zygmund decomposition on nondoubling measure spaces.

4.2. Doubling ball. The following notion of a 'doubling ball' is very important not only in the proof of Theorem 14 but also in the analysis of nondoubling measure spaces.

In the following, we always assume that μ satisfies the growth condition (14).

DEFINITION 23. We say that a ball $B(x,r)$ is a **doubling ball** if

$$\mu(B(x,2r)) \leq 3^{d+1}\mu(B(x,r)).$$

The next lemma is very important.

LEMMA 3. *For any $x \in \mathbb{R}^d$ and $r > 0$, there exist infinitely many doubling balls in the family of balls $\{B(x,2^j r)\}_{j=0}^\infty$.*

For μ-almost every $x \in \mathbb{R}^d$, there exists a series $\{r_j\}_{j=0}^\infty$ which tends to 0 such that all of the balls $B(x,r_j)$ are doubling balls.

That is, there are enough doubling balls which have a nice property. Figuratively speaking, when we compare the analysis on the Lebesgue measure spaces to walking on a concrete road, the analysis on nondoubling measure spaces is compared to walking on a muddy road. But we can go forward by making doubling balls as stepping stones.

4.3. BMO. The following definition of BMO on nondoubling measure spaces is natural:

$$f \in BMO(\mu) \Leftrightarrow \sup_{Q} \frac{1}{\mu(Q)} \int_{Q} |f(x) - m_Q(f)| d\mu(x) < \infty,$$

where $m_Q(f) = \mu(Q)^{-1} \int_Q f(x) d\mu(x)$. However, on this space, Stein's result (Proposition 5) does not hold; that is, singular integral operators are not bounded from L^∞ to $BMO(\mu)$ (see [**61**] and [**111**]). Therefore by using the idea of expanding balls (see Section 4.1), we define a new function space as follows.

DEFINITION 24 ($BMO_\rho(\mu)$).

$$BMO_\rho(\mu)(\mathbb{R}^d) = \Big\{ f \in L^1_{loc}; \|f\|_{BMO_\rho(\mu)} = \sup_{Q} \frac{1}{\mu(\rho Q)} \int_{Q} |f(x) - m_Q(f)| d\mu(x) < \infty \Big\}.$$

Then we obtain the next result.

THEOREM 15. *Let $\rho > 1$. We assume that T_μ is an n-dimensional singular integral operator and is bounded on $L^2(\mu)$. Then*

T_μ is bounded from $L^\infty(\mu)$ to $BMO_\rho(\mu)$.

We can prove this theorem by the same argument as in the classical case by virtue of the parameter ρ. The following result is important. Nazarov, Treil and Volberg [**77**] proved a $T1$ theorem (see also [**91**], [**94**], [**97**] and [**111**]). For simplicity, here again we assume the antisymmetric condition $K(x, y) = -K(y, x)$.

THEOREM 16 ($T1$ theorem, Nazarov–Treil–Volberg [**77**] (1997)). *We assume that μ satisfies the growth condition (14) and that T_μ is an antisymmetric n-dimensional singular integral operator. Then the following two statements are equivalent.*

(15) *T_μ is bounded on $L^2(\mu)$.*

(16) *There exists a $\rho > 1$ such that $T_\mu 1 \in BMO_\rho(\mu)$.*

A drawback of this theorem is that the spaces $BMO_\rho(\mu)$ depend on parameters ρ. Tolsa [**95**] defined a new function space $RBMO(\mu)$ (regular BMO) and obtained a similar result. Note that $RBMO(\mu) \subset \bigcap_{\rho>1} BMO_\rho(\mu)$ and functions in RBMO satisfy the John–Nirenberg inequality [**95**] (see also [**57**] and [**87**]). Before we define RBMO we state the result.

THEOREM 17 ($T1$ theorem, Tolsa [**95**] (2001)). *Under the same assumption as in Theorem 16, the following two statements are equivalent.*

(17) *T_μ is bounded on $L^2(\mu)$.*

(18) *$T_\mu 1 \in RBMO(\mu)$.*

Since the definition of RBMO is unnatural at a glance, we explain the reason why we need such a definition. By a simple calculation we have

$$|m_Q(f) - m_{2Q}(f)| \leq \frac{\mu(2Q)}{\mu(Q)} \|f\|_{BMO(\mu)}.$$

Therefore, when μ is doubling, for all balls such that $Q \subset R$,

$$|m_Q(f) - m_R(f)| \leq C_0 \log\Big(\frac{\mathrm{rad}(R)}{\mathrm{rad}(Q)}\Big)\|f\|_{BMO(\mu)},$$

where $\mathrm{rad}(Q)$ denotes the radius of a ball Q.

That is, the mean values of a function f are well controlled by the BMO norm of f. However, in the case of a nondoubling measure this does not hold.

The following notion of a 'Tolsa constant' plays a very important role in the analysis of nondoubling measure spaces.

DEFINITION 25 (Tolsa constant). For any balls such that $Q \subset R$, we denote the least integer k which satisfies $R \subset 2^k Q$ by $N_{Q,R}$. The Tolsa constant $K_{Q,R}$ is defined by

$$K_{Q,R} = 1 + \sum_{k=1}^{N_{Q,R}} \frac{\mu(2^k Q)}{\mathrm{rad}(2^k Q)^d}.$$

DEFINITION 26 (RBMO). We say that $f \in RBMO(\mu)(R^d)$ if there exists a positive constant C_0 such that

$$\sup_Q \frac{1}{\mu(2Q)} \int_Q |f(x) - m_Q(f)| d\mu(x) \leq C_0,$$

$$|m_Q(f) - m_R(f)| \leq C_0 K_{Q,R}\Big(\frac{\mu(2Q)}{\mu(Q)} + \frac{\mu(2R)}{\mu(R)}\Big), \quad \text{where} \quad Q \subset R.$$

REMARK. Of course, when μ is doubling, $RBMO(\mu) = BMO_\rho(\mu) = BMO(\mu)$.

Nazarov, Treil and Volberg [**79**] also proved the Tb theorem. (See also [**22**], [**64**] and [**80**].)

THEOREM 18 (Tb theorem, Nazarov–Treil–Volberg [**79**] (2003)). *We assume that μ satisfies the growth condition (14) and that T_μ is an antisymmetric n-dimensional singular integral operator. If there exists a bounded accretive function b such that*

$$Tb \in BMO_\rho(\mathbb{R}^d) \text{ for some } \rho > 1, \text{ then } T \text{ is bounded on } L^2(\mu).$$

This theorem played an most important role when Tolsa [**100**] solved the Painlevé problem.

Since RBMO is defined, we can also define the Hardy spaces ([**45**], [**46**], [**95**] and [**99**]). We give an outline of the theorem.

THEOREM 19 (Tolsa [**95**], [**99**]). *By using some maximal function correspondent to the one in Definition* 17, *we define the Hardy space $H^1(\mu)$. Then the dual space of $H^1(\mu)$ is $RBMO(\mu)$, and an atomic decomposition theorem for $H^1(\mu)$ also holds.*

One knows the atomic decomposition theorem for $H^p(R^d)$ $(p < 1)$, but one has not obtained a theory for $H^p(\mu)$ on nondoubling measure spaces. The theorem above is an existence theorem using Banach's closed range theorem; therefore one cannot apply its proof to $H^p(\mu)$, where $p < 1$. One needs a constructive proof of this theorem as in the case of $H^p(R^d)$.

4.4. Other results. By using the methods developed by Nazarov, Treil and Volberg and Tolsa, many studies have been done on the harmonic analysis of nondoubling measure spaces. For fractional integrals, see [**32**] and [**35**]. For the commutator of Coifman, Rochberg and Weiss, see [**6**], [**43**], [**44**], [**45**] and [**95**]. Lipschitz spaces are studied in [**33**], Besov spaces are in [**30**], Triebel–Lizorkin spaces are in [**42**], and Morrey spaces are in [**86**].

Many studies have been done on weighted estimates of nondoubling measure spaces (see [**34**], [**36**], [**53**], [**54**], [**60**] and [**82**]).

Recently Tolsa [**105**] defined some weight classes using another maximal operator different from the one in Section 4.1, and he obtained a necessary and sufficient condition that singular integrals are bounded on weighted L^p spaces. This result corresponds to the classical result by Coifman and Fefferman [**12**].

References

[1] L. V. Ahlfors, Bounded analytic functions, Duke Math. J., **14** (1947), 1–11. MR0021108 (9:24a)

[2] J. M. Aldaz, An example on the maximal function associated to a nondoubling measure, Publ. Mat., **49** (2005), 453–458. MR2177637 (2006d:42026)

[3] A. P. Calderón, Commutators of singular integral operators, Proc. Nat. Acad. Sci. U.S.A., **53** (1965), 1092–1099. MR0177312 (31:1575)

[4] A. P. Calderón, Cauchy integrals on Lipschitz curves and related operators, Proc. Nat. Acad. Sci. U.S.A., **74** (1977), 1324–1327. MR0466568 (57:6445)

[5] A. P. Calderón and A. Zygmund, On the existence of certain singular integrals, Acta Math., **88** (1952), 85–139. MR0052553 (14:637f)

[6] W. Chen and E. Sawyer, A note on commutators of fractional integrals with RBMO functions, Illinois J. Math., **46** (2002), 1287–1298. MR1988264 (2004c:42029)

[7] M. Christ, Lectures on singular integral operators, CBMS Regional Conf. Ser. in Math., **77**, Amer. Math. Soc., Providence, RI, 1990. MR1104656 (92f:42021)

[8] M. Christ, A $T(b)$ theorem with remarks on analytic capacity and the Cauchy integral, Colloq. Math., **60/61** (1990), 601–628. MR1096400 (92k:42020)

[9] M. Christ and J.-L. Journé, Polynomial growth estimates for multilinear singular integral operators, Acta Math., **159** (1987), 51–80. MR906525 (89a:42024)

[10] M. Christ, C. Kenig and C. Sadosky, Harmonic Analysis and Partial Differential Equations, Essays in Honor of Alberto P. Calderón, The Univ. of Chicago Press, 1999. MR1731192 (2000j:00032)

[11] R.R. Coifman, A real variable characterization of H^p, Studia Math., **51** (1974), 269–274. MR0358318 (50:10784)

[12] R.R. Coifman and C. Fefferman, Weighted norm inequalities for maximal functions and singular integrals, Studia Math., **51** (1974), 241–250. MR0358205 (50:10670)

[13] R. Coifman, P. Jones and S. Semmes, Two elementary proofs of the L^2 boundedness of Cauchy integrals on Lipschitz curves, J. Amer. Math. Soc., **2** (1989), 553–564. MR986825 (90k:42017)

[14] R.R. Coifman, A. McIntosh and Y. Meyer, L'intégrale de Cauchy définit un opérateur borné sur L^2 pour les courbes lipschitziennes, Ann. of Math. (2), **116** (1982), 361–387. MR672839 (84m:42027)

[15] R.R. Coifman and Y. Meyer, On commutators of singular integrals and bilinear singular integrals, Trans. Amer. Math. Soc., **212** (1975), 315–331. MR0380244 (52:1144)

[16] R.R. Coifman, R. Rochberg and G. Weiss, Factorization theorems for Hardy spaces in several variables, Ann. of Math. (2), **103** (1976), 611–635. MR0412721 (54:843)

[17] R.R. Coifman and G. Weiss, Analyse harmonique non-commutative sur certains espaces homogènes, Lecture Notes in Math., **242**, Springer, 1971. MR0499948 (58:17690)

[18] R.R. Coifman and G. Weiss, Extensions of Hardy spaces and their use in analysis, Bull. Amer. Math. Soc., **83** (1977), 569–645. MR0447954 (56:6264)

[19] G. David, Opérateurs intégraux singuliers sur certaines courbes du plan complexe, Ann. Sci. École Norm. Sup., **17** (1984), 157–189. MR744071 (85k:42026)

[20] G. David, Wavelets and Singular Integrals on Curves and Surfaces, Lecture Notes in Math., **1465**, Springer, 1991. MR1123480 (92k:42021)

[21] G. David, Analytic capacity, Calderón-Zygmund operators, and rectifiability, Publ. Mat., **43** (1999), 3–25. MR1697514 (2000e:30044)

[22] G. David, Unrectifiable 1-sets have vanishing analytic capacity, Rev. Mat. Iberoamericana, **14** (1998), 369–479. MR1654535 (99i:42018)

[23] G. David and J.-L. Journé, A boundedness criterion for generalized Calderón-Zygmund operators, Ann. of Math. (2), **120** (1984), 371–397. MR763911 (85k:42041)

[24] G. David, J.-L. Journé and S. Semmes, Opérateurs de Calderón-Zygmund, fonctions para-accrétives et interpolation, Rev. Mat. Iberoamericana, **1** (1985), 1–56. MR850408 (88f:47024)

[25] G. David and P. Mattila, Removable sets for Lipschitz harmonic functions in the plane, Rev. Mat. Iberoamericana, **16** (2000), 137–215. MR1768535 (2001h:31001)

[26] G. David and S. Semmes, Singular integrals and rectifiable sets in R^n: Beyond Lipschitz graphs, Astérisque, **193** (1991), 152 pp. MR1113517 (92j:42016)

[27] G. David and S. Semmes, Analysis of and on uniformly rectifiable sets, Math. Surveys Monogr., **38**, Amer. Math. Soc., Providence, RI, 1993. MR1251061 (94i:28003)

[28] A. M. Davie, Analytic capacity and approximation problems, Trans. Amer. Math. Soc., **171** (1972), 409–444. MR0350009 (50:2502)

[29] A. M. Davie and B. Øksendal, Analytic capacity and differentiability properties of finely harmonic functions, Acta Math., **149** (1982), 127–152. MR674169 (84h:31001)

[30] D. Deng, Y. Han and D. Yang, Besov spaces with non-doubling measures, Trans. Amer. Math. Soc., **358** (2006), 2965–3001. MR2216255 (2006k:42046)

[31] C. Fefferman and E.M. Stein, H^p spaces of several variables, Acta Math., **129** (1972), 137–193. MR0447953 (56:6263)

[32] J. García-Cuerva and E. Gatto, Boundedness properties of fractional integral operators associated to non-doubling measures, Studia Math., **162** (2004), 245–261. MR2047654 (2005c:42012)

[33] J. García-Cuerva and E. Gatto, Lipschitz spaces and Calderón-Zygmund operators associated to non-doubling measures, Publ. Mat. **49** (2005), 285–296. MR2177069 (2006m:42021)

[34] J. García-Cuerva and J. M. Martell, Weighted inequalities and vector-valued Calderón- Zygmund operators on non-homogeneous spaces, Publ. Mat., **44** (2000), 613–640. MR1800824 (2002k:42031)

[35] J. García-Cuerva and J. M. Martell, Two-weight norm inequalities for maximal operators and fractional integrals on non-homogeneous spaces, Indiana Univ. Math. J., **50** (2001), 1241–1280. MR1871355 (2003a:42023)

[36] J. García-Cuerva and J.M. Martell, On the existence of principal values for the Cauchy integral on weighted Lebesgue spaces for non-doubling measures, J. Fourier Anal. Appl., **7** (2001), 469–487. MR1845099 (2002i:42013)

[37] J. Garnett, Positive length but zero analytic capacity, Proc. Amer. Math. Soc., **21** (1970), 696–699. MR0276456 (43:2203)

[38] J. Garnett, Analytic capacity and measure, Lecture Notes in Math., **297**, Springer, 1972. MR0454006 (56:12257)

[39] J. Garnett, L. Prat and X. Tolsa, Lipschitz harmonic capacity and bilipschitz images of Cantor sets, Math. Res. Lett., **13** (2006), 865–884. MR2280781 (2007k:31008)

[40] J. Garnett and J. Verdera, Analytic capacity, bilipschitz maps and Cantor sets, Math. Res. Lett., **10** (2003), 515–522. MR1995790 (2004g:30038)

[41] L. Grafakos, Classical and Modern Fourier Analysis, Pearson Education, Inc., Upper Saddle River, NJ, 2004. MR2449250

[42] Y. Han and D. Yang, Triebel-Lizorkin spaces with non-doubling measures, Studia Math., **162** (2004), 105–140. MR2046565 (2005b:42020)

[43] G. Hu, Y. Meng and D. Yang, Multilinear commutators for fractional integrals in non-homogeneous spaces, Publ. Mat., **48** (2004), 335–367. MR2091009 (2005g:42031)

[44] G. Hu, Y. Meng and D. Yang, Multilinear commutators of singular integrals with non doubling measures, Integral Equations Operator Theory, **51** (2005), 235–255. MR2120079 (2005k:42040)

[45] G. Hu, Y. Meng and D. Yang, New atomic characterization of H^1 space with nondoubling measures and its applications, Math. Proc. Cambridge Philos. Soc., **138** (2005), 151–171. MR2127235 (2005m:42030)

[46] G. Hu and S. Liang, Another characterization of the Hardy space with non doubling measures, Math. Nachr., **279** (2006), 1797–1807. MR2274835 (2007h:42027)

[47] F. John and L. Nirenberg, On functions of bounded mean oscillation, Comm. Pure Appl. Math., **14** (1961), 415–426. MR0131498 (24:A1348)

[48] P. W. Jones, Square functions, Cauchy integrals, analytic capacity, and harmonic measure, In: Harmonic analysis and partial differential equations, El Escorial, 1987, Lecture Notes Math., **1384**, Springer, 1989, pp. 24-68. MR1013815 (91b:42032)

[49] P. W. Jones, Rectifiable sets and the traveling salesman problem, Invent. Math., **102** (1990), 1–15. MR1069238 (91i:26016)

[50] J.-L. Journé, Calderón-Zygmund operators, pseudodifferential operators and the Cauchy integral of Calderón, Lecture Notes in Math., **994**, Springer, 1983. MR706075 (85i:42021)

[51] S. Kass and Karl Menger, Notices Amer. Math. Soc., **43** (1996), 558–561. MR1388728

[52] C.E. Kenig, Harmonic analysis techniques for second order elliptic boundary value problems, CBMS Regional Conf. Ser. in Math., **83**, Amer. Math. Soc., Providence, RI, 1994. MR1282720 (96a:35040)

[53] V. Kokilashvili and A. Meskhi, On some weighted inequalities for fractional integrals on nonhomogeneous spaces, Z. Anal. Anwendungen, **24** (2005),871–885. MR2200116 (2006k:42025)

[54] Y. Komori, Weighted estimates for operators generated by maximal functions on nonhomogeneous spaces, Georgian Math. J., **12** (2005), 121–130. MR2136890 (2006a:42024)

[55] R. H. Latter, A decomposition of $H^p(R^n)$ in terms of atoms, Studia Math., **62** (1977), 92–101. MR0482111 (58:2198)

[56] J. C. Léger, Menger curvature and rectifiability, Ann. of Math. (2), **149** (1999), 831–869. MR1709304 (2001c:49069)

[57] A. K. Lerner, On the John-Strömberg characterization of BMO for nondoubling measures, Real Anal. Exchange, **28** (2003), 649–660. MR2010347 (2004m:42021)

[58] R. A. Macías and C. Segovia, Lipschitz functions on spaces of homogeneous type, Adv. Math., **33** (1979), 257–270. MR546295 (81c:32017a)

[59] R. A. Macías and C. Segovia, A decomposition into atoms of distributions on spaces of homogeneous type, Adv. Math., **33** (1979), 271–309. MR546296 (81c:32017b)

[60] J. Martín and M. Milman, Gehring's lemma for nondoubling measures, Michigan Math. J., **47** (2000), 559–573. MR1813544 (2002c:42026)

[61] J. Mateu, P. Mattila, A. Nicolau and J. Orobitg, BMO for nondoubling measures, Duke Math. J., **102** (2000), 533–565. MR1756109 (2001e:26019)

[62] J. Mateu, L. Prat and J. Verdera, The capacities associated to signed Riesz kernels, and Wolff potentials, J. Reine Angew. Math., **578** (2005), 201–223. MR2113895 (2005m:31012)

[63] J. Mateu and X. Tolsa, Riesz transforms and harmonic Lip_1-capacity in Cantor sets, Proc. London Math. Soc.(3), **89** (2004), 676–696. MR2107011 (2005g:42038)

[64] J. Mateu, X. Tolsa and J. Verdera, The planar Cantor sets of zero analytic capacity and the local $T(b)$-theorem, J. Amer. Math. Soc., **16** (2003), 19–28. MR1937197 (2003k:30041)

[65] J. Mateu, X. Tolsa and J. Verdera, On the semiadditivity of analytic capacity and planar Cantor sets, Contemp. Math., **320** (2003), 259–278. MR1979945 (2005f:30050)

[66] J. Mateu and J. Verdera, L^p and weak L^1 estimates for the maximal Riesz transform and the maximal Beurling transform, Math. Res. Lett., **13** (2006), 957–966. MR2280788 (2007i:42022)

[67] P. Mattila, Cauchy singular integrals and rectifiability in measures of the plane, Adv. Math., **115** (1995), 1–34. MR1351323 (96h:30078)

[68] P. Mattila, Geometry of sets and measures in Euclidean spaces, Cambridge Stud. Adv. Math., **44**, Cambridge Univ. Press, Cambridge, 1995. MR1333890 (96h:28006)

[69] P. Mattila, M.S. Melnikov and J. Verdera, The Cauchy integral, analytic capacity, and uniform rectifiability, Ann. of Math. (2), **144** (1996), 127–136. MR1405945 (97k:31004)

[70] P. Mattila and M.S. Melnikov, Existence and weak type inequalities for Cauchy integrals of general measures on rectifiable curves and sets, Proc. Amer. Math. Soc., **120** (1994), 143–149. MR1160305 (94b:30047)

[71] P. Mattila and P. V. Paramonov, On geometric properties of harmonic Lip_1-capacity, Pacific J. Math., **171** (1995), 469–490. MR1372240 (97b:31005)

[72] A. McIntosh and Y. Meyer, Algèbres d'opérateurs définis par des intégrales singulières, C.R. Acad. Sci. Paris Ser. I Math., **301** (1985), 395–397. MR808636 (87b:47053)

[73] M. S. Melnikov, Analytic capacity: Discrete approach and curvature of measure, Sb. Math., **186** (1995), 827–846. MR1349014 (96f:30020)

[74] M. S. Melnikov and J. Verdera, A geometric proof of the L^2 boundedness of the Cauchy integral on Lipschitz graphs, Internat. Math. Res. Notices, **1995**, no.7, 325–331. MR1350687 (96f:45011)

[75] Y. Meyer and R. Coifman, Wavelets: Calderón-Zygmund and multilinear operators, Cambridge Stud. Adv. Math., **48**, Cambridge Univ. Press, 1997. MR1456993 (98e:42001)

[76] T. Murai, A real variable method for the Cauchy transform, and analytic capacity, Lecture Notes in Math., **1307**, Springer, 1988. MR944308 (89k:30022)

[77] F. Nazarov, S. Treil and A. Volberg, Cauchy integral and Calderón-Zygmund operators on non-homogeneous spaces, Internat. Math. Res. Notices, **1997**, no. 15, 703–726. MR1470373 (99e:42028)

[78] F. Nazarov, S. Treil and A. Volberg, Weak type estimates and Cotlar inequalities for Calderón-Zygmund operators on nonhomogeneous spaces, Internat. Math. Res. Notices, **1998**, no. 9, 463–487. MR1626935 (99f:42035)

[79] F. Nazarov, S. Treil and A. Volberg, Tb theorem on nonhomogeneous spaces, Acta Math., **190** (2003), 151–239. MR1998349 (2005d:30053)

[80] F. Nazarov, S. Treil and A. Volberg, Accretive system Tb theorems on nonhomogeneous spaces, Duke Math. J., **113** (2002), 259–312. MR1909219 (2003g:42030)

[81] K. Okikiolu, Characterization of subsets of rectifiable curves in R^n, J. London Math. Soc. (2), **46** (1992), 336–348. MR1182488 (93m:28008)

[82] J. Orobitg and C. Pérez, A_p weights for nondoubling measures in R^n and applications, Trans. Amer. Math. Soc., **354** (2002), 2013–2033. MR1881028 (2002k:42044)

[83] H. Pajot, Analytic Capacity, Rectifiability, Menger Curvature and the Cauchy Integral, Lecture Notes in Math., **1799**, Springer, 2002. MR1952175 (2004d:28009)

[84] L. Prat, Potential theory of signed Riesz kernels: Capacity and Hausdorff measures, Internat. Math. Res. Notices, **2004**, no. 19, 937–981. MR2037051 (2004k:31012)

[85] Y. Sawano, Sharp estimates of the modified Hardy-Littlewood maximal operator on the nonhomogeneous space via covering lemmas, Hokkaido Math. J., **34** (2005), 435–458. MR2159006 (2006k:42040)

[86] Y. Sawano and H. Tanaka, Morrey spaces for non-doubling measures, Acta Math. Sin. (Engl. Ser.), **21** (2005), 1535–1544. MR2190025 (2006g:42042)

[87] Y. Sawano and H. Tanaka, The John-Nirenberg type inequality for non-doubling measures, Studia Math., **181** (2007), 153–170. MR2320693 (2008i:42049)

[88] P. Sjögren, A remark on the maximal function for measures in R^n, Amer. J. Math. **105**, (1983), 1231–1233. MR714775 (86a:28003)

[89] E. M. Stein, Singular integrals, harmonic functions, and differentiability properties of functions of several variables, Proc. Symp. Pure Math. 1967, pp. 316–335. MR0482394 (58:2467)

[90] N. Suita, On subadditivity of analytic capacity for two continua, Kodai Math. J., **7** (1984), 73–75. MR733588 (85h:30034)

[91] X. Tolsa, Cotlar's inequality and existence of principal values for the Cauchy integral without the doubling condition, J. Reine Angew. Math., **502** (1998), 199–235. MR1647575 (2000a:42030)

[92] X. Tolsa, L^2-boundedness of the Cauchy integral operator for continuous measures, Duke Math. J., **98** (1999), 269–304. MR1695200 (2000d:31001)

[93] X. Tolsa, Principal values for the Cauchy integral and rectifiability, Proc. Amer. Math. Soc., **128** (2000), 2111–2119. MR1654076 (2000m:30048)

[94] X. Tolsa, A $T(1)$ theorem for non-doubling measures with atoms, Proc. London Math. Soc. (3), **82** (2001), 195–228. MR1794262 (2002i:42019)

[95] X. Tolsa, BMO, H^1 and Calderón-Zygmund operators for non doubling measures, Math. Ann., **319** (2001), 89–149. MR1812821 (2002c:42029)

[96] X. Tolsa, A proof of the weak $(1,1)$ inequality for singular integrals with non doubling measures based on a Calderón-Zygmund decomposition, Publ. Mat., **45** (2001), 163–174. MR1829582 (2002d:42019)

[97] X. Tolsa, Littlewood-Paley theory and the $T(1)$ theorem with non-doubling measures, Adv. Math., **164** (2001), 57–116. MR1870513 (2003e:42029)

[98] X. Tolsa, On the analytic capacity γ_+, Indiana Univ. Math. J., **51** (2002), 317–343. MR1909292 (2003m:31004)
[99] X. Tolsa, The space H^1 for nondoubling measures in terms of a grand maximal operator, Trans. Amer. Math. Soc., **355** (2003), 315–348. MR1928090 (2003e:42030)
[100] X. Tolsa, Painlevé's problem and the semiadditivity of analytic capacity, Acta Math., **190** (2003), 105–149. MR1982794 (2005c:30020)
[101] X. Tolsa, The semiadditivity of continuous analytic capacity and the inner boundary conjecture, Amer. J. Math. **126** (2004), 523–567. MR2058383 (2005f:30052)
[102] X. Tolsa, Finite curvature of arc length measure implies rectifiability: A new proof, Indiana Univ. Math. J., **54** (2005), 1075–1105. MR2164419 (2006i:28003)
[103] X. Tolsa, Bilipschitz maps, analytic capacity, and the Cauchy integral, Ann. of Math. (2), **162** (2005), 1243–1304. MR2179730 (2006g:30033)
[104] X. Tolsa, Painlevé's problem and analytic capacity, Collect. Math., 2006, Vol. Extra, 89–125. MR2264206 (2007g:30042)
[105] X. Tolsa, Weighted norm inequalities for Calderón-Zygmund operators without doubling conditions, Publ. Mat., **51** (2007), 397–456. MR2334796 (2008f:42018)
[106] A. Torchinsky, Real-Variable Methods in Harmonic Analysis, Academic Press, 1986. MR869816 (88e:42001)
[107] A. Uchiyama, A maximal function characterization of H^p on the space of homogeneous type, Trans. Amer. Math. Soc., **262** (1980), 579–592. MR586737 (81j:46085)
[108] N. X. Uy, Removale sets of analytic functions satisfying a Lipschitz condition, Ark. Mat., **17** (1979), 19–27. MR543500 (81i:30076)
[109] J. Verdera. A weak type inequality for Cauchy transforms of finite measures, Publ. Mat., **36** (1992), 1029–1034. MR1210034 (94k:30099)
[110] J. Verdera. Removability, capacity and approximation, In: Complex Potential Theory, Montreal, PQ, 1993, NATO Adv. Sci. Int. Ser. C Math. Phys. Sci., **439**, Kluwer Acad. Publ., Dordrecht, 1994, pp. 419–473. MR1332967 (96b:30086)
[111] J. Verdera, On the $T(1)$-theorem for the Cauchy integral, Ark. Mat., **38** (2000), 183–199. MR1749365 (2001e:30074)
[112] J. Verdera, The fall of the doubling condition in Calderón-Zygmund theory, Proceedings of the 6th International Conference on Harmonic Analysis and Partial Differential Equations, El Escorial, 2000, Publ. Mat., 2002, Vol. Extra, 275–292. MR1964824 (2004b:42035)
[113] J. Verdera, L^2 boundedness of the Cauchy integral and Menger curvature, In: Harmonic analysis and boundary value problems, Fayetteville, AR, 2000, Contemp. Math., **277** (2001), 139–158. MR1840432 (2002f:30052)
[114] A. G. Vituškin, Example of a set of positive length but of zero analytic capacity (Russian), Dokl. Akad. Nauk SSSR, **127** (1959), 246–249. MR0118838 (22:9607)
[115] A. G. Vituškin, Estimate of the Cauchy integral (Russian), Mat. Sb. (N.S.), **71** (1966), 515–534. MR0206305 (34:6124)
[116] A. G. Vituškin, Analytic capacity of sets in problems of approximation theory (Russian), Uspehi Mat. Nauk., **22** (1967), 141–199. MR0229838 (37:5404)
[117] A. Volberg, Calderón-Zygmund capacities and operators on nonhomogeneous spaces, CBMS Regional Conf. Ser. in Math., **100** Amer. Math. Soc., Providence, RI, 2003. MR2019058 (2005c:42015)

School of High Technology for Human Welfare, Tokai University, 317 Nishino Numazu Shizuoka, 410-0395 Japan
E-mail address: komori@wing.ncc.u-tokai.ac.jp

Translated by YASUO KOMORI

Amer. Math. Soc. Transl.
(2) Vol. **230**, 2010

Theory of Reproducing Kernels: Applications to Approximate Solutions of Bounded Linear Operator Equations on Hilbert Spaces

Saburou Saitoh

ABSTRACT. This paper is a translation of a paper which was published in Sūgaku of the Mathematical Society of Japan and which was a survey paper for results obtained mainly by the author over 30 years in the theory of reproducing kernels.

1. Introduction

The theory of reproducing kernels started with the two papers of 1921 and 1922 which dealt with the typical reproducing kernels of the Szegö kernel and the Bergman kernel, respectively, and then the theory developed as a large and deep theory in complex analysis by many mathematicians. Meanwhile, the general theory of reproducing kernels was established in a complete form by N. Aronszajn ([**6**]) in 1950. Furthermore, the general theory was developed remarkably by L. Schwartz in 1964 with the paper ([**76**]) which is over 140 pages. L. Schwartz is a Field-Medalist and the founder of distribution theory.

The general theory is certainly beautiful. It seems, however, that the importance of the general theory of reproducing kernels was not realized for a long time. We were not able to find an essential reason why the theory is important. Indeed, it was an abstract theory and from the theory, we were not able to derive some definite results and any essential developments in mathematics. The great theory by Schwartz is still not realized for its importance and it has been ignored until now.

This article originally appeared in Japanese in Sūgaku **60** (3) (2008), 248–271.

2000 *Mathematics Subject Classification.* Primary 30C40, 44A05, 35A22; Secondary 44A15, 35K05, 46E22.

Key words and phrases. Reproducing kernels, complex analysis, functional analysis, partial differential equations, inverse problem, numerical analysis, Tikhonov regularization.

This work was supported in part by Research Unit Mathematics and Applications, University of Aveiro, Portugal, through FCT - Portuguese Foundation for Science and Technology. The author was also supported in part by the Grant-in-Aid for Scientific Research (C)(2)(No. 16540137, 19540164 and 21540111) from the Japan Society for the Promotion of Science and by the Mitsubishi Foundation, Natural Sciences, Vol. 36, 20(2005-2006).

The author showed clearly its importance and fundamental concept in ([**43**]) in 1983 by combining the general theory and linear mappings in the framework of Hilbert spaces; then the general theory was not restricted to the Bergman and Szegö kernels, but the general theory is important as in the concept of Hilbert spaces. The general theory is a fundamental concept and is important mathematically. In this paper, we will firstly show compactly such definite results obtained since 1983 from a wide viewpoint. Secondly, we will show the applications of the general theory of reproducing kernels to the Tikhonov regularization, which gives the approximate solutions for bounded linear operator equations on Hilbert spaces with the viewpoint of numerical solutions by computers. The general theory of reproducing kernels is based on the elementary theorems in Hilbert spaces which are the minimum core of functional analysis; however, when the general theory is combined with the linear mappings on Hilbert spaces, it will have many relations in various fields, and its fruitful applications will spread over to differential equations, integral equations, generalizations of the Pythagorean theorem, inverse problems, sampling theory, nonlinear transforms combined with linear mappings, various operators among Hilbert spaces and many other wide fields.

Meanwhile, concrete reproducing kernels such as the Bergman kernel and the Szegö kernel will have many and wide results in complex analysis, and will require surprisingly some deep theory and lead to profound results in complex analysis. We will show here some typical and beautiful results; simply, however, for their great frameworks, we did not refer to them as a global and self-contained theory, but we gave good references for them. Such a theory will, however, not have general interest for mathematicians except for complex analysists.

Therefore, in Section 2, we will give the essential parts of the general theory of reproducing kernels, in Section 3, we will refer to concrete reproducing kernels and their essential properties in one-variable complex analysis and in Section 4, we will simply introduce various and concrete applications of the general theory of reproducing kernels; for their many applications, we give their basic references. The results in Sections 3 and 4 are beautiful and basic; however, for many mathematical physicians and engineers they seem to be abstract and ideal beautiful results. So, in Section 4, we will give our recent results obtained as the concrete applications of the general theory of reproducing kernels to the Tikhonov regularization. In Sections 5 and 6, we will introduce our results for inverse problems for the heat conduction and real inversion formulas of the Laplace transform that are famous and typical ill-posed problems. In Section 8, we will state some essential and important problems when we appy our mathematics to physical and concrete problems. In the last Section 9, we shall refer to a global viewpoint for the theory of reproducing kernels. The general theory is entirely elementary it, however, will give wide applications, and furthermore, it will have a great power that creates new concepts and many basic relations. Surprisingly enough, the general theory itself will be undergraduate-level mathematics and in the theory we will be able to find many intuitive problems. In that sense, this field may still be considered still as a young mathematical research field.

2. Reproducing kernel Hilbert spaces

In this section, we will show that the general theory of reproducing kernels is fairly fundamental in mathematics by considering linear mappings in the framework of Hilbert spaces; indeed, when we consider linear mappings, we will meet naturally the concept of reproducing kernels. This will show that the theory is general and fundamental mathematics, as in the linear algebra in an undergraduate course. We will introduce its core results.

Fundamental theory for linear mappings. We will show the fundamental results for the general theory of reproducing kernels itself and its wide applications.

Identification of the image space of a linear mapping and relation of the image and domain spaces of the linear mapping. Let $\mathcal{F}(E)$ be the linear space comprising all complex-valued functions on any fixed abstract set E. Let $\mathcal{H}$ be a Hilbert space with the inner product $(\cdot,\cdot)_{\mathcal{H}}$. Let $\mathbf{h} : E \longrightarrow \mathcal{H}$ be a fixed $\mathcal{H}$-valued mapping on E. Then, we shall consider the linear mapping L from $\mathbf{f} \in \mathcal{H}$ into $\mathcal{F}(E)$ defined by the following:

$$f(p) = (L\mathbf{f})(p) = (\mathbf{f}, \mathbf{h}(p))_{\mathcal{H}}. \tag{2.1}$$

The most fundamental problem in the linear mapping (2.1) will be the characterization of the image $f(p)$ of L, and the next interesting problem will be the interrelationship between the input $\mathbf{f}$ and the output $f(p)$.

The key to considering these fundamental problems is to form the two-variable complex-valued function

$$K(p,q) = (\mathbf{h}(q), \mathbf{h}(p))_{\mathcal{H}} \tag{2.2}$$

defined on $E \times E$. We shall denote by $\mathcal{R}(L)$ the function space comprising all complex-valued functions of the images of $\mathcal{H}$ by L defined on E. In the image space, we shall introduce the norm by

$$\|f\|_{\mathcal{R}(L)} = \inf\{\|\mathbf{f}\|_{\mathcal{H}}; f = L\mathbf{f}\}. \tag{2.3}$$

Then the image space will be a Hilbert space. Indeed, we obtain precisely

THEOREM 2.1. *By (2.3) we can introduce the norm in $\mathcal{R}(L)$ and furthermore, from the norm we can introduce the inner product $[\mathcal{R}(L), (\cdot,\cdot)_{\mathcal{R}(L)}]$ induced away from the norm. Then, the image space forms a Hilbert space. The function $K(p,q)$ defined by (2.2) has the following two properties:*

(i) For any $q \in E$, $K(p,q)$ belongs to $\mathcal{R}(L)$ as a function in p.

(ii) For any function $f \in \mathcal{R}(L)$ and for any point $q \in E$,

$$f(q) = (f(\cdot), K(\cdot,q))_{\mathcal{R}(L)}.$$

Here, the function $K(p,q)$ satisfying (i) and (ii) is determined uniquely in the space $\mathcal{R}(L)$. Furthermore, L is isometric between $\mathcal{H}$ onto $\mathcal{R}(L)$ if and only if $\{\mathbf{h}(p); p \in E\}$ is complete in $\mathcal{H}$.

The properties (i) and (ii) of $K(p,q)$ are called a reproducing property in $\mathcal{R}(L)$ of $K(p,q)$, and we call $K(p,q)$ a **reproducing kernel**. This is why the function f in the inner product in (ii) comes out f after taking the inner product in the same way as in (ii). In this case, we shall call such a Hilbert space admitting a reproducing kernel **reproducing kernel Hilbert space**, and we shall write it simply by RKHS.

In fact, the reproducing kernel Hilbert space will be determined from some positive definite kernel $K(p,q)$ (this function has such a special property as is stated later) intuitively and independently of the linear mapping as in Theorem 2.1. That is, the Hilbert space is determined by $K(p,q)$, and so we shall denote its

Hilbert space by H_K. Then, the image space of $\mathcal{H}$ by (2.1) is uniquely determined by the function (2.2) as a reproducing kernel Hilbert space $\mathcal{R}(L) = H_K$. From this viewpoint, Theorem 2.1 may be stated as follows:

THEOREM 2.2. *The image space $\{f(p)\}$ by (2.1) of $\mathcal{H}$ is characterized as the reproducing kernel Hilbert space H_K determined by $K(p,q)$ in (2.2). Then, we have the inequality*

$$\|f\|_{H_K} \leq \|\mathbf{f}\|_{\mathcal{H}}.$$

Furthermore, for any $f \in H_K$, there exists a uniquely determined function $\mathbf{f}^*$ in $\mathcal{H}$ satisfying:

$$f(p) = (\mathbf{f}^*, \mathbf{h}(p))_{\mathcal{H}} \quad \text{on} \quad E \quad \text{and} \quad \|f\|_{H_K} = \|\mathbf{f}^*\|_{\mathcal{H}}.$$

Theorem 2.2 gives a method for constructing a reproducing kernel Hilbert space for the image space that is isometric to $\mathcal{H}$. Therefore, the essentially important part in Theorem 2.2 is to construct practically the space H_K from $K(p,q)$ defined by (2.2). For the construction, we have many methods depending on the kernels $K(p,q)$ and their representations.

Inverses for linear mappings. We shall consider the linear mapping defined by (2.1). In order to derive a general inversion formula for (2.1) that is widely applicable in analysis, we shall assume that both Hilbert spaces $\mathcal{H}$ and H_K are given as $\mathcal{H} = L_2(T, dm), H_K \subset L_2(E, d\mu)$, on the sets T and E, respectively. (We assume that for $dm, d\mu$ measurable sets T, E, they are the Hilbert spaces comprising $dm, d\mu - L_2$ integrable complex-valued functions, respectively.) Therefore, we shall consider the integral transform

$$f(p) = \int_T F(t)\overline{h(t,p)}dm(t). \tag{2.4}$$

Here, $h(t,p)$ is a function on $T \times E$, $h(\cdot,p) \in L_2(T,dm)$, and $F \in L_2(T,dm)$. The corresponding reproducing kernel for (2.2) is given by

$$K(p,q) = \int_T h(t,q)\overline{h(t,p)}dm(t) \quad \text{on} \quad E \times E.$$

The norm of the reproducing kernel Hilbert space H_K is represented as $L_2(E, d\mu)$. Under these situations, we obtain

THEOREM 2.3. *We assume that an approximating sequence $\{E_N\}_{N=1}^{\infty}$ of E satisfies (a) $E_1 \subset E_2 \subset \cdots \subset \cdots$, (b) $\bigcup_{N=1}^{\infty} E_N = E$, (c) $\int_{E_N} K(p,p)d\mu(p) < \infty$ $(N = 1, 2, ...)$.*

Then, for $f \in H_K$ satisfying $\int_{E_N} f(p)h(t,p)d\mu(p) \in L_2(T,dm)$ for any N, the sequence

$$\left\{\int_{E_N} f(p)h(t,p)d\mu(p)\right\}_{N=1}^{\infty} \tag{2.5}$$

converges to F^* in (2.4) in Theorem 2.2 in the sense of the $L_2(T,dm)$ norm.

Practically for many cases, the assumptions in Theorem 2.3 will be satisfied automatically, and so Theorem 2.3 may be applied in many cases. To construct the inversion formula is, in general, difficult. The formulation in Theorem 2.3 may, however, be considered as a natural one. Indeed, it was given as the strong convergence in the Hilbert space $L_2(T,dm)$ and so, it will be of interest and it may be considered as a natural one.

Identification of linear systems. In Theorem 2.2, conversely, if there exists an isometric mapping (we do not need its linearity) $\widetilde{L}$ between a reproducing kernel Hilbert space H_K admitting the kernel $K(p,q)$ on E and a Hilbert space $\mathcal{H}$, then the inverse of $\widetilde{L}$ is a linear mapping and, by using $\widetilde{L}$ and $K(p,q)$ and by using (called a **generator vector of** $\widetilde{L}$)

$$\mathbf{g}_{\widetilde{L}}(q) = \widetilde{L}K(\cdot, q), \tag{2.6}$$

it is represented as follows:

THEOREM 2.4. *For*

$$f(p) = (\mathbf{f}, \mathbf{g}_{\widetilde{L}}(p))_{\mathcal{H}}, \quad \textit{for} \quad \mathbf{f} \in \mathcal{H} \tag{2.7}$$

we obtain $\|f\|_{H_K} = \|\mathbf{f}\|_{\mathcal{H}}$. *The linear mapping (2.7) gives the inversion of* $\widetilde{L}$, *and* $\{\mathbf{g}_{\widetilde{L}}(p); p \in E\}$ *is complete in* $\mathcal{H}$.

Theorem 2.3 will give a new viewpoint and method for the Fredholm integral equation (2.4) of the first kind, which is a fundamental integral transform. Our method and solution will have the following properties:

(1) We use the naturally determined reproducing kernel Hilbert space H_K which is determined by the integral kernel.

(2) The solution is given in the sense of $\mathcal{H}$ norm convergence.

(3) Our solution (inverse) is given by $\mathbf{f}^*$ in Theorem 2.2.

(4) For the ill-posed problem in (2.4), our method gives a well-posed solution.

In general, in a linear problem, when we cannot determine the existence of a solution, or the uniqueness of a solution, or the continuity of its solution, then the linear problem is called an **ill-posed problem** in the sense of Hadamard. The linear problem (2.1) or (2.4) has been considered as a typical ill-posed problem. Our method gives a characterization of the image, that is, the solution space. For the inverse solution, we consider the minimum norm one among many solutions, and then we have the isometric relation in Theorem 2.2; that is, our solution has uniqueness and continuity properties. This viewpoint is, however, a mathematical and theoretical one. In practical and physical linear systems, the observation data will be a finite number of data containing error or noises, and so we will meet various delicate problems numerically and there will exist a field of **ill-posed problems**. In Section 5, in order to overcome such cases, we will give new methods.

We gave many applications of Theorems 2.1, 2.2 and 2.3. We did, however, not give many applications of Theorem 2.4 at this moment. These four theorems are, however, equivalent situations and so we can expect many applications of Theorem 2.4. By Theorem 2.4, we can determine the linear system from the isometric relation, and when the generator vector may be considered as a Green function, it may be applied to determine a physical law of the system. The output space of a linear system in our situation is a reproducing kernel Hilbert space and so by using typical and various reproducing kernel Hilbert spaces (as physical models of output spaces) and corresponding isometric mappings, we will be able to find corresponding laws by our method.

Fundamental properties of reproducing kernel Hilbert spaces. As we see, when we consider a linear mapping in the framework of Hilbert spaces, its image space forms naturally a reproducing kernel Hilbert space. Therefore, we shall see the fundamental properties of reproducing kernel Hilbert spaces. The following theorem shows that a reproducing kernel Hilbert space is a good function space ([**79**]).

THEOREM 2.5. *For a Hilbert space H comprising functions $\{f(p)\}$ on a set E, the space admits a reproducing kernel if and only if, for any point $q \in E$, $f \to f(q)$ is a bounded linear functional on H. If a function sequence $\{f_n\}$ converges to f in the space H, then it converges to the function pointwisely on E. Furthermore, on a subset of E where $K(p,p)$ is bounded, its convergence is uniform.*

We shall call a complex-valued function $k(p,q)$ in a set $E \times E$ a **positive definite quadratic function** on the set E when it satisfies the property: for an arbitrary function $X(p)$ on E that is zero on E except for a finite number of points of E, $\sum_{p,q} \overline{X(p)} X(q) k(p,q) \geq 0$.

As we can see simply, a reproducing kernel $K(p,q)$ on E is a positive definite quadratic function on E. Its converse statement is very important:

THEOREM 2.6. *For any positive definite quadratic function $K(p,q)$ on E, there exists a uniquely determined reproducing kernel Hilbert space H_K admitting the reproducing kernel $K(p,q)$ on E.*

For any positive definite quadratic function $K(p,q)$ on E, we can construct a Hilbert space $\mathcal{H}$ and a mapping $\mathbf{h}(p)$ from E into $\mathcal{H}$ satisfying (2.2) by using the Gaussian probability distribution with nth order co-distributions and zero mean that are given by $n \times n$ matrices $\|K(p_j, p_{j'})\|$. For this construction, we use the **Kolmogorov theorem on measures**, which is not elementary. This fact is, however, very important, because from this fact we can derive simply Theorem 2.6 from Theorem 2.1 or Theorem 2.2. Furthermore, this fact will essentially be needed when we introduce various operators on a Hilbert space and among Hilbert spaces ([**53**]).

Meanwhile, since the co-distribution on any stochastic series is a positive definite quadratic function, between both the stochastic series theory and the theory of reproducing kernels, there exists a fundamental relation ([**9**]).

It is important to know the method how to construct the reproducing kernel Hilbert space H_K from the kernel $K(p,q)$ on E.

THEOREM 2.7. *If a Hilbert space H comprising functions on a set E admits a reproducing kernel $K(p,q)$, then it is represented by any complete orthonormal system $\{v_j(p)\}_j$ of H as follows: on the set $E \times E$, in the sense of the absolute convergence $K(p,q) = \sum_j v_j(p) \overline{v_j(q)}$.*

A complete orthonormal system in H_K may be constructed by the Gram-Schmidt orthogonalization procedure from a linearly independent system in H_K. In this sense, a reproducing kernel is computational and with this basic idea, many representations of domain functions, which are defined on the domain like canonical conformal mappings of the domain, in terms of reproducing kernels have been considered. This is a theory of kernel functions whose center was Stanford University and its origins were the two papers by S. Bergman and G. Szegö that are thesis at Berlin University in 1922 and 1921, respectively, and they tried to represent the Riemann mapping function in terms of the reproducing kernels.

Generations of reproducing kernel Hilbert spaces. For any positive definite quadratic function $K(p,q)$ on a set E, there exists a uniquely determined reproducing kernel Hilbert space H_K admitting the kernel $K(p,q)$. We can consider intuitively the relations induced from relations among positive definite quadratic form functions. This problem will still be among the up-to-date problems. In this section, we shall examine such problems induced from restriction, sum and product of positive definite quadratic form functions.

For any positive definite quadratic form function $K(p,q)$ on E and for any subset E_1 of E, the restriction $K^{(1)}(p,q) = K(p,q)\big|_{E_1}$ to $E_1 \times E_1$ is, of course, a positive definite quadratic form function on E_1. Therefore, for the restriction $K^{(1)}(p,q)$, there exists a uniquely determined reproducing kernel Hilbert space $H_{K^{(1)}}$ admitting the kernel $K^{(1)}(p,q)$. Then, the relation between $H_{K^{(1)}}$ and H_K is given by the following:

THEOREM 2.8. *Any function $f^{(1)} \in H_{K^{(1)}}$ is the restriction $f^{(1)} = f\big|_{E_1}$ of a function f of H_K, and its norm in $H_{K^{(1)}}$ is given by*

$$\|f^{(1)}\|_{H_{K^{(1)}}} = \min\left\{\|f\|_{H_K} \mid f\big|_{E_1} = f^{(1)}, f \in H_K\right\}.$$

For any two positive definite quadratic form functions $K^{(1)}(p,q)$ and $K^{(2)}(p,q)$ on E, the usual sum $K(p,q) = K^{(1)}(p,q) + K^{(2)}(p,q)$ on $E \times E$ is also a positive definite quadratic form function on E. The relation among the corresponding reproducing kernel Hilbert spaces $H_K, H_{K^{(1)}}$ and $H_{K^{(2)}}$ is given by

THEOREM 2.9. *Any function $f \in H_K$ is represented as, for some functions $f^{(1)} \in H_{K^{(1)}}$ and $f^{(2)} \in H_{K^{(2)}}$, $f(p) = f^{(1)}(p) + f^{(2)}(p)$ on E. Its norm in H_K is given by*

$$\|f\|^2_{H_K} = \min\{\|f^{(1)}\|^2_{H_{K^{(1)}}} + \|f^{(2)}\|^2_{H_{K^{(2)}}} \mid$$
$$f(p) = f^{(1)}(p) + f^{(2)}(p) \quad on \quad E, \qquad f^{(1)} \in H_{K^{(1)}}, f^{(2)} \in H_{K^{(2)}}\}.$$

For any two positive definite quadratic form functions $K^{(1)}(p,q)$ and $K^{(2)}(p,q)$ on E, the usual product $K(p,q) = K^{(1)}(p,q)K^{(2)}(p,q)$ is again a positive definite quadratic form function on E by **Schur's theorem**, H_K is the restriction of the tensor product $H_{K^{(1)}} \otimes H_{K^{(2)}}$ to the diagonal set, and so, by Theorem 2.8, we obtain

THEOREM 2.10. *Let $\{f_j^{(1)}\}_j$ and $\{f_j^{(2)}\}_j$ be some complete orthonormal systems in $H_{K^{(1)}}$ and $H_{K^{(2)}}$, respectively. Then the reproducing kernel Hilbert space H_K is comprised of all functions on E which are represented as, in the sense of absolute convergence on E,*

$$f(p) = \sum_{i,j} \alpha_{i,j} f_i^{(1)}(p) f_j^{(2)}(p) \quad on \quad E, \quad \sum_{i,j} |\alpha_{i,j}|^2 < \infty \tag{2.8}$$

and its norm in H_K is given by $\|f\|^2_{H_K} = \min \sum_{i,j} |\alpha_{i,j}|^2$, where the $\{\alpha_{i,j}\}$ are considered satisfying (2.8).

This theorem is very important when we consider some general nonlinear mappings in connection with linear mappings ([**52**]).

3. Examples of reproducing kernels and RKHSs in one-variable complex analysis

Reproducing kernels will represent many important properties of their reproducing kernel Hilbert spaces, and so the concrete forms of reproducing kernels themselves will be valuable. We have many concrete reproducing kernels that are found in the solutions of many partial differential equations, in Sobolev spaces comprising real-valued functions and, furthermore, are found in many complex analysis of several variables. (When we restrict the reproducing kernels in several-variable complex analysis to the real spaces, their restrictions are also reproducing kernels, and so we have many problems in analytic continuations). In this section, we shall give typical reproducing kernels and their elementary properties. The theory of reproducing kernels might be considered as that of the Bergman kernel and the Szegö kernel. In such a way we have many references and profound results on those typical reproducing kernels, and we even have several technical books on them.

In applied mathematics we find, however, real-valued reproducing kernels ([**9**], [**15**], [**79**]), and so, we shall firstly simply introduce real-valued reproducing kernels. As we shall refer to in Section 5, when we do not know the linear systems, by using those reproducing kernels as input physical data and the observations of the outputs, we can calculate the inverses of the linear systems (see [**72**], Section 8) in the following cases:

(1) Any positive definite Hermitian matrices. (2) $\delta(x-y)$ ($\delta(0)=1$ and $\delta(x)=0$ for $x \neq 0$). (3) For any $\alpha > 0$, $\exp(-\alpha|x-y|)$. (4) For any $\alpha > 0$, $\exp(\alpha xy)$. (5) For any $\alpha > 0$, $\exp(-\alpha(x-y)^2)$. (6) $\exp(-|x-y|)(1+|x-y|)$. (7) $\min(x,y)$. (8) For any $\alpha > 0$, $\sin(\alpha(x-y))/(x-y)$ (see the sampling theorem in Section 6).

On the half-space $\{x > 0\}$, we have the additional systems:

$$(1) \quad \frac{1}{(x+y)^{2q}} \quad (q \geq \frac{1}{2}). \quad (2) \quad \frac{1}{(x^2+y^2)^{2q}} \quad (q \geq \frac{1}{2}),$$
$$(3) \quad \exp\{\min(x,y)\}.$$

On the interval $\{-1 < x < 1\}$, we also have:

$$(1) \quad \frac{1}{(1-xy)^{2q}} \quad (q \geq \frac{1}{2}), \quad (2) \quad \log\frac{1}{1-xy}.$$

For any set E and for any separable reproducing kernel Hilbert space comprising functions on the set E, its reproducing kernel $K(p,q)$ is represented by a series, for some functions $\{\varphi_j(p)\}$ on E, $K(p,q) = \sum_j \varphi_j(p)\overline{\varphi_j(q)}$ in the sense of absolutely convergence. Conversely, for any functions $\{\varphi_j(p)\}$ on E, the absolutely convergent function as in the above $K(p,q)$ is a reproducing kernel. Furthermore, many Green functions are reproducing kernels and, the Green functions and the distribution delta may be considered as a family of reproducing kernels.

Bergman kernels. Let D be a bounded regular domain on the complex $z = x+iy$ plane whose boundary is composed of a finite number of disjoint analytic Jordan curves. Let $AL_2(D)$ be a Hilbert space (Bergman space) comprising analytic functions $f(z)$ on D and with finite norms $\|f\|_{AL_2(D)} = \{\int\int_D |f(z)|^2 dxdy\}^{\frac{1}{2}}$. As we see easily, $f \to f(z)(z \in D)$ is a bounded linear functional on $AL_2(D)$. Therefore, there exists a reproducing kernel $K(z,u)$ such that for any $u \in D$ and for any function $f \in AL_2(D)$,

$$f(u) = \int\int_D f(z)\overline{K(z,u)}dxdy. \tag{3.1}$$

From $K(z,u) = \overline{K(u,z)}$, $K(z,u)$ is analytic in $\overline{u}$ (complex conjugate). So, we shall denote it as $K(z,\overline{u})$. This is the Bergman kernel of D or on D.

Let $G(z,u)$ be the Green's function for the Laplace equation on D with pole at $u \in D$. Then, we have the identity

$$K(z,\overline{u}) = -\frac{2}{\pi}\frac{\partial^2 G(z,u)}{\partial z \partial \overline{u}}. \tag{3.2}$$

Here we shall introduce the adjoint L kernel for $K(z,\overline{u})$ by

$$L(z,u) = -\frac{2}{\pi}\frac{\partial^2 G(z,u)}{\partial z \partial u}.$$

The adjoint L kernel $L(z,u)$ has one pole at u on D such that

$$L(z,u) = \frac{1}{\pi(z-u)^2} + \text{regular terms}$$

of the second order. Furthermore, along the boundary ∂D we have the identity

$$\overline{K(z,\overline{u})dz} = -L(z,u)dz.$$

This important identity shows that the Bergman kernel is indeed an analytic differential $K(z,\overline{u})dz$ and it is continued analytically to the doubly closed Riemann surface of D, and at the symmetric point u of $\overline{u}$ it has the same double pole as in the above. So, the Bergman kernel is a fundamental differential on the closed Riemann surface. Many identities are well known as in (3.2).

When D is N-ply connected, $K(z,\overline{u})$ and $L(z,u)$ have $N-1$ zeros each when u is sufficiently near to the boundary ∂D (N. Suita and A. Yamada [**75**]). Their zero points are not always the same in numbers.

We can also consider the reproducing kernel for the closed subspace of the above Bergman space whose members are comprised of the functions with single-valued integrals on the domain. For this reproducing kernel, we can find many and profound results and, in particular, the kernel is important in conformal mapping theory. Then, the reproducing kernel has always $2N-2$ zeros by counting multiplicity. This part is a beautiful part between conformal mappings and reproducing kernels and so, we shall refer to it some more.

Let $AL_2^e(D)$ be the subspace of $AL_2(D)$ whose members have single-valued integrals on D. Let $\hat{N}(z,u)$ denote the Neumann function with a pole at u for the Laplace equation. Then, the corresponding reproducing kernel and its adjoint L-kernel are given by the identities

$$K_e(z,\overline{u}) = \frac{2}{\pi}\frac{\partial^2 \hat{N}(z,u)}{\partial z \partial \overline{u}}, \quad L_e(z,u) = -\frac{2}{\pi}\frac{\partial^2 \hat{N}(z,u)}{\partial z \partial u}.$$

Furthermore, they satisfy the identity, along the boundary ∂D,

$$\overline{K_e(z,\overline{u})dz} = -L_e(z,u)dz.$$

Let $\phi(z,u)(\psi(z,u), \text{resp.})$ denote the conformal mapping, with the normalization

$$\frac{1}{z-u} + \text{regular terms}$$

around u, such that it maps D onto the extended complex plane except for N horizontal (vertical, resp.) segments, one-to-one and analytically. Then, we shall introduce two functions

$$M(z,u) = \frac{1}{2}(\phi(z,u) - \psi(z,u)), \quad N(z,u) = \frac{1}{2}(\phi(z,u) + \psi(z,u)).$$

Then, we have the identities:

$$K_e(z,\overline{u}) = \frac{1}{\pi}\frac{\partial M(z,u)}{\partial z} = \frac{1}{\pi}M'(z,u), \quad L_e(z,u) = \frac{1}{\pi}\frac{\partial N(z,u)}{\partial z} = \frac{1}{\pi}N'(z,u).$$

Here, we shall recall a beautiful result of M. Schiffer that $N(z,u)$ maps D onto an unbounded domain with a boundary of N disjoint convex closed curves. Therefore, in particular, $N'(z,u)$, that is, $L_e(z,u)$ does not have any zero points on $D \cup \partial D$. Hence, $K_e(z,\overline{u})$ has $2N-2$ zero points on D counting multiplicity.

M. Sakai [**73**] obtained the following deep result:

$$\sup_{z\in D} |M(z,u)|^2 \leq \frac{1}{\pi}\int\int_D |M'(z,u)|^2 dxdy = \pi K_e(u,\overline{u}).$$

Furthermore, $N(z,u)$ is the maximum function such that among the univalent functions $\{h(z,u)\}$ normalized at u, their image outer surface measures,

$$\frac{1}{2i}\int_{\partial D} \overline{h(z,u)}\frac{\partial h(z,u)}{\partial z}dz,$$

are maximized.

Meanwhile the reproducing property of $K_e(z,\overline{u})$ means that $M(z,u)/M'(u,u)$ is the minimum function such that among all analytic functions $\{g(z)\}$ on D normalized by $g'(u) = 1$, their image surface measures, counting multiplicity,

$$\int\int_D |g'(z)|^2 dxdy$$

are minimized.

Szegö kernel. Let $H_2(D)$ be the Hardy doubly analytic function space on D. A member of the class has nontangential boundary values belonging to $L_2(\partial D)$. Let $AL_2(\partial D)$ denote its closed subspace in $L_2(\partial D)$ and introduce the norm in $H_2(D)$ by

$$\left\{\int_{\partial D} |f(z)|^2 |dz|\right\}^{\frac{1}{2}} < \infty.$$

For a function $f \in AL_2(\partial D)$, as we see from the Cauchy integral formula, since $f \to f(u), u \in D$ is a bounded linear functional, there exists a reproducing kernel $\hat{K}(z,u)$ satisfying

$$f(u) = \int_{\partial D} f(z)\overline{\hat{K}(z,u)}|dz|.$$

As in the Bergman kernel, $\hat{K}(z,u)$ is analytic in $\overline{u}$, and so we shall denote it by $\hat{K}(z,\overline{u})$. This is the Szegö kernel of D or on D.

The important concept is the adjoint L kernel $\hat{L}(z,u)$ for $\hat{K}(z,\overline{u})$. That is, it is a meromorphic function on $D \cup \partial D$ and only at the point u, it has a simple pole

$$\hat{L}(z,u) = \frac{1}{2\pi(z-u)} + \text{regular terms}, \tag{3.3}$$

and along the boundary D it satisfies the relation

$$\overline{\hat{K}(z,\overline{u})}|dz| = \frac{1}{i}\hat{L}(z,u)dz. \tag{3.4}$$

This adjoint L kernel is uniquely determined by these properties on $D \cup \partial D$. From (3.4) we have the identity

$$\hat{L}(z,u) = -\hat{L}(u,z) \quad \text{on} \quad D \times D. \tag{3.5}$$

The adjoint L kernel $\hat{L}(z,u)$ is also characterized by the minimum property:

$$\min \int_{\partial D} |h(z,u)|^2 |dz| = \int_{\partial D} |\hat{L}(z,u)|^2 |dz|.$$

Here the minimum is considered on the meromorphic functions $\{h(z,u)\}$ on D with the singularity at u as in (3.3) and satisfying $\{\int_{\partial D} |h(z,u)|^2 |dz|\}^{\frac{1}{2}} < \infty$.

A deep and important property of the adjoint L kernel $\hat{L}(z,u)$ is that it does not have any zero point on $D \cup \partial D$. From this fact and (3.4),

$$f_0(z;u) = \frac{\hat{K}(z,\overline{u})}{\hat{L}(z,u)}$$

is the **Ahlfors function** of D with respect to u. That is, $f_0(z;u)$ is an analytic function on D. Among the analytic functions satisfying the properties

$$f(u;u) = 0, f'(u;u) \geq 0 \quad \text{and} \quad |f(z;u)| \leq 1 \quad \text{on} \quad D,$$

it satisfies the extremal property $f_0'(u;u) = \max f'(u;u)$. Then, as we see from (3.4) and the principle of argument, the function $f_0(z;u)$ maps D onto the unit disc and the disc is covered N times. Here, we assume that D is N-ply connected.

Furthermore, the adjoint L kernel $\hat{L}(z_1,z_2)$ satisfies, for any analytic function f on D satisfying $\int\int_D |f'(z)|^2 dxdy < \infty$, the identity

$$\int_{\partial D}\int_{\partial D} |(f(z_1) - f(z_2))\hat{L}(z_1,z_2)|^2 |dz_1||dz_2| = \frac{1}{\pi}\int\int_D |f'(z)|^2 dxdy,$$

and conversely the adjoint L kernel is characterized by this identity among the meromorphic functions on D with singular properties similar to the kernel. This result is not so simple and for the proof, we must examine the properties of the tensor product of the two Szegö spaces and its restriction properties to the diagonal set. Here we use the general theory of reproducing kernels essentially. This identity is the same as the Douglas integral representation, which is important in **minimum surface theory** in the case of the unit disc ([**45**]).

From (3.4) along ∂D we see that

$$\overline{\hat{K}(z,\overline{u})^2 dz} = -\hat{L}(z,u)^2 dz.$$

From this relation, we see that the square of the Szegö kernel is an analytic differential, it is continued to the double of D analytically, which is a closed Riemann surface as in the Bergman kernel. In this sense, the Szegö kernel is a half-order differential on the closed Riemann surface and so its properties are very involved for the sake of multi-valuedness on the closed Riemann surface. The profound theory for these properties was done by D. A. Hejhal [**22**] and J. D. Fay [**18**] by using the **Riemann theta functions** and the **Klein prime forms**. As we see from (3.5), the Szegö kernel is a Cauchy kernel on the Riemann surface and so it is a very important reproducing kernel. We can find three technical books for this kernel.

Hardy reproducing kernel and its conjugate kernel. Let $H_2(D)(\hat{H}_2(D)$, resp.) be the Hilbert space equipped with the norm in the space $H_2(D)$:

$$\left\{\frac{1}{2\pi}\int_{\partial D}|f(z)|^2\frac{\partial G(z,t)}{\partial \nu_z}|dz|\right\}^{\frac{1}{2}} \tag{3.6}$$

$$\left(\left\{\frac{1}{2\pi}\int_{\partial D}|f(z)|^2\left(\frac{\partial G(z,t)}{\partial \nu_z}\right)^{-1}|dz|\right\}^{\frac{1}{2}}, \text{resp.}\right). \tag{3.7}$$

Here, $\partial/\partial\nu_z$ is the inner normal derivative with respect to D. $\partial G(z,t)/\partial\nu_z$ is a positive continuous function on ∂D and so as in the Szegö space, we can consider the reproducing kernel $K_t(z,\overline{u})(\hat{K}_t(z,\overline{u})$, resp.) satisfying

$$f(u) = \frac{1}{2\pi}\int_{\partial D} f(z)\overline{K_t(z,\overline{u})}\frac{\partial G(z,t)}{\partial \nu_z}|dz|$$

$$\left(f(u) = \frac{1}{2\pi}\int_{\partial D} f(z)\overline{\hat{K}_t(z,\overline{u})}\left(\frac{\partial G(z,t)}{\partial \nu_z}\right)^{-1}|dz|, \text{resp.}\right).$$

We call $K_t(z,\overline{u})$ and $\hat{K}_t(z,\overline{u})$ the Hardy reproducing kernel and its conjugate reproducing kernel of D.

For an arbitrary open Riemann surface S, we can introduce the space $H_2(S)$ similarly.

Let $\{D_n\}_{n=1}^{\infty}$ be a regular exhaustion of S containing $t\in S$. Let $G_n(z,t)$ be the Green function of D_n with pole at t. Then

$$\left\{\frac{1}{2\pi}\int_{\partial D_n}|f(z)|^2\frac{\partial G_n(z,t)}{\partial \nu_z}|dz|\right\}_{n=1}^{\infty}$$

increases monotonically with n, and its limit is the minimum harmonic function that is not less than $|f(z)|^2$ on S. Then, the norm in $H_2(S)$ is given by

$$\|f\|_{H_2(S)} = \lim_{n\to\infty}\left\{\frac{1}{2\pi}\int_{\partial D_n}|f(z)|^2\frac{\partial G_n(z,t)}{\partial \nu_z}|dz|\right\}^{\frac{1}{2}}. \tag{3.8}$$

Therefore, there exists an H_2 kernel $K_t^{(S)}(z,\overline{u})$ satisfying

$$f(u) = \lim_{n\to\infty}\frac{1}{2\pi}\int_{\partial D_n} f(z)\overline{K_t^{(S)}(z,\overline{u})}\frac{\partial G_n(z,t)}{\partial \nu_z}|dz|.$$

Meanwhile, the norms (3.7) are, in general, not monotone with respect to n and so, we cannot consider it for some general domains. The norm on D_n is, however, conformally invariant.

Indeed, let $G_n^*(z,t)$ be the conjugate harmonic function of $G_n(z,t)$ and when we form the multi-valued meromorphic function $W_n(z,t) = G_n(z,t) + iG_n^*(z,t)$, $idW_n(z,t)$ is a single-valued meromorphic differential, and it satisfies along ∂D_n,

$$\frac{\partial G_n(z,t)}{\partial \nu_z}|dz| = idW_n(z,t).$$

Therefore, the integral (3.7) is represented by

$$\frac{1}{2\pi}\int_{\partial D_n}\frac{|f(z)dz|^2}{idW_n(z,t)}.$$

This will mean that $\hat{H}_2(D_n)$ is indeed comprised of analytic differentials $f(z)dz$. Therefore, the conjugate space may be considered on bordered Riemann surfaces with some good boundaries.

For $K_t(z,\overline{u})$ (resp. $\hat{K}_t(z,\overline{u})$), the adjoint L kernel $L_t(z,u)$ (resp. $\hat{L}_t(z,u)$), which is a meromorphic function on $D \cup \partial D$ with only one simple pole at u with residue 1, is characterized by the identity, along ∂D,

$$\overline{K_t(z,\overline{u})}idW(z,t) = \frac{1}{i}L_t(z,u)dz \tag{3.9}$$

$$\left(\overline{\hat{K}_t(z,\overline{u})dz} = \frac{1}{i}\hat{L}_t(z,u)idW(z,t) \quad \text{along} \quad \partial D, \text{resp.}\right). \tag{3.10}$$

From (3.9) and (3.10), we have a very important property:

$$L_t(z,u) = -\hat{L}_t(u,z) \quad \text{on} \quad D \times D. \tag{3.11}$$

The relations (3.9), (3.10) and (3.11) show that $L_t(z,u)dz$ is a meromorphic differential with respect to z, a meromorphic function in u, and along $z = u$ with one simple pole with residue 1, that is, the Cauchy kernel of D. Therefore, these reproducing kernels are very important as in the Bergman and Szegö kernels as the third kind of reproducing kernels in one-variable complex analysis ([**45**]). These reproducing kernels were examined deeply on Riemann surfaces by J. D. Fay [**18**] and A. Yamada [**81**] in terms of the Riemann theta functions and the Klein prime forms.

The norms of Bergman spaces are given by surface integrals on the domains and so, we can consider them on some very general domains as on regular domains; meanwhile, for the Szegö spaces, their norms are given in terms of boundary integrals and so, the Szegö spaces may be considered for some very respective domains for their boundary integrals. In this viewpoint, we will be very interested in the H_2 spaces, because the norms are given by boundary integrals and, however, for any open domains we can consider them.

For the norm (3.7), we obtain the best possible norm inequality:

$$\int\int_D |f'(z)|^2 dxdy \leq \frac{1}{2}\int_{\partial D} \frac{|f(z)dz|^2}{idW(z,t)},$$

and from this we see that the naturality of the norm (3.7) looks curious. This inequality is not so simple to derive and for its proof we must examine deeply the relations among the Hardy reproducing kernel, its conjugate kernel and the Bergman kernel ([**45**]).

Let D be an N-ply connected regular domain with disjoint analytic Jordan curves. Then, we have the basic and deep relation, for the Bergman kernel and the Szegö kernel on D,

$$K(z,\overline{u}) \gg 4\pi\hat{K}(z,\overline{u})^2.$$

The leftside minus the rightside is a positive definite quadratic function, which was given by D. A. Hejhal [**22**]. This profound result was given in the long historical lines: G.F.B. Riemann (1826-1866); F. Klein (1849-1925); S. Bergman; G. Szegö; Z. Nehari; M.M. Schiffer; P.R. Garabedian (1949 published); D.A. Hejhal (1972 published).

It seems that any elementary proof is impossible. However, the result will, in particular, mean the fairly simple inequality:

For two functions φ and ψ of $H_2(D)$, we obtain the generalized isoperimetric inequality

$$\frac{1}{\pi}\int\int_D |\varphi(z)\psi(z)|^2 dxdy \leq \frac{1}{2\pi}\int_{\partial D} |\varphi(z)|^2|dz|\frac{1}{2\pi}\int_{\partial D} |\psi(z)|^2|dz|,$$

and we can determine completely the cases attaining the equality here. In the thesis [**42**] of the author, published in 1979, the result was given. The author realized the importance of the abstract and general theory of reproducing kernels by N. Aronszajn. In the paper, the core part was to determine the equality statement in the above inequality. Surprisingly enough, some deep and general independent proof appeared 26 years later in A. Yamada ([**82**]). Here, of course, in the thesis we can find some fundamental ideas for nonlinear transforms. In particular, for the special case $\varphi \equiv \psi \equiv 1$, for the plane measure $m(D)$ of D and the length ℓ of the boundary we have the isoperimetric inequality

$$4\pi m(D) \leq \ell^2.$$

Meanwhile, for the Hardy reproducing kernel we can give the conjecture in the converse direction, from the case of doubly connected domains

$$K(z,\overline{u}) \ll K_t(z,\overline{u})\hat{K}_t(z,\overline{u}) \tag{3.12}$$

([**45**]). From this relation we can derive various results. The conjecture will be, however, very difficult to solve.

The author attacked this problem for several years. It was, however, impossible to solve it and in those days, the author derived the above result by applying the result of Hejhal conversely to the general theory of reproducing kernels by Aronszajn. The author thinks for the conjecture we must use the deep theory of J. D. Fay; however, his profound theory seems to be too deep and great for the author. So, from the viewpoints of fundamental results and applicable analysis, the author turned his research interest to the following applications of the general theory of reproducing kernels.

4. Applications of the general theory of reproducing kernels

In this section, we shall state some general applications of the general theory of reproducing kernels in Section 3.

(1) Integral transforms ([**48**]). By using the fundamental Theorems 2.1–2.3, we derive many concrete results for typical integral transforms.
(2) Inverse problems ([**60**]). By using Theorem 2.3, we derived concrete results for various inverse problems. For numerical solutions, however, we shall need the methods in the following Section 5.
(3) Integral transforms for some smooth functions ([**56**]). We can introduce a method deriving inversion formulas for some smooth functions.
(4) Nonharmonic integral transforms ([**45**]). When an integral kernel moves a little, we cannot calculate the kernel form (2.2). However, with some conditions for the images, we can derive the inversion formula, from the idea of Paley-Wiener.
(5) Various norm inequalities ([**49**], [**52**]). From many relations among reproducing kernels, we derived many and beautiful norm inequalites.

(6) Nonlinear transforms ([**52**]). In very general nonlinear transforms, we derived many natural norm inequalities. There, we state basic relations between nonlinear transforms and analytic functions. The important meaning of these results was given as the identification problems of nonlinear transforms in ([**91**]).
(7) Integral equations ([**57**]). We gave methods for solving various linear integral equations in the framework of Hilbert spaces. Recently we gave general and numerical methods for solving linear and singular integral equations ([**25**]).
(8) Linear differential equations with variable coefficients ([**57**]). We gave general methods on how to solve general linear differential equations with arbitrary variable coefficients; however, the methods in Section 5 will be practical.
(9) Approximations of functions ([**12**], [**13**]). By using Theorem 2.2, we gave various approximations of functions for concrete cases.
(10) Representations of inverse mappings ([**50**]). From the transform of reproducing kernels, we considered the inverse mapping for an arbitrary mapping and considered its representation in terms of the direct mapping. Furthermore, we derived concrete results. See also [**64**]. Recently we developed furthermore concrete results ([**90**] and [**92**]).
(11) Various operators on Hilbert spaces ([**53**]). Among Hilbert spaces, we introduced various operators such as sum, product, differential, and so on. Furthermore, we gave some concrete meanings and examples. For example, we gave a general concept of convolution. The derived inequalities there were applied for stability of the solutions with some conditions in inverse problems ([**58, 63, 65**]).
(12) Sampling theory ([**51**], Chapter 4, Section 2). From the general theory of reproducing kernels, we can see the essence of the sampling theory and furthermore, we derived error estimates in general situations. The results are extended in various ways.
(13) Interpolation problems of Pick-Nevanlinna type ([**46**]). The problem is to consider the functions that take given values for given points and are bounded on the domain. A general and abstract theory of interpolation problems of Pick-Nevannlinna type was considered. See also the recent book ([**4**]) on this topic.
(14) Analytic continuation ([**47**], [**59**], [**62**]). From the isometric identities derived from restrictions and transforms of reproducing kernels, we derived many analytic continuation formulas in some concrete forms. For their applications to nonlinear partial differential equations, see ([**21**]).
(15) Inversions for a family of bounded linear operators ([**66**]). From some generalization of the Pythagorean theorem, we derived such a concept. In particular, from different kinds of information we derived the inversion formula.
(16) Principle of telescope ([**61**]). For analytic functions and for many solutions of linear partial differential equations, we gave the methods representing their global data or their boundary values from some local data. For the case of time-dependent problems, we shall need some long time data.

5. Approximate solution methods for bounded linear operator equations on Hilbert spaces

Best approximations of functions. Let L be any bounded linear operator from a reproducing kernel Hilbert space H_K into a Hilbert space $\mathcal{H}$. Then the following problem is a well-known classical and fundamental problem, which is known as the best approximate mean square norm problem. For any member $\mathbf{d}$ of $\mathcal{H}$,

$$\inf_{f \in H_K} \|Lf - \mathbf{d}\|_{\mathcal{H}}. \tag{5.1}$$

This problem has, however, a complicated structure, when the Hilbert spaces are infinite dimensional and the problem leads to the generalized inverse in the sense of Moore-Penrose. By Theorem 2.1, we see its complicated structure in the following theorem:

THEOREM 5.1 ([**13**]). *For any member* $\mathbf{d}$ *of* $\mathcal{H}$*, there exists a function* $\tilde{f}$ *in* H_K *satisfying*

$$\inf_{f \in H_K} \|Lf - \mathbf{d}\|_{\mathcal{H}} = \|L\tilde{f} - \mathbf{d}\|_{\mathcal{H}} \tag{5.2}$$

if and only if, for the reproducing kernel Hilbert space H_k *admitting the kernel defined by* $k(p,q) = (L^*LK(\cdot,q), L^*LK(\cdot,p))_{H_K}$,

$$L^*\mathbf{d} \in H_k. \tag{5.3}$$

Furthermore, when there exists a function $\tilde{f}$ *satisfying (5.2), there exists a uniquely determined function that minimizes the norms in* H_K *among the functions satisfying the equality, and its function* $f_{\mathbf{d}}$ *is represented as follows:*

$$f_{\mathbf{d}}(p) = (L^*\mathbf{d}, L^*LK(\cdot,p))_{H_k} \quad on \quad E. \tag{5.4}$$

Here, the adjoint operator L^* of L, as we see from

$$(L^*\mathbf{d})(p) = (L^*\mathbf{d}, K(\cdot,p))_{H_K} = (\mathbf{d}, LK(\cdot,p))_{\mathcal{H}},$$

is represented by known $\mathbf{d}, L, K(p,q)$, and $\mathcal{H}$. From this theorem, we see that the problem is well-established by the theory of reproducing kernels; that is, the existence, uniqueness and representation of the solutions in the problem are well-formulated. In particular, note that the adjoint operator is represented in a good way; this fact will be very important in our framework. The extremal function $f_{\mathbf{d}}$ is the **Moore-Penrose generalized inverse** $L^{\dagger}\mathbf{d}$ of the equation $Lf = \mathbf{d}$. The criteria (5.3) is involved and so the Moore-Penrose generalized inverse $f_{\mathbf{d}}$ is not good when the data contain error or noises as in some practical cases. So, we shall introduce the idea of the Tikhonov regularization.

Spectoral analysis. We shall set some needed important spectoral and Tikhonov regularization properties following [**17**]. For the case of compact operators L, see [**20**].

Let $\{E_\lambda\}$ be the spectoral family for the selfadjoint operator L^*L. When L^*L has a continuous inverse,

$$(L^*L)^{-1} = \int \frac{1}{\lambda} dE_\lambda.$$

Then, the Moore-Penrose generalized inverse (5.3) is represented by

$$f_{\mathbf{d}}(p) = \int \frac{1}{\lambda} dE_\lambda L^* \mathbf{d}.$$

When $\mathcal{R}(L)$ is not closed and $\mathbf{d} \notin \mathcal{D}(L^\dagger)$, that is, $Lf = \mathbf{d}$ does not have the Moore-Penrose generalized inverse, by taking $\alpha > 0$, we define

$$f_{\mathbf{d},\alpha}(p) = \int \frac{1}{\lambda + \alpha} dE_\lambda L^* \mathbf{d}.$$

Then, for $\mathbf{d}^\delta$ satisfying $\|\mathbf{d} - \mathbf{d}^\delta\|_{\mathcal{H}} \leq \delta$, we can estimate $f_{\mathbf{d}\alpha}$ and

$$f^\delta_{\mathbf{d},\alpha}(p) = \int \frac{1}{\lambda + \alpha} dE_\lambda L^* \mathbf{d}^\delta$$

in the following way:

THEOREM 5.2. *For a member* $\mathbf{d}$ *of* $\mathcal{D}(L^\dagger)$,

$$\lim_{\alpha \to 0} (L^* L + \alpha I)^{-1} L^* \mathbf{d} = \lim_{\alpha \to 0} f_{\mathbf{d},\alpha} = f_{\mathbf{d}}. \tag{5.5}$$

Furthermore,

$$\|L f_{\mathbf{d},\alpha} - L f^\delta_{\mathbf{d},\alpha}\|_{\mathcal{H}} \leq \delta, \quad \|f_{\mathbf{d},\alpha} - f^\delta_{\mathbf{d},\alpha}\|_{H_K} \leq \frac{\delta}{\sqrt{\alpha}}.$$

THEOREM 5.3. *For a member* $\mathbf{d}^\delta$ *of* $\mathcal{D}(L^\dagger)$ *satisfying* $\|\mathbf{d} - \mathbf{d}^\delta\|_{\mathcal{H}} \leq \delta$, *the function* $f^\delta_{\mathbf{d},\alpha}$ *is a uniquely determined solution of the extremal problem:*

$$\inf_{f \in H_K} \{\alpha \|f\|^2_{H_K} + \|\mathbf{d}^\delta - Lf\|^2_{\mathcal{H}}\}. \tag{5.6}$$

If $\alpha = \alpha(\delta)$ *satisfies*

$$\lim_{\delta \to 0} \alpha(\delta) = 0, \quad \lim_{\delta \to 0} \frac{\delta^2}{\alpha(\delta)} = 0,$$

then

$$\lim_{\delta \to 0} f^\delta_{\mathbf{d},\alpha} = f_{\mathbf{d}} = L^\dagger(\mathbf{d}). \tag{5.7}$$

Our practical observations contain error and noises, and so these ideas and results are very important.

Representations of the extremal functions in the Tikhonov regularization. Our main interest is in some concrete representation of the abstract extremal functions $f_{\mathbf{d},\alpha}$ and $f^\delta_{\mathbf{d},\alpha}$ in the above Tikhonov regularization. Only for the case of compact operators do we know its representation in terms of spectoral functions and values. For the above general case for bounded linear operators, we know only the above abstract representation.

Therefore, we shall give some practical and more concrete representation by using the theory of reproducing kernels. For this purpose, we shall introduce the function

$$K_L(\cdot, p; \alpha) = (L^* L + \alpha I)^{-1} K(\cdot, p)$$

and furthermore, for any $\alpha > 0$ we introduce the following inner product:

$$(f, g)_{H_K(L;\alpha)} = \alpha (f, g)_{H_K} + (Lf, Lg)_{\mathcal{H}}. \tag{5.8}$$

Under this inner product, the members of H_K form a new reproducing kernel Hilbert space $H_K(L; \alpha)$ and we can say furthermore the following:

THEOREM 5.4 ([68]). *In the Tikhonov functional*

$$\inf_{f\in H_K} \{\alpha\|f\|_{H_K}^2 + \|\mathbf{d} - Lf\|_{\mathcal{H}}^2\}, \tag{5.9}$$

the extremal function $f_{\mathbf{d},\alpha}(p)$ *is represented, by using the kernel* $K_L(p,q;\alpha)$, *as follows:*

$$f_{\mathbf{d},\alpha}(p) = (\mathbf{d}, LK_L(\cdot,p;\alpha))_{\mathcal{H}}\,. \tag{5.10}$$

Here, $K_L(p,q;\alpha)$ *is the reproducing kernel for the space* $H_K(L;\alpha)$, *and it is the solution* $\tilde{K}(p,q;\alpha)$ *of the functional equation*

$$\tilde{K}(p,q;\alpha) + \frac{1}{\alpha}(L\tilde{K}_q, LK_p)_{\mathcal{H}} = \frac{1}{\alpha}K(p,q), \tag{5.11}$$

which corresponds to the Fredholm integral equation of the second kind for many concrete cases. Here,

$$\tilde{K}_q = \tilde{K}(\cdot,q;\alpha) \in H_K \quad \textit{for} \quad q\in E, \qquad K_p = K(\cdot,p) \quad \textit{for} \quad p\in E.$$

In (5.10) we shall need error estimates, when $\mathbf{d}$ contains error or noises. For this, we obtain

THEOREM 5.5 ([35], [25]). *For* (5.10) *we have the estimate*

$$|f_{\mathbf{d},\alpha}(p)| \le \frac{1}{\sqrt{\alpha}}\sqrt{K(p,p)}\|\mathbf{d}\|_{\mathcal{H}}.$$

This theorem means that in order to obtain good approximate solutions, we must take sufficiently small α; however, here we have restrictions for them, as we see, when $\mathbf{d}$ moves to $\mathbf{d}'$, by considering $f_{\mathbf{d},\alpha}(p) - f_{\mathbf{d}',\alpha}(p)$ in connection with the relation of the difference $\|\mathbf{d}-\mathbf{d}'\|_{\mathcal{H}}$. This fact is a very natural one, because we cannot obtain good solutions from the data with error. Here we wish to know how to take a small α a prior and what is the bound for it. These problems are very important practically and are delicate ones, and we have many methods. The basic idea may be given as follows: We examine for various α tending to zero, the corresponding extremal functions. By examining the sequence of the extremal functions, when it converges to some function numerically and after that, when the sequence diverges numerically, it will give the bound for α numerically (see [17], [20]).

For concrete examples, see ([7], [29]–[37], [66]–[71]).

Inverse problem for heat conduction. We shall consider the integral transform for the $L_2(\mathbf{R}^n)$ functions F:

$$u_F(x,t) = (L_tF)(x) = \frac{1}{(4\pi t)^{n/2}}\int_{\mathbf{R}^n} F(\xi)\exp\left\{-\frac{|\xi-x|^2}{4t}\right\}d\xi. \tag{5.12}$$

This function gives, in general, the solution $u(x,t)$ for the heat equation $u_t(x,t) = \Delta u(x,t)$ satisfying the initial condition $u(x,0) = F(x)$. This integral is very fundamental and important in mathematical physics; it is also applied in medical science (see, for example, [77] and its references). For fixed t, from $u_F(x,t)$ we wish to obtain the initial data F; this is a well-known difficult and inverse problem in heat conduction. As a typical example for our methods, we shall state our results.

For the case $n = 1$ we discussed this problem 24 years ago in ([44]), by the idea and methods in Sections 2 and 3. We gave the characterization of the heat distribution $u_F(x,t)$ in terms of a complex-valued analytic function, and from the result and idea, we obtained many and concrete analytic continuation formulas. We published, furthermore, the proceedings book ([**61**]) and also ([**62**]). As an analytic function, its characterization is simple; however, our information $u_F(x,t)$ is real-valued. It is a very difficult problem numerically to obtain its analytic continuation onto the whole complex plane.

For example, the following well-known formula will show the difficulty of the real-valued inversion formula for heat conduction, containing infinite order differentials: For a bounded continuous function F for $t = 1$,

$$e^{-D^2}[(L_1F)(x)] = F(x) \quad \text{pointwisely} \quad \text{on} \quad \mathbf{R}$$

([**26**, p. 182]). We derived quite good results by the methods in Section 5, by using the Sobolev spaces in the whole spaces as reproducing kernel Hilbert spaces in [**68, 33**]. Surprisingly enough, in [**35**] we used the **Paley-Wiener reproducing kernel** Hilbert spaces that are restrictive function spaces and we, however, obtained much better results numerically.

For the Paley-Wiener spaces, see Stenger [**74**], and as **sampling theory** we have great references. The method using the Paley-Wiener spaces is called the **sinc method**.

We shall consider the Fourier transform: for functions g in

$$L_2(\mathbf{R}^n, (-\pi/h, +\pi/h)^n)(h > 0),$$

$$f(z) = \frac{1}{(2\pi)^n}\int_{\mathbf{R}^n} \chi_h(t)g(t)e^{-iz\cdot t}dt.$$

Here, $z = (z_1, z_2, ..., z_n), t = (t_1, t_2, ..., t_n), dt = dt_1 \cdot dt_2 \cdots dt_n, z \cdot t = z_1t_1 + \cdots + z_nt_n$ and for the characteristic function χ of $(-\pi/h, +\pi/h)$, $\chi_h(t) = \prod_{\nu=1}^n \chi(t_\nu)$. Its image space admits the reproducing kernel

$$K_h(z, \overline{u}) = \frac{1}{(2\pi)^n}\int_{\mathbf{R}^n} \chi_h(t)e^{-iz\cdot t}\overline{e^{-iu\cdot t}}dt = \prod_{\nu}^{n} \frac{1}{\pi(z_\nu - \overline{u}_\nu)} \sin\frac{\pi}{h}(z_\nu - \overline{u}_\nu),$$

and it forms the Paley-Wiener space W_h. Its member f satisfies that for each ν, with some constant C_ν as $z_\nu \to \infty$,

$$|f(z_1, ..., z_\nu, z_{\nu+1}, ..., z_n)| \le C_\nu \exp\left(\frac{\pi|z_\nu|}{h}\right), \quad \int_{\mathbf{R}^n} |f(x)|^2 dx < \infty.$$

Then, for $j = (j_1, j_2, ..., j_n) \in \mathcal{Z}^n$, we have the identity

$$\frac{1}{(2\pi)^n}\int_{\mathbf{R}^n} |g(t)|^2 dt = h^n \sum_j |f(jh)|^2 = \int_{\mathbf{R}^n} |f(x)|^2 dx.$$

This means that

$$f(x) = (f(\cdot), K_h(\cdot, x))_{H_{K_h}} = h^n \sum_j f(jh)K_h(jh, x) = \int_{\mathbf{R}^n} f(\xi)K_h(\xi, x)d\xi$$

and the whole data $f(x)$ is represented by the discrete data $\{f(jh)\}_j$ (sampling theorem). For the error estimates, when we consider a finite number of data $\{hj\}_j$, we refer to [**23, 45**]. When we use this space, we obtain the following theorem.

THEOREM 5.6 ([**35**]). *For any $L_2(\mathbf{R}^n)$ function g and for any $\alpha > 0$, in the sense of*

$$\inf_{F \in H_{K_h}} \left\{\alpha\|F\|_{H_{K_h}}^2 + \|g - u_F(\cdot, t)\|_{L_2(\mathbf{R}^n)}^2\right\} \tag{5.13}$$

$$= \alpha \|F^*_{t,\alpha,h,g}\|^2_{H_{K_h}} + \|g - u_{F^*_{t,\alpha,h,g}}(\cdot, t)\|^2_{L_2(\mathbf{R}^n)},$$

there exists a uniquely determined best approximate function $F^*_{t,\alpha,h,g}$, *and it is represented by*

$$F^*_{t,\alpha,h,g}(x) = \int_{\mathbf{R}^n} g(\xi) Q_{t,\alpha,h}(\xi - x) d\xi. \tag{5.14}$$

Here,

$$Q_{t,\alpha,h}(\xi - x) = \frac{1}{(2\pi)^n} \int_{\mathbf{R}^n} \frac{\chi_h(p) e^{-ip\cdot(\xi - x)} dp}{\alpha e^{|p|^2 t} + e^{-|p|^2 t}}.$$

For a function F *in* H_{K_h}, *we consider the function* $u_F(x,t)$ *and we take* $u_F(\xi, t)$ *as* g. *Then for* $\alpha \to 0$,

$$F^*_{t,\alpha,h,g} \to F$$

uniformly.

When we use the Sobolev space, for $\alpha = 0$, the corresponding (5.14) does not exist ([**68**], [**33**]); however, in these Paley-Wiener space W_h cases, for the case $\alpha = 0$, the representations (5.14) are valid. That is, in Theorem 5.6, the results are valid for $\alpha = 0$. That is, in the present case, we do not need the Tikhonov regularization. Furthermore, from $(L_t F^*_{t,0,h,g})(x) = (g(\cdot), K_h(\cdot, x))_{L_2(\mathbf{R}^n)}$, $(L_t F^*_{t,0,h,g})(x)$ is the orthogonal projection of g to the Paley-Wiener space W_h, and we can estimate the difference $\|L_t F^*_{t,0,h,g} - g\|_{L_2(\mathbf{R}^n)}$ of $F^*_{t,0,h,g}$ and g.

Of course, $F^*_{t,0,h,g}$ is, for any $L_2(\mathbf{R}^n)$ function g and for the functions F of W_h, the Moore-Penrose generalized inverse of the equation $L_t F = g$.

When we use the Paley-Wiener space W_h, we do not need the Tikhonov regularization method. However, the construction of the Moore-Penrose generalized inverses is involved, and we had better consider $F^*_{t,\alpha,h,g}$ in the Tikhonov regularization method above and consider the case $\alpha = 0$.

In both cases of the Sobolev spaces H_S and Paley-Wiener spaces W_h, we are interested in the behaviours of the cases $\alpha \to 0$ and $h \to 0$, respectively; in our numerical experiments, about 30 times $h \to 0$ is better. See the numerical experiments in [**33, 35**]. For a general $L_2(\mathbf{R}^n)$ function g, we established its error estimates, and for our numerical experiments we were able to obtain good results, even for the hardest case such as the characteristic functions for closed intervals for g.

Real inversion formulas of the Laplace transform. We shall consider the inversion formula of the Laplace transform:

$$(\mathcal{L}F)(p) = f(p) = \int_0^\infty e^{-pt} F(t) dt, \quad p > 0$$

for some natural function spaces F. For more general functions, we shall apply their transforms suitably in order to apply the results. We shall consider, in general, the complex inversion formulas, because the images of the Laplace transform are analytic functions; however, we are requested to use only real and discrete data only to obtain the inversion formula. This is the **real inversion formula** of the Laplace transform, and we must represent the analytic function of the image in terms of the data on the positive real line. This problem is a very famous difficult one. Its difficulty is seen from the well-known formulas

$$\lim_{n \to \infty} \frac{(-1)^n}{n!} \left(\frac{n}{t}\right)^{n+1} f^{(n)}\left(\frac{n}{t}\right) = F(t),$$

$$\lim_{n\to\infty} \Pi_{k=1}^{n} \left(1 + \frac{t}{k}\frac{d}{dt}\right)\left[\frac{n}{t} f\left(\frac{n}{t}\right)\right] = F(t)$$

([**40, 80**]). Further, see [**11, 36**]. We have great references in [**84, 85**]. In particular, in connection with analytic extension problems, see also [**27, 28**].

In order to consider a bounded linear operator by the Laplace transform, we shall recall a natural function space in [**36**].

On the positive real line $\mathbf{R}^+$, we shall consider the norm

$$\left\{\int_0^\infty |F'(t)|^2 \frac{1}{t} e^t dt\right\}^{1/2}$$

for absolutely continuous functions F satisfying $F(0) = 0$. This space H_K admits the reproducing kernel

$$K(t,t') = \int_0^{\min(t,t')} \xi e^{-\xi} d\xi. \tag{5.15}$$

Then, we have

$$\int_0^\infty |(\mathcal{L}F)(p)p|^2 dp \leq \frac{1}{2}\|F\|_{H_K}^2; \tag{5.16}$$

that is, $(\mathcal{L}F)(p)p$ is a bounded linear operator from H_K into $L_2(\mathbf{R}^+, dp) = L_2(\mathbf{R}^+)$. Recently, Y. Sawano proved that this operator is compact for the conjecture by H. Fujiwara ([**86**]). By using this reproducing kernel Hilbert space, we obtain, following our general method:

THEOREM 5.7 ([**36**]). *For any $g \in L_2(\mathbf{R}^+)$ and for any $\alpha > 0$, in the sense that*

$$\inf_{F\in H_K}\left\{\alpha\int_0^\infty |F'(t)|^2 \frac{1}{t} e^t dt + \|(\mathcal{L}F)(p)p - g\|_{L_2(\mathbf{R}^+)}^2\right\} \tag{5.17}$$

$$= \alpha\int_0^\infty |F_{\alpha,g}^{*\prime}(t)|^2 \frac{1}{t} e^t dt + \|(\mathcal{L}F_{\alpha,g}^*)(p)p - g\|_{L_2(\mathbf{R}^+)}^2,$$

there exists a uniquely determined best approximate function $F_{\alpha,g}^*$ and it is represented by

$$F_{\alpha,g}^*(t) = \int_0^\infty g(\xi)\left(\mathcal{L}K_\alpha(\cdot,t)\right)(\xi)\xi d\xi. \tag{5.18}$$

Here, $K_\alpha(\cdot,t)$ is determined by the functional equations for $K_{\alpha,t'} = K_\alpha(\cdot,t')$, $K_t = K(\cdot,t)$,

$$K_\alpha(t,t') = \frac{1}{\alpha}K(t,t') - \frac{1}{\alpha}((\mathcal{L}K_{\alpha,t'})(p)p, (\mathcal{L}K_t)(p)p)_{L_2(\mathbf{R}^+)}. \tag{5.19}$$

We calculate the approximate inverse $F_{\alpha,g}^*(t)$ by using (5.18). By taking the Laplace transform of (5.19) with respect to t, by changing the variables t and t',

(5.20)

$$(\mathcal{L}K_\alpha(\cdot,t))(\xi) = \frac{1}{\alpha}(\mathcal{L}K(\cdot,t))(\xi) - \frac{1}{\alpha}((\mathcal{L}K_{\alpha,t})(p)p, ((\mathcal{L}(\mathcal{L}K_\cdot)(p)p))(\xi))_{L_2(\mathbf{R}^+)}.$$

Here,

$$K(t,t') = \begin{cases} -te^{-t} - e^{-t} + 1 & \text{for} \quad t \le t', \\ -t'e^{-t'} - e^{-t'} + 1 & \text{for} \quad t \ge t', \end{cases}$$

$$(\mathcal{L}K(\cdot,t'))(p) = e^{-t'p}e^{-t'}\left[\frac{-t'}{p(p+1)} + \frac{-1}{p(p+1)^2}\right] + \frac{1}{p(p+1)^2}.$$

$$\int_0^\infty e^{-qt'}(\mathcal{L}K(\cdot,t'))(p)dt' = \frac{1}{pq(p+q+1)^2}.$$

Therefore, by setting $(\mathcal{L}K_\alpha(\cdot,t))(\xi)\xi = H_\alpha(\xi,t)$, we obtain the **Fredholm integral equation of the second kind**:

$$\alpha H_\alpha(\xi,t) + \int_0^\infty H_\alpha(p,t)\frac{1}{(p+\xi+1)^2}dp = -\frac{e^{-t\xi}e^{-t}}{\xi+1}\left(t + \frac{1}{\xi+1}\right) + \frac{1}{(\xi+1)^2}.$$

We did numerical experiments by using Mathematica for the integral equation ([**37**]) and in ([**38**]) we used the sinc methods by M. Sugihara and his group. We were able to obtain some good numerical results for the real inversion formula. Meanwhile, we have recently used a powerful numerical method based on the concept of **infinite precision** based on **multiple-precision arithmetic** ([**3, 24**]), and H. Fujiwara tried numerical simulation for our formulas. He obtained surprisingly good results for the numerical and real inversion for the Laplace transform in ([**19**]). For the integral equation, we used the **DE formula** by H. Takahashi and M. Mori, using double exponential transforms ([**89**]). Surprisingly enough, he was able to derive numerically the inversion for the Laplace transform of the distribution delta which was proposed by V. V. Kryzhniy as a difficult case. This fact will mean that our results are valid for very general functions approximated by the functions of the reproducing kernel Hilbert space H_K, as we see from the formula (5.17).

The inverse problems in heat conduction and in the real inversion of the Laplace transform are famous as difficult ones essentially. However, by our mathematical methods and Fujiwara's powerful computer systems, we were able to solve them practically and numerically. In the case of heat conduction, we calculate our formulas for 10^{-22} as α for using the Sobolev spaces and $1/300$ for h for using the sinc functions. For the case of the Laplace transform, Fujiwara calculates surprisingly our formula for 10^{-400} as α. Such values will mean how difficult it is to solve the inverse problems from the viewpoint of Theorem 5.3. These results will show some good prospects that we can get analytic functions by computers, and our mathematical theory may be applied to wide problems practically. However, we still have difficult problems, for example, to catch the global data from some local data for analytic functions.

The central problem in our method is now to solve effectively the functional equation (5.11), and there we need practically some discretization of the equation to solve it by using computers. The practical treatments are dependent on the concrete equations, case by case. When a complete orthonormal system of $\mathcal{H}$ is simply constructed, we can discretize it by using the Fourier coefficients ([**36**]); however, in the worst case, by taking points in the space, by the induction method, we can discretize it ([**34**]). In the near future, for the sake of the great power of computers, we will be able to solve effectively and sufficiently many linear problems by our methods.

The computer simulation method of H. Imai and H. Fujiwara will be one of the top methods in the world based on the concept ([**3, 24**]) of infinite precison using multiple-precision arithmetic. We started only 4 years ago our co-research in our numerical challenges, and we expect further fruitful developments in those practical problems and simulation.

Physical problems and mathematics. We were able to solve the very difficult and famous ill-posed problems practically by using computers. We therefore, by using our powerful method, would like to contribute to human beings practically in some concrete problem. With our colleagues in the Faculty of Engineering of Gunma University, we attacked the problem that from some vague image taken by a camera we reproduce a clear original image. We soon find, however, a serious problem as in the following ([**72**]):

Inverse problems in mathematics which are expected to be applied to practical problems will, sometimes, have weak points from the viewpoint that the background theories are not faithful for practical and physical problems. For example, formulated equations are the representations of some ideal models and are not those of faithful models in the real physical world. Sometimes, boundary conditions for the equations are involved in physical units and sometimes their physical realizations and observations are very difficult. Here, we shall give a new inversion formula for a linear system based on physical experimental data and by using reproducing kernels and the Tikhonov regularization based on the formulas (5.10) and (5.11). In particular, we will not assume any analytical assumption on the linear system, but we use physical experimental data for obtaining an approximate inversion formula for the linear system.

Our new method is a general one; however, we will need practical physical experiments for the problems chosen on a case-by-case basis. So we need many collabourations for some practical problems, as in the real inversion problem of the Laplace transform.

Overview on the theory of reproducing kernels. Sections 2, 3 and 4 are based on the research books [**2, 45, 51**]; for the detailed sources, contents and references, see these books. In this paper, therefore we listed only the restricted references.

For the theory of reproducing kernels in one-variable complex analysis, see the fundamental books [**8, 18, 22**]. For the advanced and profound theory, see the books D. A. Hejhal [**22**] and J. D. Fay [**18**] in connection with the Riemann theta functions and the Klein prime form. For their applications, see A. Yamada [**81**]. Their theory now seems to be, however, too deep to deal with for any mathematicians. For the old history of reproducing kernels, see [**6, 45**]. In the Proceedings [**55**] of an international conference, we can find various recent results on the theory of reproducing kernels. We find many results in the **learning theory** where the applications of the theory of reproducing kernels are important; that is, the estimation of covering numbers by the discs of reproducing kernel Hilbert spaces as subspaces of a family of continuous functions, some detailed smoothness relationships between reproducing kernels and reproducing kernel Hilbert spaces, and approximations of functions by Sobolev spaces. See, for example, S. Smale [**15**] and ([**83**]). Indeed, we have many references, and for the learning theory, see [**78**]. For **Support Vector Machines** that are favourable for engineers, see [**14**] whose research center is MIT.

We would like to note the active research results of K. Fukumizu ([**87**],[**88**]). [**5**] summarizes various applications of the theory in connection with operator theory on Hilbert spaces whose research center is in Israeru with a great group. For the connection with the stochastic theory and the theory, see Berlinet ([**9, 10**]) who was very active in the 5th ISAAC Catania Congress with his colleagues and we see many results in this field.

Acknowlegements

The author wishes to express his sincere thanks to the editors for giving him the chance for a publication in Sūgaku. For the referees of the paper the author wishes to express deep thanks for their detailed and valuable comments and suggestions. For Section 5 and later, for his colleague T. Matsuura, he expresses his thanks for his great contributions to the computer simulation and in Section 7, for H. Fujiwara and M. Sugihara for their contributions to numerical experiments and simulation.

The readers may be surprised by the elementary and fundamental results in this survey paper. The author, indeed, thinks the theory of reproducing kernels is fundamental and applicable in mathematics as in vector analysis and complex analysis, at the undergraduate level. In order to realize this idea in the mathematical world, the author founded the **Institute of Reproducing Kernels** (kbdmm360@yahoo.co.jp).

References

1. N. Suita, *Modern Function Theory II–Conformal Mapping Theory* (in Japanese), Morikita Co., (1976).
2. S. Saitoh, *Introduction to Theory of Reproducing Kernels* (in Japanese), Makino-Shoten Co., (2002).
3. H. Imai, *Multiple-Precision Arithmetic* (in Japanese), Sūgaku, Mathematical Society of Japan, Iwanami-Shoten Co., **55** (2003), 316-325.
4. J. Agler and J. E. McCarthy, *Pick Interpolation and Hilbert Function Spaces,* Graduate Studies in Mathematics, Vol. **44** (2002). MR1882259 (2003b:47001)
5. D. Alpay, *The Schur Algorithm, Reproducing Kernel Spaces and System Theory,* American Mathematical Society, 1998. MR1839648 (2002b:47144)
6. N. Aronszajn, *Theory of reproducing kernels,* Trans. Amer. Math. Soc., **68** (1950), 337–404. MR0051437 (14:479c)
7. M. Asaduzzaman, T. Matsuura, and S. Saitoh, *Constructions of approximate solutions for linear differential equations by reproducing kernels and inverse problems,* Advances in Analysis, Proceedings of the 4th International ISAAC Congress (2003), 335-344. (World Scientific, 2005). MR2184949 (2006h:65208)
8. S. Bergman, *The kernel function and conformal mapping,* Amer. Math. Soc., Providence, R. I. (1950, 1970). MR0038439 (12:402a)
9. A. Berlinet and C. Thomas-Agnan, *Reproducing kernel Hilbert spaces in probability and statistics,* Kluwer Akademic Publishers, Boston/Dordrecht/London, 2004. MR2239907 (2007b:62006)
10. A. Berlinet, *Reproducing kernels in probability and statistics,* More Progress in Analysis, Proceedings of the 5th International ISAAC Congress (2005), 153-162. (World Scientific, 2009)
11. D.-W. Byun and S. Saitoh, *A real inversion formula for the Laplace transform,* Z. Anal. Anwendungen, **12**(1993), 597-603. MR1264853 (95d:44002)
12. D.-W. Byun and S. Saitoh, *Approximation by the solutions of the heat equation,* J. Approximation Theory, **78** (1994), 226–238. MR1285260 (95j:35093)
13. D.-W. Byun and S. Saitoh, *Best approximation in reproducing kernel Hilbert spaces,* Proc. of the Second International Colloquium on Numerical Analysis, VSP-Holland, 1994, 55–61. MR1455929

14. N. Cristianini and J. Shawe-Taylor, *An Introduction to Support Vector Machines and other Kernel-based Learning Methods,* Cambridge Univ. Press, Cambridge, 2001.
15. F. Cucker and S. Smale, *On the mathematical foundations of learning,* Bull. Amer. Math. Soc., **39** (2001), 1–49. MR1864085 (2003a:68118)
16. G. Doetsch, *Handbuch der Laplace Transformation,* Vol. 1., Birkhäuser Verlag, Basel, 1950. MR0344808 (49:9547)
17. H. W. Engl, M. Hanke and A. Neubauer, *Regularization of Inverse Problems,* Mathematics and Its Applications, **376**(2000), Kluwer Academic Publishers. MR1408680 (97k:65145)
18. J.D. Fay, *Theta functions on Riemann surfaces,* Lecture Notes in Math., Vol. 352, Springer-Verlag, 1973. MR0335789 (49:569)
19. H. Fujiwara, T. Matsuura and S. Saitoh (T. Matsuura, A. Al-Shuaibi, H. Fujiwara and S. Saitoh), *Numerical real inversion formulas of the Laplace transform by using a Fredholm integral equation of the second kind,* J. Anal. Appl., **5** (2007), 123–136. MR2314784 (2009a:30018)
20. C. W. Groetsch, *Inverse Problems in the Mathematical Sciences,* Vieweg & Sohn Verlagsgesellschaft mbH, Braunschweig/Wiesbaden, 1993. MR1247696 (94m:00008)
21. N. Hayashi, *Analytic function spaces and their applications to nonlinear evolution equations,* Analytic Extension Formulas and their Applications, Kluwer Academic Publishers, 2001, 59–86. MR1830377 (2002c:35061)
22. D.A. Hejhal, *Theta functions, kernel functions and Abel integrals,* Memoirs of Amer. Math. Soc., Vol. 129, Providence, R.I., 1972. MR0372187 (51:8403)
23. J. R. Higgins, *A sampling principle associated with Saitoh's fundamental theory of linear transformations,* Analytic Extension Formulas and their Applications, Kluwer Academic Publishers, 2001, 73–86. MR1830378 (2002c:46053)
24. H. Imai, T. Takeuchi and M. Kushida, *On numerical simulation of partial differential equations in infinite precision,* Advances in Mathematical Sciences and Applications, **9** (1999), 1007-1016. MR1725697
25. H. Itou and S. Saitoh, *Analytical and numerical solutions of linear singular integral equations,* Int. J. Appl. Math. Stat., **12** (2007), 76–89. MR2374503 (2009a:45011)
26. I.I. Hirschman and D.V. Widder, *The Convolution Transform,* Princeton University Press, Princeton, New Jersey, 1955. MR0073746 (17:479c)
27. V.V. Kryzhniy, *Regularized inversion of integral transformations of Mellin convolution type,* Inverse Problems, **19** (2003), 573-583. MR2024697 (2004k:44005)
28. V. V. Kryzhniy, *Numerical inversion of the Laplace transform: analysis via regularized analytic continuation,* Inverse Problems, **22** (2006), 579-597. MR2216416 (2006k:65380)
29. T. Matsuura and S. Saitoh, *Analytical and Numerical Solutions of the Inhomogeneous Wave Equation,* Australian J. of Math. Anal. and Appl., **1** (2004), Volume 1, Issue 1, Article 7. MR2077664
30. T. Matsuura, S. Saitoh and D.D. Trong, *Numerical solutions of the Poisson equation,* Applicable Analysis, **83** (2004), 1037-1051. MR2096664 (2005k:31009)
31. T. Matsuura and S. Saitoh, *Analytical and numerical solutions of linear ordinary differential equations with constant coefficients,* J. Anal. Appl., **3** (2005), 1-17. MR2110496 (2005h:34018)
32. T. Matsuura and S. Saitoh, *Numerical inversion formulas in the wave equation,* J. Comput. Math. Optim., **1** (2005), 1-19. MR2128936
33. T. Matsuura, S. Saitoh and D.D. Trong, *Approximate and analytical inversion formulas in heat conduction on multidimensional spaces,* J. of Inverse and Ill-posed Problems, **13** (2005), 479-493. MR2188625 (2006h:80002)
34. T. Matsuura and S. Saitoh, *Dirichlet's Principle Using Computers,* Applicable Analysis, **84** (2005), 989-1003. MR2177850 (2006e:31002)
35. T. Matsuura and S. Saitoh, *Analytical and numerical inversion formulas in the Gaussian convolution by using the Paley-Wiener spaces,* Applicable Analysis, **85** (2006), 901-915. MR2250776 (2007d:44003)
36. T. Matsuura and S. Saitoh, *Analytical and numerical real inversion formulas of the Laplace transform,* More Progress in Analysis, Proceedings of the 5th International ISAAC Congress (2005), 191-200. (World Scientific 2009)
37. T. Matsuura, A. Al-Shuaibi, H. Fujiwara and S. Saitoh, *Numerical real inversion formulas of thc Laplacc transform by using a Fredholm integral equation of the second kind,* J. Anal. Appl., **5** (2007), 123-136. MR2314784 (2009a:30018)

38. T. Matsuura, A. Al-Shuaibi, H. Fuijiwara, S. Saitoh and M. Sugihara, *Numerical real inversion formulas of the Laplace transform by a sinc method,* Far East J. Math. Sci., **27** (2007), 1–14. MR2402538 (2009c:44003)
39. Z. Nehari, *Conformal mapping,* McGraw-Hill Book Company Inc., New York, 1952. MR0045823 (13:640h)
40. E. L. Post, *Generalized differentiation,* Trans. Amer. Math. Soc., **32** (1930), 723-781. MR1501560
41. Th. M. Rassias and S. Saitoh, *The Pythagorean theorem and linear mappings,* PanAmerican Math. J., **12** (2002), 1–10. MR1880156 (2002m:46033)
42. S. Saitoh, *The Bergman norm and the Szegö norm,* Trans. Amer. Math. Soc., **249** (1979), 261–279. MR525673 (80h:30011)
43. S. Saitoh, *Hilbert spaces induced by Hilbert space valued functions,* Proc. Amer. Math. Soc., **89** (1983), 74–78. MR706514 (84h:44001)
44. S. Saitoh, *The Weierstrass transform and an isometry in the heat equation,* Applicable Analysis, **16** (1983), 1-6. MR705341 (84k:44008)
45. S. Saitoh, *Theory of Reproducing Kernels and its Applications,* Pitman Research Notes in Mathematics Series, **189** (1988), Longman Scientific & Technical, UK. MR983117 (90f:46045)
46. S. Saitoh, *Interpolation problems of Pick-Nevanlinna type,* Pitman Research Notes in Mathematics Series, **212** (1989), 253–262. MR1048142 (91d:46028)
47. S. Saitoh, *Representations of the norms in Bergman-Selberg spaces on strips and half planes,* Complex Variables, **19** (1992), 231–241. MR1284115 (95b:30011)
48. S. Saitoh, *One approach to some general integral transforms and its applications,* Integral Transforms and Special Functions, **3** (1995), 49–84. MR1410596 (97j:44002)
49. S. Saitoh, *Natural norm inequalities in nonlinear transforms,* General Inequalities **7** (1997), 39–52. Birkhäuser Verlag, Basel, Boston. MR1457267 (98j:46023)
50. S. Saitoh, *Representations of inverse functions,* Proc. Amer. Math. Soc., **125** (1997), 3633–3639. MR1423330 (98b:46033)
51. S. Saitoh, *Integral Transforms, Reproducing Kernels and their Applications,* Pitman Research Notes in Mathematics Series, **369** (1997), Addison Wesley Longman, UK. MR1478165 (98k:46041)
52. S. Saitoh, *Nonlinear transforms and analyticity of functions,* Nonlinear Mathematical Analysis and Applications (1998), 223–234. Hadronic Press, Palm Harbor.
53. S. Saitoh, *Various operators in Hilbert space induced by transforms,* International J. of Applied Math., **1** (1999), 111–126. MR1689721 (2000g:46033)
54. S. Saitoh, *Applications of the general theory of reproducing kernels,* Reproducing Kernels and their Applications, Kluwer Academic Publishers, 1999, 165–188. MR1789761
55. S. Saitoh, D. Alpay, J.A. Ball and T. Ohsawa (eds), *Reproducing Kernels and their Applications,* Kluwer Academic Publishers, 1999. MR1789746 (2001d:00046)
56. S. Saitoh and M. Yamamoto, *Integral transforms involving smooth functions,* Reproducing Kernels and their Applications, Kluwer Academic Publishers, 1999, 149–164. MR1789760 (2001f:44001)
57. S. Saitoh, *Linear integro-differential equations and the theory of reproducing kernels,* Volterra Equations and Applications, C. Corduneanu and I.W. Sandberg (eds), Gordon and Breach Science Publishers, Amsterdam, 2000.
58. S. Saitoh, *Weighted L_p-norm inequalities in convolutions,* Survey on Classical Inequalities (T. M. Rassias, ed.), Kluwer Academic Publishers, 2000, 225-234. MR1894722 (2003d:26034)
59. S. Saitoh, *Analytic extension formulas, integral transforms and reproducing kernels,* Analytic Extension Formulas and their Applications, Kluwer Academic Publishers, 2001, 207–232. MR1830384 (2002d:30011)
60. S. Saitoh, *Applications of the reproducing kernel theory to inverse problems,* Comm. Korean Math. Soc., **16** (2001), 371–383. MR1843521 (2002i:30011)
61. S. Saitoh, *Principle of telethoscope,* Functional-Analytic and Complex Methods, their Interaction and Applications to Partial Differential Equations, Proceedings of the International Graz Workshop, Graz, Austria, 12-16 February 2001. World Scientific (2001), 101–117. MR1893252 (2003b:35034)
62. S. Saitoh, N. Hayashi and M. Yamamoto (eds.), *Analytic Extension Formulas and their Applications,* Kluwer Academic Publishers, 2001. MR1830372 (2001k:00017)

63. S. Saitoh, Vu Kim Tuan and M. Yamamoto, *Conditional Stability of a Real Inverse Formula for the Laplace Transform,* Z. Anal. Anwendungen **20** (2001), 193-202. MR1826326 (2002b:44001)
64. S. Saitoh and M. Mori, *Representations of analytic functions in terms of local values by means of the Riemann mapping function,* Complex Variables, **45** (2001), 387–393. MR1866503 (2002j:30006)
65. S. Saitoh, Vu Kim Tuan and M. Yamamoto, *Reverse convolution inequalities and applications to inverse heat source problems,* J. of Inequalities in Pure and Applied Mathematics, **3** (2002), Article 80. MR1966515 (2004b:26032)
66. S. Saitoh, T. Matsuura and M. Asaduzzaman, *Operator Equations and Best Approximation Problems in Reproducing Kernel Hilbert Spaces,* J. Anal. Appl., **1** (2003), 131-142. MR2013271 (2004i:46040)
67. S. Saitoh, *Constructions by Reproducing Kernels of Approximate Solutions for Linear Differential Equations with L_2 Integrable Coefficients,* International J. of Math. Sci., **2** (2003), 261-273. MR2061513 (2005i:35014)
68. S. Saitoh, *Approximate Real Inversion Formulas of the Gaussian Convolution,* Applicable Analysis, **83** (2004), 727-733. MR2072423 (2005f:44006)
69. S. Saitoh, *Best approximation, Tikhonov regularization and reproducing kernels,* Kodai. Math. J., **28** (2005), 359-367. MR2153923 (2006f:41039)
70. S. Saitoh, *Tikhonov regularization and the theory of reproducing kernels,* Finite or Infinite Dimensional Complex Analysis and Applications (Proceedings of the 12th ICFIDCAA), Kyushu University Press, 2005, 291-298. MR2359707 (2009b:47017)
71. S. Saitoh, *Applications of reproducing kernels to best approximations, Tikhonov regularizations and inverse problems,* Advances in Analysis, Proceedings of the 4th International ISAAC Congress (2003), 439-445. (World Scientific, 2005) MR2184959
72. S. Saitoh and M. Yamada, *Inversion formulas for a linear system determined by input and response relations, by using suitable function spaces,* Hokkaido University Technical Report Series in Mathematics, **118** (2007), 18-21.
73. M. Sakai, *Analytic functions with finite Dirichlet integrals on Riemann surfaces,* Acta. Math., **142** (1979), 199–229. MR521461 (80d:30042)
74. F. Stenger, *Numerical Methods Based on Sinc and Analytic Functions,* Springer Series in Computational Mathematics, **20** (1993). MR1226236 (94k:65003)
75. N. Suita and A. Yamada, *On the Lu-Qi Keng conjecture,* Proc. Amer. Math. Soc., **59** (1976), 222–224. MR0425185 (54:13142)
76. L. Schwartz, *Sous-espaces hilbertiens d'espaces vectoriels topologiques et noyaux associés (noyaux reproduisants),* J. Analyse Math., **13** (1964), 115–256. MR0179587 (31:3835)
77. W. Ulmer and W. Kaissl, *The inverse problem of a Gaussian convolution and its application to the finite size of measurement chambers/detectors in photon and proton dosimetry,* Phys. Med. Biol., **48**(2003), 707-727.
78. V. Vapnik, *Statistical Learning Theory,* John Wiley & Sons, 1998. MR1641250 (99h:62052)
79. G. Wahba, *Spline Models for Observational Data,* CBMS-NSF Regional Conference Series in Applied Mathematics, **59**, SIAM, Philadelphia, 1990. MR1045442 (91g:62028)
80. D. V. Widder, *The Laplace Transform,* Princeton University Press, Princeton, 1972. MR0005923 (3:232d)
81. A. Yamada, *Fay's trisecant formula and Hardy H^2 reproducing kernels,* Reproducing Kernels and their Applications, Kluwer Academic Publishers, 1999, 165–188. MR1789764 (2001h:30012)
82. A. Yamada, *Equality conditions for general norm inequalities in reproducing kernel Hilbert spaces,* Advances in Analysis, World Scientific, 2005, 447–455. MR2184960 (2006g:46049)
83. D. -X Zhou, *Capacity of reproducing kernel spaces in learning theory,* IEEE Trans. Inform. Theory, **49** (2003), 1743–1752. MR1985575 (2004c:62095)
84. *http://library.wolfram.com/inforcenter/MathSource/4738/*
85. *http://www.columbia.edu/ ww2040/abate.html*
86. H. Fujiwara, T. Matsuura, S. Saitoh and Y. Sawano, *The real inversion of the Laplace transform by numerical singular value decomposition,* J. Anal. Appl., **6** (2008), 55-68. MR2394888 (2009b.35151)

87. K. Fukumizu, F.R. Bach and M.I. Jordan, *Dimensionality reduction for supervised learning with reproducing kernel Hilbert spaces,* Journal of Machine Learning Research, **5** 2003/04, 73–99. MR2247974
88. M. Cuturi, K. Fukumizu and J.P. Vert, *Semigroup kernels on measures,* Journal of Machine Learning Research, **6** (2005), 1169–1198. MR2249852
89. H. Takahasi and M. Mori, *Double exponential formulas for numerical integration,* Publ. RIMS. Kyoto Univ., **10** (1974), 721–741. MR0347061 (49:11781)

Added in Translation

90. M. Yamada, T. Matsuura and S. Saitoh, *Representations of inverse functions by the integral transform with the sign kernel,* Frac. Calc. Appl. Anal., **10** (2007), 161–168. MR2351657 (2008i:44001)
91. M. Yamada and S. Saitoh, *Identification of non-linear systems,* J. Computational, Math. Optimization, **4** (2008), 47–60. MR2394377 (2009b:93051)
92. M. Yamada and S. Saitoh, *Numerical solutions of two non-linear simultaneous equations,* Applicable Analysis, **88** (2009), 151–160. MR2536026 (2010f:65104)

Department of Mathematics, University of Aveiro, 3810-193 Aveiro, Portugal
E-mail address: saburou.saitoh@gmail.com

Translated by SABUROU SAITOH

Amer. Math. Soc. Transl.
(2) Vol. **230**, 2010

An Approach to Superrigidity and Fixed-point Theorems via Harmonic Maps

Hiroyasu Izeki and Shin Nayatani

Introduction

In the middle of the 1970s, Margulis [**31**] proved a rigidity theorem for lattices in semisimple Lie groups of rank not less than two (Theorem 1.3) as an important step for the solution of the Selberg conjecture (cf. Theorem 1.4). The theorem asserted that rigidity stronger than the Mostow strong rigidity known until then held for such lattices, and it was named the superrigidity theorem by Mostow. While Margulis' proof of Theorem 1.3 was based on ergodic theory, in the early 1990s, Corlette [**5**] proved a superrigidity theorem (Theorem 2.1) for lattices in simple Lie groups of rank one by a new method using harmonic maps. By this accomplishment of Corlette and the succeeding work of Jost-Yau [**25**], Mok-Siu-Yeung [**35**] and Gromov-Schoen [**15**] and so on, the usefulness and importance of harmonic maps in the rigidity problem became widely recognized. Also, this stream of study has significantly motivated the recent progress in the theory of harmonic maps (in particular, the study of harmonic maps between spaces with singularities).

In this article, we discuss an approach to superrigidity and fixed-point theorems using harmonic maps, focusing upon the fixed-point theorem for discrete groups recently obtained by Gromov and the present authors independently. The study of superrigidity and fixed-point theorems has progressed not only because of geometric interest in themselves but also because of the important motivation of applying those theorems to algebraic problems such as arithmeticity and nonlinearity of discrete groups. We have, therefore, tried to discuss these aspects also as long as the authors' knowledge reaches. This policy is reflected in Theorem 1.4 and Proposition 4.4, but the readers may skip the surrounding algebraic arguments, as they are not necessary to understand the main body of this article.

1. Superrigidity and fixed-point theorems

Formulating geometrically, we can state the phenomenon of superrigidity as follows.

Let X and Y be spaces with some geometric structures (which are usually metrics or distances), and let Γ be a finitely generated group. We suppose that Γ

This article originally appeared in Japanese in Sūgaku **58** (3) (2006), 239–262.
2000 *Mathematics Subject Classification.* Primary 58E20; Secondary 20F65, 22E40 .

acts on X properly discontinuously, preserving its geometric structures, and that the quotient $\Gamma\backslash X$ has some sort of finiteness (e.g. compactness). Then it may happen that for any action of Γ on Y preserving its geometric structures either of the following necessarily holds:

(i) Γ has a (global) fixed point in Y;
(ii) Γ preserves a geometric copy $X' \subset Y$ of X, and the action of Γ on X' is essentially the same as that of Γ on X.

In this case Γ is said to have superrigidity (for Y) and an assertion that a certain finitely generated group Γ has superrigidity is called a superrigidity theorem. In particular, if the geometric nature of Y is very different from that of X and therefore Y cannot contain a geometric copy of X, then the assertion (i) above necessarily holds. We distinguish the assertion in such a situation and call it a fixed-point theorem.

As examples of fixed-point theorems, those related to Kazhdan's property (T) are well known. Kazhdan's property (T) was introduced for locally compact topological groups in [**27**]. Here we state the usual definition of this property, restricting ourselves to the case of discrete groups (that is, countable groups endowed with the discrete topology).

DEFINITION 1.1. Let $\mathcal{H}$ be a Hilbert space, and let $\mathcal{U}(\mathcal{H})$ be the group of all unitary transformations of $\mathcal{H}$. Let Γ be a discrete group, and let $\rho : \Gamma \longrightarrow \mathcal{U}(\mathcal{H})$ be a homomorphism, that is, a unitary representation of Γ. We say that ρ has an *almost invariant vector* if for any $\varepsilon > 0$ and any finite subset $K \subset \Gamma$, there exists $v \in \mathcal{H}$ of length one satisfying $\max_{\gamma \in K} \|\rho(\gamma)v - v\| < \varepsilon$. Also, we say that Γ has *Kazhdan's property (T)* if any unitary representation ρ with an almost invariant vector necessarily has an invariant vector. Here, an invariant vector means an element (other than the zero vector) of $\mathcal{H}$ fixed by $\rho(\Gamma)$.

Lattices in simple Lie groups of rank not less than two are typical examples of discrete groups with the Kazhdan's property (T). Here, letting G be a locally compact topological group, we say that Γ is a *lattice* of G if Γ is a discrete subgroup of G and the measure of $\Gamma\backslash G$ is finite. In particular, if $\Gamma\backslash G$ is compact, we call Γ a *uniform lattice.*

A discrete group with Kazhdan's property (T) is finitely generated.

THEOREM 1.2. (i) *A discrete group Γ has Kazhdan's property* (T) *if and only if any isometric action of Γ on a Hilbert space has a fixed point* ([**6**], [**16**]).

(ii) *If a discrete group Γ has Kazhdan's property* (T), *then any isometric action of Γ on an $\mathbb{R}$-tree has a fixed point* ([**49**]).

(iii) *If a discrete group Γ has Kazhdan's property* (T), *then any isometric action of Γ on real or complex hyperbolic spaces* (*possibly of infinite dimensions*) *has a fixed point* (*cf.* [**17**]).

The following theorem due to Margulis is the prototype of superrigidity theorems, and it will be cited repeatedly throughout this article.

THEOREM 1.3 (Margulis superrigidity theorem [**31**]). *Let* $\mathbf{G}$ *be a connected semisimple algebraic group defined over $\mathbb{R}$ of $\mathbb{R}$-rank not less than two such that $(\mathbf{G}_{\mathbb{R}})^0$ has no compact factor, and let $\Gamma \subset (\mathbf{G}_{\mathbb{R}})^0$ be an irreducible lattice. Let* $\mathbf{H}$ *be a connected algebraic group defined over a local field k of characteristic zero which* *is simple over k, and let $\rho : \Gamma \longrightarrow \mathbf{H}_k$ be a homomorphism such that the image $\rho(\Gamma)$ is Zariski dense. Then either of the following holds:*

(i) *$\rho(\Gamma)$ is relatively compact (with respect to the Hausdorff topology);*
(ii) *$k = \mathbb{R}$ or $\mathbb{C}$, and ρ extends to a rational homomorphism $\overline{\rho} : \mathbf{G} \longrightarrow \mathbf{H}$ defined over k.*

We omit various definitions related to algebraic groups and refer the reader to the concise description in section 3.1 of [**50**]. An example of the algebraic group $\mathbf{G}$ in the theorem is $\mathbf{PGL}_n$ ($n \geq 3$), and we have $(\mathbf{G}_{\mathbb{R}})^0 = PGL_n(\mathbb{R})$ in this case. ($\mathbf{PGL}_n$ is a typical example of $\mathbf{H}$ also.) The assertion of the theorem is sometimes called an *Archimedean superrigidity* when k is $\mathbb{R}$ or $\mathbb{C}$, and a *non-Archimedean superrigidity* when k is a totally-disconnected field such as the r-adic field $\mathbb{Q}_r$.

It will be natural to expect that because of discreteness, representations of discrete groups should have more freedom than those of continuous groups. Theorem 1.3, however, asserts that such a naive feeling does not apply to a certain kind of discrete group at all. Note that Theorem 1.3 does not assume that $\rho(\Gamma)$ is a lattice in $\mathbf{H}_k$, and at this point it is further stronger than the strong rigidity theorem (due to Mostow).

To restate the conclusion of Theorem 1.3 in geometric form, we begin with setting up relevant spaces. Let K and L be maximal compact subgroups of $(\mathbf{G}_{\mathbb{R}})^0$ and $\mathbf{H}_k$, respectively, and consider the homogeneous spaces $X = (\mathbf{G}_{\mathbb{R}})^0/K$ and $Y = \mathbf{H}_k/L$. (Maximal compact subgroups of $PGL_n(k)$ for $k = \mathbb{C}, \mathbb{R}, \mathbb{Q}_r$ are given by $PU_n, PO_n, PGL_n(\mathbb{Z}_r)$, respectively, where $\mathbb{Z}_r$ is the r-adic integer ring.) The space X is a Riemannian manifold referred to as a symmetric space of noncompact type, and the left action of $(\mathbf{G}_{\mathbb{R}})^0$ on X preserves the metric. The space Y is also a symmetric space of noncompact type when $k = \mathbb{R}, \mathbb{C}$, while it is a metric simplicial complex called a Bruhat-Tits building when k is totally disconnected. In any case, Y is a metric space, and the left action of $\mathbf{H}_k$ on Y preserves the distance. Moreover, when $k = \mathbb{R}, \mathbb{C}$, Y has nonpositive sectional curvature. When k is totally disconnected, Y has nonpositive curvature in the sense of Alexandroff though the curvature itself is not defined for Y.

We now restate in this setting the conclusion (i) of Theorem 1.3. First, since $\rho(\Gamma)$ is a subgroup of $\mathbf{H}_k$, it acts on Y isometrically. Now if (i) holds, every orbit of $\rho(\Gamma)$ is bounded. Since Y has nonpositive curvature, uniquely determined is a point called the circumcenter of the orbit ([**2**, p. 179]), and it is fixed by $\rho(\Gamma)$. Conversely, if $\rho(\Gamma)$ has a fixed point in Y, then since the stabilizer of any point in Y is a compact subgroup of $\mathbf{H}_k$, it follows that $\rho(\Gamma)$ is relatively compact in $\mathbf{H}_k$. In this way, the conclusion (i) is equivalent to the assertion that $\rho(\Gamma)$ has a fixed point in Y. On the other hand, corresponding to the conclusion (ii) is the assertion that there exists an isometric totally geodesic ρ-equivariant embedding of X into Y. However, this last assertion is substantially stronger than the conclusion (ii), and it cannot be read directly from (ii).

Theorem 1.3 was proved by Margulis around 1974, and his proof was based on ergodic theory on the whole. On the other hand, beginning in the early 1990s, various superrigidity theorems, including Theorem 1.3, have been proved by using the differential-geometric instruments of harmonic maps. We shall discuss the details of this issue in §2. Here we just remark that the existence of the embedding map corresponding to the conclusion (ii) can be verified only through the proof using harmonic maps.

Now the common feature of Theorem 1.2 and Theorem 1.3 is that the spaces Y are all CAT(0) spaces, that is, metric spaces with nonpositive curvature in the sense of Alexandroff. On the other hand, whereas in Theorem 1.3 the space $Y = \mathbf{H}_k/L$ is locally compact, in Theorem 1.2 the space Y is not necessarily locally compact. Also, whereas in Theorem 1.3 $\rho(\Gamma)$ is assumed to be Zariski dense, in Theorem 1.2 the isometric action of Γ on Y is totally arbitrary. At this point naturally arises the problem of searching for sufficient conditions for any isometric action of a discrete group on a CAT(0) space to have a fixed point, by comprehending the situations where Theorem 1.2 and the conclusion of Theorem 1.3 hold. In fact, Gromov [**14**] and the authors [**21**] have formulated, for a discrete goup given as a subgroup of the automorphim group of a simplicial complex, a criterion for any isometric action of the group on a CAT(0) space to have a fixed point. The details will be given in §3, where we shall use combinatorial harmonic maps (precisely, gradient flows of the corresponding energy functional) obtained by replacing the source spaces from manifolds to simplicial complexes in the definition of harmonic maps. Ballmann-Świątkowski [**1**] and Żuk [**51**] gave, for the same class of discrete groups, a sufficient condition for any isometric action of the group on a Hilbert space to have a fixed point (hence, the group has Kazhdan's property (T)). The result of Gromov and the authors given in §3 generalizes that of Ballmann-Świątkowski and Żuk, replacing the Hilbert space by a general CAT(0) space.

We conclude this section with the discussion of applications of (geometric) superrigidity theorems and fixed-point theorems. As a classical application, we have the following.

THEOREM 1.4 (Margulis arithmeticity theorem [**31**]). *Let* $\mathbf{G}$ *be a connected semisimple Lie group with trivial center and no compact factor, and let* $\Gamma \subset G$ *be an irreducible lattice. If the* $\mathbb{R}$*-rank of* G *is not less than two, then* Γ *is arithmetic; that is, there exist a semisimple algebraic group* $\mathbf{H}$ *defined over* $\mathbb{Q}$ *and a surjective continuous homomorphism* $\varphi : (\mathbf{H}_{\mathbb{R}})^0 \longrightarrow G$ *with* $\operatorname{Ker}\varphi$ *compact such that* $\varphi(\mathbf{H}_{\mathbb{Z}} \cap (\mathbf{H}_{\mathbb{R}})^0)$ *and* Γ *are commensurable.*

Here, we say that $\varphi(\mathbf{H}_{\mathbb{Z}}\cap(\mathbf{H}_{\mathbb{R}})^0)$ and Γ are commensurable if $\varphi(\mathbf{H}_{\mathbb{Z}}\cap(\mathbf{H}_{\mathbb{R}})^0)\cap\Gamma$ has finite index both in $\varphi(\mathbf{H}_{\mathbb{Z}} \cap (\mathbf{H}_{\mathbb{R}})^0)$ and Γ.

Selberg and Piatetski-Shapiro conjectured the assertion of the theorem in the 1960s, and then Margulis proved it around 1974 by using his superrigidity theorem. We refer the reader to [**32**], [**50**] for the details of the proof of Theorem 1.4. Here we only outline the proof, following [**50**]. First, by a well-known argument, we may assume that the group G in the theorem is defined over $\mathbb{Q}$. Namely, there exists a connected semisimple algebraic group $\mathbf{G} \subset GL_n(\mathbb{C})$ defined over $\mathbb{Q}$ (with trivial center and of $\mathbb{R}$-rank not less than two) such that $G = (\mathbf{G}_{\mathbb{R}})^0$. If we denote the Lie algebra of $\mathbf{G}$ by $\mathfrak{g} \subset M(n, \mathbb{C})$, then $\mathfrak{g}$ and the adjoint representation $\mathrm{Ad} : \mathbf{G} \longrightarrow GL(\mathfrak{g})$ are also defined over $\mathbb{Q}$. Based upon the above, we may divide the proof of Theorem 1.4 into four steps:

(i) Show that $\mathrm{tr}(\mathrm{Ad}(\Gamma)) \subset K$, where K denotes the field of algebraic real numbers.
(ii) Show that $\mathbf{G}$ is defined over K and we may assume that $\Gamma \subset \mathbf{G}_K$.
(iii) Show that there are a semisimple algebraic group $\mathbf{H}$ defined over $\mathbb{Q}$ and a rational homomorphism $p : \mathbf{H} \longrightarrow \mathbf{G}$ defined over $\mathbb{R}$ with $(\operatorname{Ker} p)_{\mathbb{R}}$ compact.

(iv) Show that Γ and $p(\mathbf{H}_{\mathbb{Z}} \cap (\mathbf{H}_{\mathbb{R}})^0)$ are commensurable.

The superrigidity theorem involves the steps (i), (iii), (iv). In what follows, since the arguments for (i), (iv) are necessary in later applications, we outline the proof with focusing on these two steps. As for (i), noting that the automorphism group $\mathrm{Aut}(\mathbb{C})$ of the field $\mathbb{C}$ acts on the set of all transcendental numbers transitively, it is sufficient to prove that $\{\sigma(\mathrm{tr}(\mathrm{Ad}(\gamma))) \mid \sigma \in \mathrm{Aut}(\mathbb{C})\}$ is a bounded set of $\mathbb{C}$. Note that $\mathrm{Aut}(\mathbb{C})$ acts naturally on $M(n, \mathbb{C})$ and preserves $\mathbf{G}$, and we have $\sigma(\mathrm{tr}(\mathrm{Ad}(\gamma))) = \mathrm{tr}(\mathrm{Ad}(\sigma(\gamma)))$. Therefore, it is enough to prove that $|\mathrm{tr}(\mathrm{Ad}(\sigma(\gamma)))|$ is bounded uniformly independently of σ and this is done by the superrigidity theorem. Assuming for simplicity that $\mathbf{G}$ is simple and applying Theorem 1.3 to $\sigma|_\Gamma : \Gamma \longrightarrow \mathbf{G}$, we conclude that either $\sigma(\Gamma)$ is contained in a compact subgroup of $\mathbf{G}$ or $\sigma|_\Gamma$ extends to a rational isomorphism $\pi_\sigma : \mathbf{G} \longrightarrow \mathbf{G}$. The eigenvalues of $\mathrm{Ad}(\sigma(\gamma))$ all have absolute value one in the former case and coincide with the eigenvalues of $\mathrm{Ad}(\gamma)$ in the latter. Hence, in either case, $|\mathrm{tr}(\mathrm{Ad}(\sigma(\gamma)))|$ is bounded uniformly independently of σ. (When $\mathbf{G}$ is not simple, one applies Theorem 1.3 to each simple factor of $\mathbf{G}$.)

The argument for (ii) is algebraic on the whole and does not need the superrigidity theorem. The proof of this step is due to Vinberg [**46**]. In proceeding from (ii) to (iii), we first use the fact that Γ is finitely generated. (When G is simple, this follows from the fact that Γ has Kazhdan's property (T). For the general case, see [**42**].) If we denote by k the field generated by the matrix components of finite generators of Γ, then by the assertion of (ii), k is a real algebraic number field (a finite extension of $\mathbb{Q}$), and it is clear that the matrix components of every element of Γ belong to k. From this, it follows that $\mathbf{G}$ is defined over k and $\Gamma \subset \mathbf{G}_k$. Then applying further the operation called "restriction of scalar" to $\mathbf{G}$, we obtain an algebraic group $\mathbf{H}'$ defined over $\mathbb{Q}$ and a surjective rational homomorphism $\mathbf{H}' \longrightarrow \mathbf{G}$ which induces the isomorphism between $\mathbf{H}'_{\mathbb{Q}}$ and $\mathbf{G}_k$. Taking the Zariski closure of the image of the inclusion map $\Gamma \hookrightarrow \mathbf{H}'$ via the above isomorphism and dividing by its nilpotent radical gives $\mathbf{H}$ in the assertion of (iii), and a surjective rational homomorphism $p : \mathbf{H} \longrightarrow \mathbf{G}$ is induced. Clearly, a homomorphism $\alpha : \Gamma \longrightarrow \mathbf{H}_{\mathbb{Q}}$ exists and satisfies $p \circ \alpha = \mathrm{id}_\Gamma$, and $\alpha(\Gamma)$ is Zariski dense in $\mathbf{H}$. Note that in verifying that $(\mathrm{Ker}\, p)_{\mathbb{R}}$ is compact, one uses the Archimedean superrigidity again.

As for (iv), it is sufficient to show that Γ and $p(\mathbf{H}_{\mathbb{Z}})$ are commensurable, and in proving that $\Gamma \cap p(\mathbf{H}_{\mathbb{Z}})$ has finite index in Γ, one uses the non-Archimedean superrigidity. We now present this argument. If r is an arbitrary prime, then through the natural embedding $\mathbf{H}_{\mathbb{Q}} \longrightarrow \mathbf{H}_{\mathbb{Q}_r}$, we may regard $\alpha : \Gamma \longrightarrow \mathbf{H}_{\mathbb{Q}_r}$. Since Γ is finitely generated, there exist finitely many primes $r_1, \ldots, r_n$ such that for any $\gamma \in \Gamma$, the denominator of every matrix component of $\alpha(\gamma) \in \mathbf{H}_{\mathbb{Q}}$ is a product of powers of r_i. Let $\gamma, \gamma' \in \Gamma$ and assume that $\alpha(\gamma), \alpha(\gamma') \in \mathbf{H}_{\mathbb{Q}_{r_i}}$ are equal modulo $\mathbf{H}_{\mathbb{Z}_{r_i}}$ for each i. Write $\alpha(\gamma) = \alpha(\gamma')h$ in $\mathbf{H}_{\mathbb{Q}}$ and look at this equation in $\mathbf{H}_{\mathbb{Q}_{r_i}}$. Then the assumption and the above fact imply that in fact $h \in \mathbf{H}_{\mathbb{Z}}$. This means that the natural map $\alpha(\Gamma)/\alpha(\Gamma) \cap \mathbf{H}_{\mathbb{Z}} \longrightarrow \prod_{i=1}^n \mathbf{H}_{\mathbb{Q}_{r_i}}/\mathbf{H}_{\mathbb{Z}_{r_i}}$ is injective. On the other hand, by the non-Archimedean superrigidity, for each $1 \leq i \leq n$, $\alpha(\Gamma)$ is relatively compact in $\mathbf{H}_{\mathbb{Q}_{r_i}}$; in other words, the image of $\alpha(\Gamma)$ in $\mathbf{H}_{\mathbb{Q}_{r_i}}/\mathbf{H}_{\mathbb{Z}_{r_i}}$ is finite. Hence, $\alpha(\Gamma) \cap \mathbf{H}_{\mathbb{Z}}$ has finite index in $\alpha(\Gamma)$, and by applying p it follows that $\Gamma \cap p(\mathbf{H}_{\mathbb{Z}})$ has finite index in Γ. In particular, $\Gamma \cap p(\mathbf{H}_{\mathbb{Z}}) \subset p(\mathbf{H}_{\mathbb{Z}})$ is an inclusion between lattices in $\mathbf{G}_{\mathbb{R}}$, and therefore $\Gamma \cap p(\mathbf{H}_{\mathbb{Z}})$ has finite index in $p(\mathbf{H}_{\mathbb{Z}})$ also.

Thus, Γ and $p(\mathbf{H}_{\mathbb{Z}})$ are commensurable. This concludes the outline of the proof of Theorem 1.4.

Finally, as a more recent problem, Ghys [**11**] and Pansu [**41**] point out the possibility of applying fixed-point theorems for isometric actions of discrete groups on Hadamard manifolds and Bruhat-Tits buildings to the problem of nonlinearity of random groups. The argument therein is similar on the whole to the proof of the arithmeticity theorem outlined above. The details will be discussed in §4.

2. Harmonic maps and superrigidity

Application of the harmonic map theory to the superrigidity of lattices started with the work of Corlette [**5**], in which he proved the following Archimedean superrigidity theorem:

THEOREM 2.1. *Let G be either $Sp_{n,1}$ $(n \geq 2)$ or F_4^{-20}, and let Γ be a lattice in G. Suppose that H is a noncompact simple Lie group with trivial center, and that $\rho : \Gamma \longrightarrow H$ is a homomorphism whose image $\rho(\Gamma)$ is Zariski dense in H. Then ρ extends to a continuous homomorphism from G onto H.*

This theorem asserts the superrigidity of lattices in rank one simple Lie groups $Sp(n,1)$ and F^4_{-20}, which is not covered by Theorem 1.3. This theorem, together with the non-Archimedean superrigidity theorem due to Gromov-Schoen, which was proved also by using harmonic maps, implies the arithmeticity of the lattices as in the preceding section.

These works were followed by those of Mok-Siu-Yeung [**35**] and Jost-Yau [**25**], both of which, as a special case, gave a proof of the Margulis superrigidity (Theorem 1.3) for cocompact lattices in the case of $k = \mathbb{R}, \mathbb{C}$ by means of harmonic maps. We note that their method allows us to replace the algebraic group H in Theorem 1.3 by the isometry group of a simply connected complete Riemannian manifold with nonpositive curvature.

In this section we review the theorem of Mok-Siu-Yeung and Jost-Yau and its proof. Let G be a noncompact simple Lie group, and K its maximal compact subgroup (e.g. $G = SL_n(\mathbb{R})$ and $K = SO_n$). Then $X = G/K$ is a manifold diffeomorphic to $\mathbb{R}^n$, and X admits a Riemannian metric g invariant under the left action of G on X. The metric g is uniquely determined up to a constant multiple, and with this metric g, X becomes a Riemannian symmetric space of noncompact type. Throughout this section, we always assume that X is equipped with this metric g. Then X has nonpositive sectional curvature, and is irreducible since G is simple. Here a Riemannian manifold is called irreducible if the universal cover of X does not split as a Riemannian product. (Note that our X is simply connected, and hence the irreducibility means the indecomposability of X into a Riemannian product.) The dimension of the maximal totally geodesic flat submanifold in X is called the *rank* of X. For example, the rank of X is $n-1$ when $G = SL_n(\mathbb{R})$.

A simply connected complete Riemannian manifold with nonpositive sectional curvature (e.g. $X = G/K$ as above) is often called a *Hadamard manifold.* A Hadamard manifold is diffeomorphic to $\mathbb{R}^n$ by the Cartan-Hadamard theorem, thus in particular is contractible. Let Y be a Hadamard manifold. Two geodesic rays $c_1, c_2 : [0, \infty) \longrightarrow Y$ parametrized by the arclength are called *asymptotic* if $d_Y(c_1(t), c_2(t))$ is uniformly bounded independent of t, where $d_Y(\cdot, \cdot)$ denotes the distance of Y defined by the Riemannian metric of Y. This relation becomes an equivalence relation on the set of all geodesic rays on Y. The set of equivalence classes (asymptotic classes of geodesic rays) is denoted by $Y(\infty)$ and called the geometric boundary of Y. For any point $p \in Y$, there exists a unique geodesic ray in each asymptotic class that starts from p. Therefore, $Y(\infty)$ can be identified with the set of all the directions at p. Moreover we may regard a point ξ in $Y(\infty)$ as the point at $t = \infty$ on the geodesic $c(t)$ representing ξ and starting from p. Thus $Y(\infty)$ can be understood as the set of points at infinity of Y. There is a natural way to define a topology on $Y(\infty)$, and the identification with the set of directions as above gives a homeomorphism between $Y(\infty)$ and S^{n-1}. Thus $Y \cup Y(\infty)$ is homeomorphic to a closed ball $\overline{B^n}$ ($n = \dim Y$). It is clear that the asymptotic classes of geodesic rays are preserved under the action of isometries of Y. In fact, isometries of Y extend to homeomorphisms of $Y \cup Y(\infty)$.

THEOREM 2.2 (Jost-Yau [**25**], Mok-Siu-Yeung [**35**]). *Let $X = G/K$ be an irreducible noncompact symmetric space of rank not less than two, and let Γ be a cocompact lattice in G. Let Y be a Hadamard manifold, and $\rho : \Gamma \longrightarrow \mathrm{Isom}(Y)$ a homomorphism, where* $\mathrm{Isom}(Y)$ *denotes the isometry groups of Y. Suppose that $\rho(\Gamma)$ does not fix a point in $Y(\infty)$. Then either of the following holds:*

(i) *$\rho(\Gamma)$ fixes a point in Y.*

(ii) *There exists a ρ-equivariant totally geodesic embedding $f : X \longrightarrow Y$, which is an isometric embedding up to scaling of the metrics.*

Here a map $f : X \longrightarrow Y$ is called *ρ-equivariant* if $f(\gamma x) = \rho(\gamma) f(x)$ holds for any $\gamma \in \Gamma$ and $x \in X$, while f is called *totally geodesic* if f sends the geodesics in X to those in Y. In case (ii) in Theorem 2.2, the action of $\rho(\Gamma)$ on $f(X)$ agrees with that of Γ on X through $f : X \longrightarrow Y$. Theorem 2.2 asserts that the action of Γ on Y is superrigid in the sense we have explained at the begining of §1 as long as the action of Γ does not fix a point in $Y(\infty)$.

Let us check that Theorem 2.2 covers the cases of $k = \mathbb{R}, \mathbb{C}$ in Theorem 1.3. Suppose Y is given as a noncompact symmetric space H/L. By the assumption of Theorem 1.3, which says that $\rho(\Gamma)$ is Zariski dense in H, we conclude that $\rho(\Gamma)$ does not fix a point in $Y(\infty)$. In fact, if $\rho(\Gamma)$ fixes a point ξ in $Y(\infty)$, then $\rho(\Gamma)$ must lie in the isotropy group of ξ, which is an algebraic subgroup (called the parabolic subgroup) of H. Thus, by Theorem 2.2, either (i) or (ii) hold under the assumption of Theorem 1.3. If $\rho(\Gamma)$ fixes a point in Y as claimed in case (i) of Theorem 2.2, then $\rho(\Gamma)$ is relatively compact in H with respect to the Hausdorff topology as we have explained in the preceding section. Note that any isometry of a complete, totally geodesic submanifold in a symmetric space Y extends to that of the whole Y. This shows that ρ extends to a continuous homomorphism $\overline{\rho} : G \longrightarrow H$ in case (ii) in Theorem 2.2.

Now we sketch the proof of Theorem 2.2 following [**35**]. For the details, see the original paper [**35**] or the article [**37**] (written in Japanese). Let Γ, X, Y, ρ be as in Theorem 2.2. Since Y is a Hadamard manifold, it is contractible. Hence there exists a ρ-equivariant map $f : X \longrightarrow Y$, and its equivariant homotopy class is uniquely determined. We first show that there is a good representative in this equivariant homotopy class, which is given as a harmonic map, and then show that the harmonic map from X into Y is totally geodesic.

DEFINITION 2.3. Let F be a fundamental domain for the action of Γ on X, and $f : X \longrightarrow Y$ a ρ-equivariant map. Denote by dv_g the volume form of X. We regard the differential df of f as a section of the bundle $T^*X \otimes f^{-1}TY$, and denote by $\| \cdot \|$ the norm naturally defined from the metric on $T^*X \otimes f^{-1}TY$. The *energy* $E(f)$ of a ρ-equivariant map f is defined by

$$E(f) = \frac{1}{2} \int_F \|df\|^2 \, dv_g.$$

Since the actions of Γ and $\rho(\Gamma)$ are isometric, $E(f)$ is independent of the choice of F. The critical point of the functional E is called a *harmonic map*.

In [**8**], Eells-Sampson introduced the notion of (nonequivariant) harmonic maps, and proved the existence of a harmonic map in each homotopy class of maps from a compact Riemannian manifold M into a compact Riemannian manifold N with nonpositive curvature. They constructed a deformation of a given map $\varphi : M \longrightarrow N$ so that the energy decreases efficiently along the deformation and showed that the deformation converges to a harmonic map at the limit. The energy is a function on the space of maps, and its differential at a map f is given by

$$\Delta f(x) = -\sum_{i=1}^{m} (\nabla_{e_i} df)(e_i),$$

where $x \in M$, $\{e_i\}$ is an orthonormal basis of T_xM, $m = \dim M$, and ∇ denotes the connection on $T^*M \otimes f^{-1}TN$ defined from the Levi-Civita connections on M and N. Thus ∇df is a section of the bundle $T^*M \otimes T^*M \otimes f^{-1}TN$, and Δf is the negative of its trace. Since a harmonic map f is a critical point of the energy, Δf must be 0. To deform a given map φ so that the energy decreases most efficiently, it is appropriate to consider the gradient flow of the energy, namely

$$\frac{\partial f_t}{\partial t}(x) = -\Delta f_t(x) \quad (t \geq 0), \quad f_0 = \varphi.$$

If f_t converges to a map as $t \to \infty$, then the limit should be a critical point of the energy. This is obviously true for a gradient flow of a function on a compact manifold. However, for the gradient flow on the space of maps, even the existence of the gradient flow is nontrivial. Eells-Sampson overcome these difficulties by using the compactness of manifolds and the nonpositivity of the sectional curvature of the target manifold N, and show the existence of a harmonic map. Because of the equation involved, their method is often called the heat flow method. It is possible to follow their arugument in our setting up to some point, and we obtain the following (cf. [**7**], [**29**], [**5**], and [**24**]).

PROPOSITION 2.4. *If $\rho(\Gamma)$ does not fix a point in $Y(\infty)$, then there exists a ρ-equivariant harmonic map $f : X \longrightarrow Y$.*

When $\rho(\Gamma)$ fixes a point $\xi \in Y(\infty)$, then the image $f_t(F)$ of the fundamental domain F under the gradient flow f_t ($t \in [0, \infty)$) may approach ξ. Since ξ is a point at infinity of Y, f_t cannot converge to a map into Y in this case.

In the proof of the proposition above, the nonpositivity of the curvature of Y is essential. However, we do not need to assume X to be an irreducible symmetric space. It suffices to assume that X is a universal covering of a compact manifold, and we can take Γ to be its covering transformation group. On the other hand, for the following proposition, the assumption on X as well as that on Y is crucial.

PROPOSITION 2.5. *Every ρ-equivariant map $f : X \longrightarrow Y$ is totally geodesic.*

These propositions lead us to Theorem 2.2: If $\rho(\Gamma)$ does not fix a point in $Y(\infty)$, then, by Proposition 2.4, there exists a ρ-equivariant harmonic map $f : X \longrightarrow Y$. According to Proposition 2.5, f is a totally geodesic map. Noting that X is irreducible, one sees that f must be either a constant map or an isometric embedding (up to scaling of the metrics). When f is a constant map, then $\rho(\Gamma)$ leaves the point $f(Y)$ invariant by the equivariance, that is, $\rho(\Gamma)$ fixes the point. Therefore, if $\rho(\Gamma)$ does not fix a point in $Y(\infty)$, either $\rho(\Gamma)$ fixes a point in Y or there exists a totally geodesic ρ-equivariant embedding $f : X \longrightarrow Y$.

The key ingredient of the proof of Proposition 2.5 is a certain integral formula involving a tensor of curvature type. Here a $(0,4)$-tensor field Q on a Riemannian manifold is called a tensor of *curvature type* if Q satisfies

$$Q(X,Y,Z,W) = -Q(Y,X,Z,W),$$
$$Q(X,Y,Z,W) = Q(Z,W,X,Y),$$
$$Q(X,Y,Z,W) + Q(X,Z,W,Y) + Q(X,W,Y,Z) = 0,$$

for arbitrary tangent vectors X, Y, Z, and W. We say that Q *has nonpositive curvature* if $Q(X,Y,Y,X) \leq 0$ holds for arbitrary tangent vectors X, Y.

LEMMA 2.6. *Let X be an irreducible noncompact symmetric space of rank greater than one. Then X admits a parallel tensor Q of curvature type satisfying the following.*

(i) *Q is positive definite on the linear space of trace-free $(0,2)$-tensors. Here we define $Q(S,S) = Q_{ijkl}S^{ik}S^{jl}$ for a $(0,2)$-tensor $S = S_{ij}$.*
(ii) *For any tensor T of curvature type with nonpositive curvature, $g(Q,T) \leq 0$ holds. Here the metric on the vector bundle of $(0,4)$-tensors induced from the Riemannian metric g on X is denoted by the same symbol g.*
(iii) *For any ρ-equivariant map $f : X \longrightarrow Y$,*

$$\int_F Q \otimes f^{-1}h^*(\nabla df, \nabla df)\, dv_g = \frac{1}{2}\int_F g(Q, f^{-1}R^Y)\, dv_g$$

holds, where h^* is a metric on the cotangent bundle T^*Y of Y induced from the Riemmanian metric of Y, R^Y is the Riemannian curvature tensor of Y, and f^{-1} is the pull-back of tensor fields on Y by f.

Proposition 2.5 follows from the lemma above: Let f be a ρ-equivariant harmonic map and apply the integral formula in (iii) to f. Since Y has nonpositive curvature, $f^{-1}R^Y$ is a tensor of curvature type with nonpositive curvature on X. By (ii), the right-hand side of the formula in (iii) is nonpositive. On the other hand, since f is a harmonic map, f satisfies $\Delta f = 0$, which means that the trace of ∇df vanishes. Noting this and (i), we see that the left-hand side of the formula in (iii) must be nonnegative. Therefore the left-hand side of the formula in (iii) vanishes. Again by the positive-definiteness (i) of Q, we obtain $\nabla df = 0$, that is, f is totally geodesic.

In the case of $Y = \mathbb{R}$, the argument above gives the vanishing of a de Rham cohomology group as follows. Let $\rho : \Gamma \longrightarrow \mathbb{R}$ be a homomorphism. Then $\rho(\Gamma)$ acts on Y by translations. Assume that $M = \Gamma\backslash X$ is a manifold. A ρ-equivariant map f is a function on X, and the function defined by $x \mapsto f(x) - f(\gamma x)$ is equal to $\rho(\gamma)$ and independent of x. Therefore df is a Γ-invariant 1-form on X. Hence it descends to M and defines a closed 1-form α_f on M. On the other hand, since $H^1(X, \mathbb{R}) = 0$, the lift of any closed 1-form on M to X is exact. From these observations one easily sees that the map that sends ρ to the cohomology class of α_f gives the isomorphism between the linear space of homomorphisms from Γ into $\mathbb{R}$ and $H^1(M,\mathbb{R})$. Finding a ρ-equivariant map minimizing the energy is equivalent to finding the 1-form that minimizes the L^2-norm among the cohomology class corresponding to ρ, that is, finding the harmonic form in the cohomology class. Since M is compact, in any cohomology class, there is a harmonic form α_0. Regarding the integral formula in (iii) of Lemma 2.6 as an integral formula on M, we can apply it to α_0 and conclude that α_0 is a parallel 1-form. Since X is irreducible, a parallel form must be trivial, and hence $H^1(M,\mathbb{R})=0$. This result is the so-called Matsushima vanishing theorem [**33**], and the integral formula in (iii) of Lemma 2.6 is sometimes called the Matsushima formula. In [**33**], the vanishing of the cohomology was proved directly; there was no reason to consider equivariant maps. However, rewriting the proof of Matsushima's vanishing theorem as we have just sketched and replacing the homomorphism into $\mathbb{R}$ by a homomorphism into the isometry group of a Hadamard manifold, we obtain a proof of Theorem 2.2. In this sense, Matsushima's vanishing theorem and its proof should be regarded as an origin of the harmonic map approach to superrigidity theorems.

In the discussion above, we have assumed that X and Y are both manifolds. However, about the same time that [**25**] and [**35**] appeared, Gromov-Schoen [**15**] generalized Corlette's approach to the case that Y is a Bruhat-Tits building. As we have mentioned in the preceding section, a Bruhat-Tits building is a metric space with nonpositive curvature (see the next section), and it arises as the quotient space of an algebraic group over a totally disconnected local field by its maximal compact subgroup. Gromov and Schoen considered a harmonic map into such a singular space and proved that Theorem 2.1 is true even if H in the theorem is replaced by an algebraic group over a p-adic number field:

THEOREM 2.7. *Let G be either $Sp_{n,1}$ $(n\geq 2)$ or F_4^{-20}, and let Γ be a lattice in G. Suppose that H is a simple algebraic group over a p-adic number field, and that $\rho:\Gamma\longrightarrow H$ is a homomorphism whose image $\rho(\Gamma)$ is Zariski dense in H. Then $\rho(\Gamma)$ is relatively compact with respect to the Hausdorff topology.*

Combining this with Theorem 2.1, and repeating the argument mentioned in the preceding section, we see that the lattices in $Sp_{n,1}(n\geq 2)$ and F_4^{-20} are arithmetic.

In the next section, we introduce the notion of the combinatorial harmonic map. Even in the combinatorial setting, an argument analogous to what we have seen above works up to some point. This enables us to show, for a large class of discrete groups including certain lattices, a fixed-point theorem for an isometric action on a metric space of nonpositive curvature.

3. Combinatorial harmonic maps and fixed-point theorems

In this section, we discuss combinatorial harmonic maps and their application to fixed-point theorems. Combinatorial harmonic maps refer to the maps obtained by replacing the source spaces from Riemannian manifolds to (weighted) simplicial complexes in the definition of harmonic maps. As the target spaces, we consider metric spaces called CAT(0) spaces, generalizing Hadamard manifolds.

Let us briefly recall the basics of CAT(0) spaces. A metric space (Y,d) is said to be a *geodesic space* if for any two points $p,q\in Y$ there exists a geodesic joining

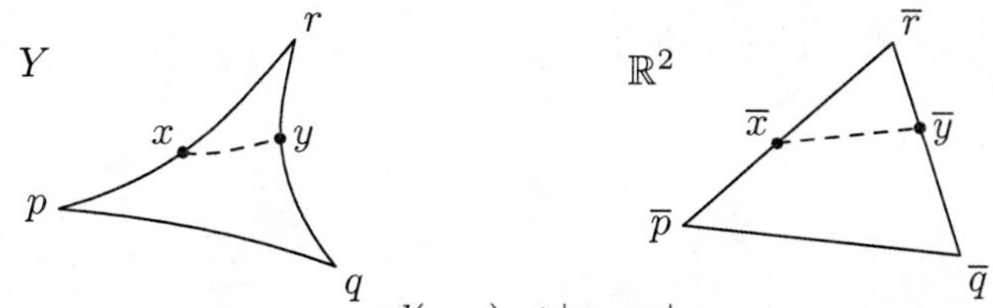

FIGURE 1

p and q, that is, a curve $c : [a,b] \longrightarrow Y$ satisfying $c(a) = p$, $c(b) = q$ and

$$d(c(t), c(t')) = |t - t'|, \quad t, t' \in [a,b].$$

Let $\triangle$ be a geodesic triangle in Y; that is, $\triangle$ is a triple of points $p, q, r \in Y$ with a triple of geodesic segments joining them specified (see Figure 1). Corresponding to $\triangle$, take a (possibly degenerate) triangle $\overline{\triangle} = \overline{p}\,\overline{q}\,\overline{r}$ in the plane $\mathbb{R}^2$, called a comparison triangle, so that

$$d(p,q) = |\overline{p} - \overline{q}|, \quad d(q,r) = |\overline{q} - \overline{r}|, \quad d(r,p) = |\overline{r} - \overline{p}|.$$

Then we say that $\triangle$ satisfies the CAT(0) condition if for any two points x, y on the circumference of $\triangle$ and the corresponding two points $\overline{x}, \overline{y} \in \overline{\triangle}$, the inequality

$$d(x,y) \leq |\overline{x} - \overline{y}|$$

holds. A geodesic space (Y, d) is said to be a *CAT*(0) *space* if any geodesic triangle in Y satisfies the CAT(0) condition. In a CAT(0) space, a geodesic joining any two points is unique. Moreover, a CAT(0) space is contractible. Throughout this section, we assume that a CAT(0) space is always complete.

Henceforth, let (Y, d) be a CAT(0) space.

The notion of barycenter in a CAT(0) space is fundamental, and it plays an important role also in the theory of harmonic maps.

DEFINITION 3.1. Let Y be a CAT(0) space, and let $p_1, \ldots, p_m \in Y$ and $t_1, \ldots, t_m \in \mathbb{R}_{>0}$ with $\sum_{i=1}^m t_i = 1$ be given. Then the function

$$Y \ni q \;\mapsto\; F(q) = \sum_{i=1}^{m} t_i\, d(p_i, q)^2 \in \mathbb{R}$$

is strictly convex, and therefore there exists a unique point $p \in Y$ minimizing F. We call the point p the *barycenter* of $\{p_1, \ldots, p_m\}$ with weight $\{t_1, \ldots, t_m\}$.

For a point $p \in Y$, one can define the tangent cone TC_pY of Y at p as well as the "inner product" $\langle \cdot, \cdot \rangle_p$ and the distance d_p on it. We outline the definitions of these (for the details, see [**2**], [**4**]). First, define the angle $\angle_p(c, c')$ formed by nontrivial geodesics c, c' emanating at $p \in Y$ as the limit of the interior angle at the vertex $\overline{p}$ of the comparison triangle $\overline{p}\,\overline{c(t)}\,\overline{c(t')}$ as $t, t' \to +0$. Then equip an equivalence relation on the set of all nontrivial geodesics emanating at p, by defining two such geodesics to be equivalent if the angle formed by them at p is 0. We denote the completion of the set of all equivalence classes with respect to the distance induced from $\angle_p(\cdot, \cdot)$ by S_pY, and call it the *space of directions* at p. We call the cone

$$TC_pY = S_pY \times [0, \infty)/S_pY \times \{0\}$$

over S_pY the *tangent cone* of Y at p. The "inner product" $\langle \cdot, \cdot \rangle_p$ and the distance d_p on TC_pY are defined by

$$\langle ([c], t), ([c'], t') \rangle_p = tt' \cos \angle_p(c, c')$$

and

$$d_p(([c], t), ([c'], t'))^2 = t^2 + t'^2 - 2\langle ([c], t), ([c'], t') \rangle_p$$

respectively, where $[c]$ denotes the equivalence class containing the geodesic c. (Therefore, precisely speaking, these expressions make sense only on a dense subset of TC_pY.) Note that (TC_pY, d_p) is a CAT(0) space again. Finally, define a map $\pi_p : Y \longrightarrow TC_pY$ by $\pi_p(q) = ([c_{p,q}], d_Y(p, q))$, where $c_{p,q}$ denotes the geodesic joining p and q. Note that π_p is distance-decreasing.

We now give some typical examples of CAT(0) spaces.

EXAMPLE 1. (1) (Hadamard manifolds) Hadamard manifolds provide the most fundamental examples of CAT(0) spaces. (The underlying manifolds may be Hilbert manifolds. In this case, completeness should be understood to mean complete as metric spaces.)

(2) (Trees) Let Y be a locally finite tree and assume that every vertex is the common endpoint of at least three edges. With the distance such that the length of each edge is say one, Y becomes a CAT(0) space. The tangent cone at an interior point of an edge is isometric to a line. On the other hand, if p is a vertex, then TC_pY is isometric to the union of finitely many half-lines with all endpoints identified, and the angle between $v, w \in TC_pY$ is given by

$$\angle_p(v, w) = \begin{cases} 0, & \text{if } v,\ w \text{ belong to the same half-line,} \\ \pi, & \text{otherwise.} \end{cases}$$

(3) (Bruhat-Tits buildings) A building is a simplicial complex for which a family of subcomplexes with certain properties, called apartments, is specified, and each apartment is required to be isomorphic to a Coxeter complex, the simplicial complex associated with a Coxeter group (cf. [**3**], [**43**]). A building is called a Bruhat-Tits building if the corresponding Coxeter complex is a triangulation of a Euclidean space. A Bruhat-Tits building becomes a CAT(0) space if one defines a distance by transplanting the Euclidean distance of the corresponding Coxeter complex to each apartment.

A one-dimensional Bruhat-Tits building is nothing but a tree such that every vertex is the common endpoint of at least two edges.

As examples of two-dimensional Bruhat-Tits buildings, the "r-adic symmetric spaces"

$$Y = PGL_3(\mathbb{Q}_r)/PGL_3(\mathbb{Z}_r)$$

are well known, where r runs over prime numbers. The corresponding Coxeter complex is the triangulation of the plane as in Figure 2, and the Coxeter group is the Euclidean reflection group, called a group of type $\widetilde{A}_2$, generated by the reflections with respect to the lines in the picture.

When $r = 2$, the link of each vertex of Y is a bipartite graph as in Figure 3.

One finds various hexagons in Figure 3. These hexagons represent the sections of the link by the apartments, and correspond to the regular hexagon which is the link of each vertex in the Coxeter complex.

FIGURE 2

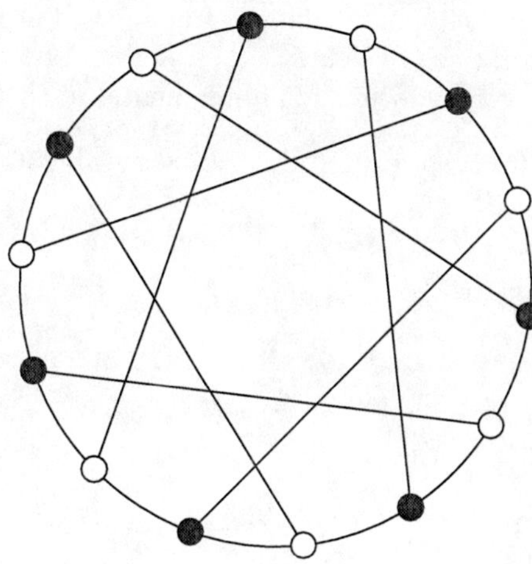

FIGURE 3

If one defines a distance on the graph by specifying the length of each edge to be $\pi/3$, the tangent cone of a vertex of Y is isometric to the metric cone over this metric graph.

We now define the energy of an equivariant map from a simplicial complex into a CAT(0) space. Let X be a locally finite simplicial complex, Γ a finitely generated group, and suppose that Γ acts simplicially, properly continuously and cofinitely on X. For each $r \geq 0$, we denote the set of all r simplices by $X(r)$ and denote a representative system of the Γ-action on $X(r)$ by $\mathcal{F}(r)$. In what follows, we assume for the sake of simplicity that the quotient space $\Gamma\backslash X$ is a finite simplicial complex such that the projection $X \longrightarrow \Gamma\backslash X$ is a local isomorphism (in particular, the action of Γ on X is free). Let Y be a CAT(0) space, and let $\rho : \Gamma \longrightarrow \mathrm{Isom}(Y)$ be a homomorphism, where $\mathrm{Isom}(Y)$ denotes the isometry group of Y. Then we define the *energy* $E(f)$ of a ρ-equivariant map $f : X(0) \longrightarrow Y$ by

$$E(f) = \sum_{\{x,y\}\in\mathcal{F}(1)} m(x,y) d_Y(f(x), f(y))^2,$$

where $m(x,y)$ denotes the number of 2-simplices containing the edge $\{x,y\}$ (cf. [**47**], [**48**], [**21**], [**28**], [**30**]).

Let $f : X(0) \longrightarrow Y$ be a ρ-equivariant map. For each $x \in X(0)$, we define $F_x : (\mathrm{Lk}\,x)(0) \longrightarrow TC_{f(x)}Y$ by $F_x(y) = \pi_{f(x)}(f(y))$. If f is a map minimizing the functional E, then it can be shown that for each $x \in X(0)$ the barycenter of $\{F_x(y) \mid y \in (\mathrm{Lk}\,x)(0)\}$ with weight $\{m(x,y)/m(x) \mid y \in (\mathrm{Lk}\,x)(0)\}$ coincides with the origin $0_{f(x)}$ of $TC_{f(x)}Y$, where $m(x) = \sum_{y\in(\mathrm{Lk}\,x)(0)} m(x,y)$. We then define a harmonic map as follows.

DEFINITION 3.2. Let $f : X(0) \longrightarrow Y$ be a ρ-equivariant map, and denote the barycenter of $\{F_x(y) \mid y \in (\mathrm{Lk}\,x)(0)\}$ with weight $\{m(x,y)/m(x) \mid y \in (\mathrm{Lk}\,x)(0)\}$ by $-\Delta f(x)$ $(\in TC_{f(x)}Y)$. We say that f is *harmonic* if $-\Delta f(x) = 0_{f(x)}$ holds for all $x \in \mathcal{F}(0)$.

Gromov's definition of harmonicity of an equivariant map [**14**] does not refer to the tangent cone, and so it appears different from ours, but it can be shown that the two definitions are in fact equivalent.

The following Bochner-type formula is fundamental in the argument hereafter.

PROPOSITION 3.3. *Let $f : X(0) \longrightarrow Y$ be a ρ-equivariant map. Then the following formula holds:*

(3.1)

$$\begin{aligned}
&\sum_{x\in\mathcal{F}(0)} m(x)\,|-\Delta f(x)|^2_{f(x)} \\
&\quad= \sum_{x\in\mathcal{F}(0)} \Bigg[2 \sum_{\{y,y'\}\in(\mathrm{Lk}\,x)(1)} d_{f(x)}(F_x(y),F_x(y'))^2 \\
&\qquad\qquad - \sum_{y\in(\mathrm{Lk}\,x)(0)} m(x,y)\, d_{f(x)}(-\Delta f(x),F_x(y))^2 \\
&\qquad\qquad +2 \sum_{\{y,y'\}\in(\mathrm{Lk}\,x)(1)} \left\{ d_Y(f(y),f(y'))^2 - d_{f(x)}(F_x(y),F_x(y'))^2 \right\} \Bigg].
\end{aligned}$$

Note that since the map $\pi_{f(x)} : Y \longrightarrow TC_{f(x)}Y$ is distance-decreasing, we have

$$d_Y(f(y),f(y')) \geq d_{f(x)}(F_x(y),F_x(y')), \quad \{y,y'\} \in (\mathrm{Lk}\,x)(1),$$

and hence the third term in [ ] of (3.1) is nonnegative.

Garland [**10**] carried out a computation equivalent to the derivation of the linear version (the $Y = \mathbb{R}$ case) of (3.1) when X is a Bruhat-Tits building and states that the resulting formula is a discrete analogue of the Matsushima formula mentioned in §2 (see also [**1**], [**40**], [**21**]).

To formulate a fixed-point theorem, we introduce the following invariant.

DEFINITION 3.4. For $x \in X(0)$ and a CAT(0) space T (it is typically a tangent cone of a CAT(0) space), define

$$\lambda_1(\mathrm{Lk}\,x,T) = \inf_{\varphi} \frac{\displaystyle\sum_{\{y,z\}\in(\mathrm{Lk}\,x)(1)} d_T(\varphi(y),\varphi(z))^2}{\displaystyle\sum_{y\in(\mathrm{Lk}\,x)(0)} m(x,y)\, d_T(\overline{\varphi},\varphi(y))^2},$$

where $\overline{\varphi}$ denotes the barycenter of $\{\varphi(y) \mid y \in (\mathrm{Lk}\, x)(0)\}$ with weight $\{m(x,y)/m(x) \mid y \in (\mathrm{Lk}\, x)(0)\}$, and the infimum is taken over all nonconstant maps $\varphi : (\mathrm{Lk}\, x)(0) \longrightarrow T$.

This invariant was introduced by Wang [**48**]. For $x \in X(0)$, the discrete Laplacian Δ on the 1-skeleton (which is a graph) of $\mathrm{Lk}\, x$ is defined by

$$(\Delta\varphi)(y) = \varphi(y) - \frac{1}{\deg(y)} \sum_{z \sim y} \varphi(z)$$

for functions $\varphi : (\mathrm{Lk}\, x)(0) \longrightarrow \mathbb{R}$. The first eigenvalue of Δ is zero, and we therefore denote the second eigenvalue by $\mu_1(\mathrm{Lk}\, x)$. Clearly, $\lambda_1(\mathrm{Lk}\, x, \mathbb{R}) = \mu_1(\mathrm{Lk}\, x)$, and more generally we have $\lambda_1(\mathrm{Lk}\, x, \mathcal{H}) = \mu_1(\mathrm{Lk}\, x)$ for any Hilbert space $\mathcal{H}$.

The following is the main theorem of this section, and it was proved by Gromov [**14**] and the authors [**21**] independently.

THEOREM 3.5. *Let X be a locally finite simplicial complex. Let Γ be a finitely generated group, and suppose that it acts simplicially, properly discontinuously and cofinitely on X. Let Y be a* CAT(0) *space. Assume that there exists a constant $C > 1/2$ such that $\lambda_1(\mathrm{Lk}\, x, TC_pY) \geq C$ holds for all $x \in X(0)$ and $p \in Y$. Then for any homomorphism $\rho : \Gamma \longrightarrow \mathrm{Isom}(Y)$, $\rho(\Gamma)$ has a fixed point in Y.*

Note that the assumption on λ_1 in the theorem involves the local properties of the spaces X, Y only, and it is totally independent of the homomorphism ρ.

The theorem was proved by Wang [**48**] also under the assumption that Y is locally compact and ρ is reductive (in the sense of Jost [**23**]). Indeed, the argument in that case is rather similar to that in the proof of Theorem 2.2. Under these assumptions, the existence of a map minimizing E can be easily verified, and therefore the assertion of the theorem follows by applying the Bochner-type formula to this map.

We shall now outline the proof of Theorem 3.5 following [**21**]. First, if we set $\mathcal{M} = \{f : X(0) \longrightarrow Y \mid \rho\text{-equivariant maps}\}$, then $\mathcal{M}$ is a CAT(0) space, and E is a convex function on $\mathcal{M}$. Jost [**22**] and Mayer [**34**] defined a gradient flow for a convex function on a CAT(0) space, and in the case of our E it is described as follows. For any $f_0 \in \mathcal{M}$ and $h > 0$, there exists a unique point which attains the minimum of the function

$$\mathcal{M} \ni f \mapsto E(f) + \frac{1}{2h} d_{\mathcal{M}}(f, f_0)^2 \in \mathbb{R},$$

and we denote this point by $J_h(f_0)$. The gradient flow of E starting from f_0 is defined by

$$f(0) = f_0, \quad f(t) = \lim_{n \to \infty} \left(J_{t/n}\right)^n (f_0) \quad (t > 0).$$

For $f \in \mathcal{M}$, it is appropriate to interpret $(-\mathrm{grad}E)(f) = (2(-\Delta f(x)))_{x \in \mathcal{F}(0)}$. Then by the Bochner-type formula (3.1) and the assumption on λ_1, we obtain a constant $C > 0$ such that $|(-\mathrm{grad}E)(f)|^2 \geq CE(f)$ holds for all $f \in \mathcal{M}$. If Y is smooth, then so are $\mathcal{M}$ and E, and the usual gradient flow $f(t)$ of E satisfies the differential equation $f'(t) = (-\mathrm{grad}E)(f(t))$. Hence,

$$\begin{aligned} -\frac{d}{dt} E(f(t)) &= \langle (-\mathrm{grad}E)(f(t)), f'(t) \rangle \\ &= |(-\mathrm{grad}E)(f(t))|^2 \\ &\geq CE(f(t)); \end{aligned}$$

namely, $E(t) = E(f(t))$ satisfies the differential inequality

$$E'(t) \leq -CE(t). \tag{3.2}$$

From this, we obtain the exponential decay estimate of the energy

$$E(t) \leq C'e^{-Ct}. \tag{3.3}$$

Moreover, noting that

$$|f'(\tau)| = |(-\mathrm{grad}E)(\tau)| = \sqrt{-E'(\tau)},$$

we obtain for $t < t'$,

$$\begin{aligned} d_{\mathcal{M}}(f(t), f(t')) &\leq \int_t^{t'} |f'(\tau)|\, d\tau \\ &= \int_t^{t'} \sqrt{-E'(\tau)}\, d\tau \\ &= -2\int_t^{t'} \frac{\left(\sqrt{E(\tau)}\right)'}{\sqrt{-E'(\tau)/E(\tau)}}\, d\tau \\ &\leq C''e^{-Ct/2}. \end{aligned} \tag{3.4}$$

If Y is a general CAT(0) space, it can be proved that $E(t)$ is Lipschitz continuous and satisfies the differential inequality (3.2) almost everywhere. From this, (3.3) follows, and we can verify (3.4) by a slight modification of the above argument. Hence, for any divergent sequence of times $\{t_j\}_{j=1}^{\infty}$, $\{f(t_j)\}_{j=1}^{\infty}$ is a Cauchy sequence in $\mathcal{M}$, and therefore it has a limit f_∞. By (3.3), we obtain $E(f_\infty) = 0$, that is, f_∞ is a constant map. By the equivariance of f_∞, $\rho(\Gamma)$ fixes the image of f_∞, which is a single point of Y. This concludes the outline of the proof of Theorem 3.5.

Now, in order to derive more concrete fixed-point theorems from Theorem 3.5, we need to compute or estimate from below the invariant λ_1. For this purpose, we first introduce an invariant for a CAT(0) space (particularly for its tangent cones) which measures the difference of the barycenter therein and that in Euclidean spaces.

Definition 3.6. Let T be a CAT(0) space. Suppose that collections of points $\{v_1, \dots, v_m\} \subset T$ and positive real numbers $\{t_1, \dots, t_m\}$ with $\sum_{i=1}^m t_i = 1$ are given, and let $\overline{v}$ denote the barycenter of $\{v_i\}$ with weight $\{t_i\}$. A *realization* of $\{v_i\}$ is a collection of m vectors $\{\mathbf{v}_1, \dots, \mathbf{v}_m\}$ in the Euclidean space $\mathbb{R}^m$ such that

$$||\mathbf{v}_i|| = d_T(\overline{v}, v_i), \quad ||\mathbf{v}_i - \mathbf{v}_j|| \leq d_T(v_i, v_j).$$

We set

$$\delta(\{v_i\}, \{t_i\}) = \inf \left[\left|\left| \sum_{i=1}^m t_i \mathbf{v}_i \right|\right|^2 \Big/ \sum_{i=1}^m t_i ||\mathbf{v}_i||^2 \right] \in [0, 1],$$

where the infimum is taken over all realizations $\{\mathbf{v}_i\}$ of $\{v_i\}$. We then define

$$\delta(T) = \sup\ \delta(\{v_i\}, \{t_i\}),$$

where the supremum is taken over all collections $\{v_i\}$, $\{t_i\}$. Here, if we restrict the choices of $\{v_i\}$, $\{t_i\}$ to those with barycenter at a given $v \in T$, we denote the corresponding number by $\delta(T, v)$. Clearly $0 \leq \delta(T, v) \leq \delta(T) \leq 1$ and $\delta(T) = \sup_{v \in T} \delta(T, v)$. We say that T is *flexible* if $\delta(T) = 0$.

PROPOSITION 3.7. *Let T be a* CAT(0) *sapce. Then we have*

$$(1-\delta(T))\mu_1(\mathrm{Lk}\,x) \le \lambda_1(\mathrm{Lk}\,x, T) \le \mu_1(\mathrm{Lk}\,x).$$

In paticular, if T is flexible, then we have $\lambda_1(\mathrm{Lk}\,x, T) = \mu_1(\mathrm{Lk}\,x)$.

By using this proposition, we obtain the following as a special case of Theorem 3.5.

THEOREM 3.8. *Let X be a locally finite simplicial complex, and assume that $\mu_1(\mathrm{Lk}\,x) > 1/2$ holds for all $x \in X(0)$. Let Γ be a finitely generated group, and suppose that it acts simplicially, properly discontinuously and cofinitely on X. Let Y be a* CAT(0) *space, and assume that its tangent cones are all flexible. Then for any homomorphism $\rho : \Gamma \longrightarrow \mathrm{Isom}(Y)$, $\rho(\Gamma)$ has a fixed point in Y.*

There are plenty of simplicial complexes which satisfy the assumptions of the theorem (cf. [**1**], [**52**]). For example, if $X = PGL_3(\mathbb{Q}_p)/PGL_3(\mathbb{Z}_p)$, the "$p$-adic symmetric space", then we have $\mu_1(\mathrm{Lk}\,x) = 1 - \sqrt{p}/(p+1) > 1/2$ [**9**].

Clearly, a tangent cone of a CAT(0) space is flexible if it is isometric to a Hilbert space, and in particular, if Y is a Hadamard manifold, then any isometric action of Γ satisfying the conditions of Theorem 3.8 on Y has a fixed point. This result (and Gromov's result to be mentioned later) gives a negative answer to the following question of Gromov [**13**, p. 16]:

> Is a finitely generated group with finite cohomological dimension necessarily the fundamental group of a complete Riemannian manifold with nonpositive sectional curvature?

In fact, as Γ, it is sufficient to take, for example, a uniform lattice in $PGL_n(\mathbb{Q}_p)$ which contains no element of finite order other than the identity element.

As a further special case, if we choose Y to be a Hilbert space, then any isometric action of Γ satisfying the conditions of Theorem 3.8 on a Hilbert space has a fixed point, and therefore Γ has Kazhdan's property (T). This fact was proved by Ballmann-Świątkowski [**1**, Corollary 1] and Żuk [**51**, Theorem 1].

As an interesting example of an infinite-dimensional Hadamard manifold, the universal Teichmüller space is known. The universal Teichmüller space had been traditionally treated as a Banach manifold; Takhtajan-Teo [**45**] defined a Hilbert-manifold structure and a Riemannian metric on the universal Teichmüller space. They computed the curvature of the metric and verified that the sectional curvature is negative. As a Hilbert manifold, the universal Teichmüller space has infinitely many connected components, and therefore it cannot be a geodesic space. However, each connected component is a geodesic space, and therefore it is a CAT(0) space. So by applying the theorem to the connected component containing $\mathrm{Diff}_+(S^1)/\mathrm{M\ddot{o}b}(S^1)$ and supplementing some more arguments, one can show that any homomorphism from Γ satisfying the conditions of Theorem 3.8 to $\mathrm{Diff}_+(S^1)$ has a finite image. On the other hand, Navas [**36**] proved more generally that any homomorphism from a group with Kazhdan's property (T) to $\mathrm{Diff}_+(S^1)$ has a finite image.

Now, as the following example shows, a tangent cone of CAT(0) space can be flexible if it is not isometric to a Hilbert space.

EXAMPLE 2. (Trees) Let Y be the tree as in Example 1. The tangent cone at an interior point of an edge is isometric to the line, and therefore it is flexible. On the other hand, the tangent cone TC_pY at a vertex p is isometric to the union of finitely many half-lines with endpoints identified, and one can show that this also is flexible. For the sake of simplicity, suppose that the number of half-lines is three, and denote them by H_1, H_2, H_3. Furthermore, suppose that $m = 3$, $v_1 \in H_1$, $v_2 \in H_2$, $v_3 \in H_3$, $t_1 = t_2 = t_3 = 1/3$, and assume that $\overline{v} = 0$ (0 is the origin of TC_pY). We now show that $\delta(\{v_i\}, \{t_i\}) = 0$ in this case. If $e \in H_1$ is a unit vector, then for $s > 0$ we have

$$\begin{aligned} F(se) &= \sum_{i=1}^{3} \frac{1}{3} d(se, v_i)^2 \\ &= \frac{1}{3}\{(|v_1| - s)^2 + (|v_2| + s)^2 + (|v_3| + s)^2\} \\ &= F(0) + \frac{2s}{3}(-|v_1| + |v_2| + |v_3|) + s^2, \end{aligned}$$

and since 0 is the barycenter, we obtain $|v_1| \leq |v_2| + |v_3|$ and two similar inequalities. These inequalities mean that there exists a (possibly degenerate) triangle on $\mathbb{R}^2$ such that the lengths of three sides are $|v_1|$, $|v_2|$, $|v_3|$, and this means furthermore that there exists a realization $\{\mathbf{v}_i\}$ of $\{v_i\}$ satisfying $\sum_{i=1}^{3} \mathbf{v}_i = \mathbf{0}$. Thus $\delta(\{v_i\}, \{t_i\}) = 0$.

Since only collections of finitely many points of T are considered in Definition 3.6, the last argument holds true if the number of half-lines is infinite. Therefore, the tangent cones of an $\mathbb{R}$-tree are flexible also.

$\mathbb{R}$-trees are obtained also as the ultralimits of a sequence of negatively curved manifolds (cf. [**26**]). For example, let $(\mathbb{H}^2, d_{\mathbb{H}^2})$ be the hyperbolic plane ($d_{\mathbb{H}^2}$ denotes the Poincaré distance), and set $Y_j = (\mathbb{H}^2, \frac{1}{j} d_{\mathbb{H}^2})$ for each positive integer j. Then for any nonprincipal ultrafilter ω on $\mathbb{N}$, the ultralimit Y_ω of $\{Y_j\}_{j \in \mathbb{N}}$ with respect to ω becomes an $\mathbb{R}$-tree. The flexibility of the tangent cones of an $\mathbb{R}$-tree follows also from this fact together with the following proposition.

PROPOSITION 3.9. *Let $\{Y_j\}_{j \in \mathbb{N}}$ be a sequence of* CAT(0) *spaces. Let ω be a nonprincipal ultrafilter on $\mathbb{N}$, and let Y_ω be the ultralimit of $\{Y_j\}_{j \in \mathbb{N}}$ with respect to ω. Then we have $\delta(Y_\omega) \leq \limsup_{j \to \infty} \delta(Y_j)$.*

It follows from this proposition, in particular, that if each Y_j is a Hadamard manifold, then for any $p \in Y_\omega$, TC_pY_ω is flexible. Moreover, using this, one can prove the following result of Gromov [**14**]. Gromov defined a class $\overline{\mathcal{C}}_{\text{reg}}$ of CAT(0) spaces as the minimal class which contains all Hadamard manifolds and is closed under taking ultralimits as well as closed convex subsets, and obtained the following.

THEOREM 3.10 (Gromov [**14**]). *Let X be a locally finite simplicial complex, and assume that $\mu_1(\mathrm{Lk}\, x) > 1/2$ holds for all $x \in X(0)$. Let Γ be a finitely generated group, and suppose that it acts simplicially, properly discontinuously and cofinitely on X. Let Y be a* CAT(0) *space belonging to the class $\overline{\mathcal{C}}_{\text{reg}}$. Then for any homomorphism $\rho : \Gamma \longrightarrow \mathrm{Isom}(Y)$, $\rho(\Gamma)$ has a fixed point in Y.*

The class $\overline{\mathcal{C}}_{\text{reg}}$ as well as the class of CAT(0) spaces all of whose tangent cones are flexible are nice classes, as they contain singular or non-locally-compact CAT(0) spaces and still one can formulate a fixed-point theorem of a general form for these classes. On the other hand, these classes should be very small in the class of all CAT(0) spaces, as for example, most Bruhat-Tits buildings do not belong to these classes.

EXAMPLE 3. Let Y be the "r-adic symmetric space" of Example 1. If $p \in Y$ is not a vertex, TC_pY is flexible. So in what follows let $p \in Y$ be a vertex. Recall that TC_pY is isometric to the cone over the metric graph obtained by assigning length $\pi/3$ to each edge of the link of p. Then one can show that

$$\delta(TC_pY, 0_p) \geq 1 - \frac{1}{2(1-\sqrt{r}/(r+1))} = \frac{(\sqrt{r}-1)^2}{2(r-\sqrt{r}+1)}, \tag{3.5}$$

and in particular, TC_pY is not flexible. Note that since $\delta(TC_pY, v) = 0$ when $v \neq 0_p$, we have $\delta(TC_pY) = \delta(TC_pY, 0_p)$.

We conjecture that the equality holds in (3.5), and we did a related investigation in [**21**]. Also, we gave a meaningful upper estimate of $\delta(TC_pY)$ when $r = 2$, and using this together with Proposition 3.7 and Theorem 3.5, we obtained a fixed-point theorem for isometric actions of a certain class of finitely generated groups on the "2-adic symmetric space".

4. A fixed-point theorem for random groups

As we have seen, the proof of Theorem 3.5 was given by tracing over a part of the proof of Theorem 2.2 in the combinatorial setting. We note here that, however, the integral formula in Lemma 2.6, which is a key ingredient of the proof of Theorem 2.2, is a formula for the symmetric space associated to the Lie group containing the given lattice Γ. In other words, the superrigidity of a lattice is deduced from the geometry of a continuous object associated to the lattice. On the other hand, in the fixed-point theorem described in the preceding section, we do not assume the existence of a continuous geometric object which contains the given discrete group Γ. Although we assume the existence of a simplicial complex on which Γ acts in a convenient way, we can easily construct such a complex for a finitely presented group from a suitably chosen presentation of the group. Moreover, by generalizing the argument presented in the preceding section, we are able to treat equivariant maps from a given finitely generated group into a CAT(0) space. In this case, the quantity corresponding to $\mu_1(\mathrm{Lk}\, x)$ is determined completely by the presentation of the discrete group. This suggests that our combinatorial harmonic map may be a useful tool for investigating finitely generated groups. In this section, we present a result on random groups as an example of the study in this direction.

Let $S = \{s_1, \ldots, s_m, s_1^{-1}, \ldots, s_m^{-1}\}$ be a set of $2m$ elements such that if $s \in S$, then $s^{-1} \in S$. A sequence $r = s_{i_1}^{\pm 1} \cdots s_{i_l}^{\pm 1}$ consisting of elements of S is called a *word*, where $i_j \in \{1, \ldots, m\}$ for $j = 1, \ldots, l$, and l is called the *length* of the word r. A word r is said to be *reduced* if r does not contain a sequence of the form $s_k s_k^{-1}$ or $s_k^{-1} s_k$. Let R be a subset of the set of reduced words. A pair $P = (S, R)$ is called a *presentation*, where S is a set of *generators* and R is a set of *relations*. We denote by $\Gamma(P)$ the group defined by $P = (S, R)$, that is, the quotient of the free group $F_S = \langle s_1 \rangle * \cdots * \langle s_m \rangle$ generated from S by the normal closure $\overline{R}$ of R. Here the normal closure of R is the minimal normal subgroup containing R.

Let R_3 denote the subset of R consisting of words with length 3. Note that R_3 is a set with only finitely many elements since S is a finite set. We define a graph $L(P)$ from a given presentation P as follows:

$$\begin{cases} \{\text{vertices of } L(P)\} = S, \\ \{\text{edges of } L(P)\} = \{\{s_a^{-1}, s_b\}, \{s_b^{-1}, s_c\}, \{s_c^{-1}, s_a\} \mid s_a s_b s_c \in R_3\}. \end{cases}$$

Since $L(P)$ may have loops and multiple edges, $L(P)$ may not be a simplicial complex. However, we can define $\lambda_1(L(P), TC_pY)$ for $L(P)$ in the same manner as we did for a link of a simplicial complex. Then we have

THEOREM 4.1 ([**18**]). *Let $\Gamma(P)$ be a finitely generated group defined by a presentation P, and let Y be a* CAT(0) *space. Suppose there exists a constant $C > 1/2$ such that $\lambda_1(L(P), TC_pY) \geq C$ holds for every $p \in Y$. Then for any homomorphism $\rho : \Gamma(P) \longrightarrow \mathrm{Isom}(Y)$, $\rho(\Gamma(P))$ has a fixed point in Y.*

If there is no element $s \in S$ such that s becomes the identity in $\Gamma(P)$, then it is easy to construct a simplicial complex X having $\Gamma(P)$ as the set of vertices. Then $\Gamma(P)$ acts on X simplicially, properly discontinuously, and cofinitely. Using the notation in the preceding section, we have $\mathcal{F}(0) = \{e\}$ and $L(P)$ turns out to be $\mathrm{Lk}\, e$ essentially. Therefore, for a presentation P which does not contain an element corresponding to the identity, Theorem 4.1 essentially follows from Theorem 3.5. On the other hand, when S contains an element corresponding to the identity, then $L(P)$ is not a simplicial complex, and it seems quite difficult to construct a simplicial complex on which $\Gamma(P)$ acts directly from P. However, when a finitely presented group acts on a countable set equipped with a certain weight function compatible with the action, which is similar to the admissible weight m in the preceding section, we can trace over the argument we have explained in the preceding section, and prove a result analogous to Theorem 3.5 under this setting. The proof of Theorem 4.1 is given by taking $\Gamma(P)$ itself as X and defining a weight function from P suitably.

According to Theorem 4.1, we see that if $\mu_1(L(P)) > 1/2$, then $\Gamma(P)$ has the fixed-point property for Hilbert spaces, that is, $\Gamma(P)$ has Kazhdan's property (T). Żuk proved this result by a different argument. (See [**52**, Theorem 1, Proposition 6]. Our $L(P)$ is denoted by $L(S)$ in [**52**].) Also Ballmann-Świątkowski constructed plenty of groups having Kazhdan's property (T) as an application of [**1**, Corollary 1], which we have mentioned in the preceding section.

Before these works appeared, most of the known examples of discrete groups with Kazhdan's property (T) were lattices in Lie groups or algebraic groups, and the property (T) of these lattices is inherited from the ambient Lie groups (or algebraic groups). This fact, together with the definition and fixed-point properties following from Kazhdan's property (T), might make one suspect that a group with Kazhdan's property (T) is exotic. However, some recent results, such as [**52**, Theorem 1, Proposition 6] that we have just mentioned, give sufficient conditions for a finitely generated group to have Kazhdan's property (T), which can be checked from the presentation of a group. This enables us to show that groups with Kazhdan's property (T) are, in a sense, generic; Żuk proved a result, stated using the notion of random groups, which says that, in a certain class of finitely presented groups, the probability of the occurrence of groups with Kazhdan's property (T) is considerably large.

Let us consider a presentation $P = (S, R)$ that satisfies $R = R_3$, that is, all the relations of P have length 3. If the number of the generators is $2m$, then the number of reduced words with length 3 equals $2m(2m-1)^2$, and almost $(2m-1)^3$ as m gets large. Let us denote by $P_{\mathcal{M}}(m, d)$ the set of presentations such that the logarithmic density of R in the set of all length-three words is about $0 < d < 1$:

$$P_{\mathcal{M}}(m, d) = \{P = (S, R) \mid \#S = 2m, c^{-1}(2m-1)^{3d} \leq \#R \leq c(2m-1)^{3d}\},$$

where $c > 1$ is an arbitrary constant, and the result stated below does not depend on the choice of c.

If the set of relations R is empty in a given presentation $P = (S, R)$, then the group we obtain from P is a free group F_S. In this case, for any given group G, determining the image of each element of S produces a homomorphism of F_S into G. Thus there are plenty of homomorphisms from F_S into G. On the other hand, if R contains many elements relative to the cardinality of S, then we do not obtain a homomorphism from $\Gamma(P)$ into G so easily since the images of the elements of S must satisfy all the relations. Therefore, for a group defined by $P \in P_{\mathcal{M}}(m, d)$, the set of homomorphisms from $\Gamma(P)$ into G should be small when d is close to 1. Noting that the fixed-point property as well as Kazhdan's property (T) asserts that homomorphisms from the group into the isometry group must be restrictive, and thinking naively, we expect that, by taking d suitably large, the groups defined by $P \in P_{\mathcal{M}}(m, d)$ have the fixed-point property with a large probability. On the other hand, if we take d too large, the group defined by $P \in P_{\mathcal{M}}(m, d)$ might get too small, for instance, might be a finite group. It is easy to see, by an argument using the barycenter, that any isometric action of a finite group on any CAT(0) space fixes a point. Therefore we are interested only in the case that P defines an infinite group. As a matter of fact, Żuk proved:

THEOREM 4.2 ([**52**, Theorems 3, 4]). *Suppose* $1/3 < d < 1/2$. *Then we have*

$$\lim_{m\to\infty} \frac{\#\left\{ P \in P_{\mathcal{M}}(m, d) \;\middle|\; \begin{array}{l} P \text{ defines an infinite hyperbolic group} \\ \text{with Kazhdan's property (T)} \end{array} \right\}}{\#P_{\mathcal{M}}(m, d)} = 1.$$

Here the hyperbolicity in the statement means hyperbolicity in the sense of Gromov [**12**]. Theorem 4.2 says that if we choose a group randomly from a certain class of finitely presented groups, then we get a group with Kazhdan's property (T) with a large probability. We often read the statement simply as: when the density d satisfies $1/3 < d < 1/2$, random groups have Kazhdan's property (T). In what follows, we often use a phrase such as "random groups with density d have a property that....". The precise meaning of this phrase is "if we take a presentation randomly from $P_{\mathcal{M}}(m, d)$, then the group obtained from the chosen presentation has a property that....with a large probability so that it tends to 1 when letting $m \to \infty$. Note that "random groups" is the name of the formulation introduced in order to express the genericity of groups with certain properties, and that we do not mean that there is a class of groups called random groups. The set of presentations $P_{\mathcal{M}}(m, d)$ is called the *triangular model* of random groups. There are other models of random groups, which have different sets of presentations as their populations.

Although Theorem 4.2 does not give us new concrete examples of groups with Kazhdan's property (T), it claims that there are plenty of such groups. We remark that Theorem 4.2 has some meaning even for a group with relations of length other than 3, since the quotient of a group with Kazhdan's property (T) is again a group with the property.

Such a viewpoint for the study of finitely generated groups has its origin in the theorem of Gromov claiming that most of the groups defined by long enough relations with a fixed number of generators are infinite hyperbolic groups ([**12**]). Although the proof of this statement was not seen in [**12**], Ol'shanskii gave a proof in [**39**]. In [**13**], Gromov proposed the so-called *density model* of random groups and proved that, for this model of random groups, random groups with density $d < 1/2$ are infinite hyperbolic groups. The results in this direction assert not only the richness of the set of groups with a certain property but also the existence of a group with the property, even when it is not easy to construct a concrete example with the property. (For example, a counterexample to the Baum-Conne conjecture with coefficient was given in [**14**].) For expositions on random groups we refer the reader to [**11**] and [**41**]. Also, the reader will find a detailed exposition on recent results in [**38**].

When we deal with random groups, we regard a set of presentations to be a set of groups and examine the probability of the occurrence of groups with the property under consideration. This fits very well into the framework we have discussed so far. In fact, Theorem 4.2 is obtained by proving that the probability of $P \in P_{\mathcal{M}}(m,d)$ satisfying $\mu_1(L(P)) > 1/2$ approaches 1 when letting $m \to \infty$. Looking at the details of the proof of Theorem 4.2, we notice that, for any given $\varepsilon > 0$ the probability of the occurrence of $P \in P_{\mathcal{M}}(m,d)$ with $\mu_1(L(P)) > 1 - \varepsilon$ converges to 1 when letting $m \to \infty$. Therefore, by Proposition 3.7 and Theorem 4.1, we get the following:

THEOREM 4.3 ([**18**]). *Fix a positive constant $\delta_0 < 1/2$. Let $\mathcal{Y}_{\delta_0}$ be the class of* CAT(0) *spaces whose tangent cone TC_pY satisfies $\delta(TC_pY) \leq \delta_0$ at each point $p \in Y$. Suppose $1/3 < d < 1/2$. Then we have*

$$\lim_{m\to\infty} \frac{\#\left\{ P \in P_{\mathcal{M}}(m,d) \;\middle|\; \begin{array}{l} P \text{ defines an infinite hyperbolic group with} \\ \text{the fixed-point property for all } Y \in \mathcal{Y}_{\delta_0} \end{array} \right\}}{\# P_{\mathcal{M}}(m,d)} = 1.$$

Here we say that Γ has the fixed-point property for Y if every isometric action of Γ fixes a point in Y.

There is an interesting (possibility of) application of fixed-point theorems to random groups. Familiar examples of infinite groups are often picked up from infinite subgroups in $GL_n(\mathbb{C})$ such as lattices of algebraic groups. Gromov suggests that such groups are far from generic in the set of infinite groups; he claims that, by taking an appropriate model of random groups, any homomorphism of random groups into $GL_n(\mathbb{C})$ has finite image. Ghys[**11**] and Pansu[**41**] pointed out that Gromov's claim follows from a certain fixed-point theorem for random groups. Actually, the next proposition tells us that a certain fixed-point property implies the finiteness of the image of a linear representation.

PROPOSITION 4.4. *Let k be either $\mathbb{R}$, $\mathbb{C}$, or an r-adic field $\mathbb{Q}_r$, and denote by K the maximal compact subgroups of $PGL_m(k)$. Suppose that a finitely generated group Γ has the fixed-point property for Hilbert spaces and $PGL_m(k)/K$ for any m and k. Then, for any n, the image of any homomorphism from Γ into $GL_n(\mathbb{C})$ is finite.*

Suppose the following is true: For suitable $0 < d < 1/2$, random groups with density d satisfy $\lambda_1(L(P), TC_pY) > 1/2$ for $Y = PGL_m(k)/K$ (note that $\lambda_1(L(P), Y) = \mu_1(L(P))$ when $k = \mathbb{R}, \mathbb{C}$). Then random groups with density d satisfy the assumption of Proposition 4.4 by Theorem 4.1. Thus the image of any homomorphism from random groups with density d into $GL_n(\mathbb{C})$ must be finite. On the other hand, since random groups with density $d < 1/2$ are infinite hyperbolic groups, they cannot be subgroups of $GL_n(\mathbb{C})$.

As we have just seen, Gromov's claim can be reduced to the estimate of λ_1 of random groups. (Here we have to mention that, from Example 3 in the preceding section, we know that $\liminf_{r\to\infty} \delta(PGL_3(\mathbb{Q}_r)/K) \geq 1/2$, and hence it is impossible to take δ_0 so that $\delta(TC_pY) \leq \delta_0 < 1/2$ for all $Y = PGL_m(\mathbb{Q}_r)/K$. In particular, we cannot deduce Gromov's claim for random groups of the triangular model from Theorem 4.3.)

The outline of the proof of Proposition 4.4 is as follows. Take any homomorphism $\rho : \Gamma \longrightarrow GL_n(\mathbb{C})$. Since Γ has the fixed-point property for Hilbert spaces, that is, Γ has Kazhdan's property (T), we see that any finite index subgroup of $\rho(\Gamma)$ cannot be solvable. Noting this, we can show that there exists a semisimple algebraic group G and a homomorphism $\rho_1 : \Gamma \longrightarrow G$ such that $\rho_1(\Gamma)$ is infinite whenever $\rho(\Gamma)$ is. Applying the arguments of (i), (ii), (iv) in the proof of the arithmeticity we have explained in §1, we see that the fixed-point property for $PGL_m(\mathbb{Q}_r)/K$ and $PGL_m(\mathbb{C})/K$ implies that there exists a finite index subgroup Γ_0 of Γ, $N \in \mathbb{N}$, and $\rho_2 : \Gamma_0 \longrightarrow SL_N(\mathbb{Z})$ such that $\rho_2(\Gamma_0)$ is infinite whenever $\rho_1(\Gamma)$ is. In particular, $\rho_2(\Gamma_0)$ is a discrete subgroup of $SL_N(\mathbb{R})$. Moreover, by the assumption in the case $k = \mathbb{R}$, namely the fixed-point property for $PGL_N(\mathbb{R})/K = SL_N(\mathbb{R})/SO_N$, we conclude that $\rho_2(\Gamma_0)$ must lie in a maximal compact subgroup K', which is a conjugate of SO_N. Therefore $\rho_2(\Gamma_0) \subset SL_N(\mathbb{Z}) \cap K'$, namely $\rho_2(\Gamma_0)$ is a discrete compact subgroup of $SL_N(\mathbb{R})$. Hence $\rho_2(\Gamma_0)$ must be a finite group, and so is $\rho(\Gamma)$ by our choice of ρ_1 and ρ_2.

5. Some problems

At present the results we have obtained as applications of combinatorial harmonic maps are fixed-point theorems for the discrete-group actions on CAT(0) spaces, and we have not succeeded in proving a superrigidity result. The original motivation of introducing combinatorial harmonic maps was to give a geometric proof of the following superrigidity theorem for lattices in p-adic semisimple algebraic groups.

THEOREM 5.1 (**[31]**). *Suppose that $\mathbf{G}$ is a connected, (algebraically) simply connected, semisimple algebraic group defined over $\mathbb{Q}_p$ with $\mathbb{Q}_p$ rank greater than one, and that $\mathbf{G}_{\mathbb{Q}_p}$ does not have a compact factor. Let $\Gamma \subset \mathbf{G}_{\mathbb{Q}_p}$ be an irreducible lattice. Let k be either $\mathbb{R}$, $\mathbb{C}$, or $\mathbb{Q}_r$, and let $\mathbf{H}$ be a connected algebraic group over k that is k-simple. Assume that $\rho : \Gamma \longrightarrow \mathbf{H}_k$ is a homomorphism and that the image $\rho(\Gamma)$ is Zariski dense in $\mathbf{H}$. Then one of the following holds.*

(i) *$\rho(\Gamma)$ is relatively compact with respect to the Hausdorff topology.*
(ii) *$k = \mathbb{Q}_p$ and ρ extends to a rational homomorphism $\overline{\rho} : \mathbf{G} \longrightarrow \mathbf{H}$ defined over k.*

In order to give a proof of this theorem via combinatorial harmonic maps, we have to handle the case (ii) in the combinatorial setting. In the proof of Theorem 2.2, this part follows from the existence of a totally geodesic embedding. Is it possible to transplant this argument to the combinatorial setting so that it gives a geometric proof of Theorem 5.1?

In order to achieve the theorem above by using combinatorial harmonic maps, we must give a good estimate of $\lambda_1(\mathrm{Lk}\, x, TC_pY)$ for Y with nonflexible tangent cones such as r-adic symmetric spaces. One way to do this is to give an estimate of

$\lambda_1(\mathrm{Lk}\,x, TC_pY)$ by using the invariant δ of the tangent cone introduced in §3. The conjecture mentioned in §3 is related to this problem.

The fixed-point theorem for random groups discussed in the preceding section seems to be a quite interesting subject. We note that a good estimate of λ_1 of r-adic symmetric spaces should again be necessary in order to show that random groups satisfy the assumption of Proposition 4.4.

References

[1] W. Ballmann and J. Świątkowski, *On L^2-cohomology and property* (T) *for automorphism groups of polyhedral cell complexes*, Geom. Funct. Anal. **7** (1997), 615–645. MR1465598 (98m:20043)

[2] M. R. Bridson and A. Haefliger, Metric Spaces of Non-positive Curvature, Springer-Verlag, Berlin, Heidelberg, 1999. MR1744486 (2000k:53038)

[3] K. S. Brown, Buildings, Springer-Verlag, New York, Berlin, Heidelberg, 1988. MR969123 (90e:20001)

[4] D. Burago, Y. Burago, and S. Ivanov, A Course in Metric Geometry, Graduate Studies in Mathematics 33, American Mathematical Society, 2001. MR1835418 (2002e:53053)

[5] K. Corlette, *Archimedean superrigidity and hyperbolic geometry,* Ann. of Math. (2) **135** (1992), 165–182. MR1147961 (92m:57048)

[6] P. Delorme, 1-*cohomologie des représentations unitaires des groupes de Lie semi-simples et résolubles. Produit tensoriels continus et représentations*, Bull. Soc. Math. France **105** (1977), 281–336. MR0578893 (58:28272)

[7] S. K. Donaldson, *Twisted harmonic maps and the self-duality equations*, Proc. London Math. Soc. **55** (1987), 127–131. MR887285 (88g:58040)

[8] J. Eells and J. H. Sampson, *Harmonic maps of Riemannian manifolds*, Amer. J. Math. **86** (1964), 109–160. MR0164306 (29:1603)

[9] W. Feit and G. Higman, *The nonexistence of certain generalized polygons*, J. Alg. **1** (1964), 114–131. MR0170955 (30:1189)

[10] H. Garland, *p-adic curvature and the cohomology of discrete subgroups of p-adic groups*, Ann. of Math. (2) **97** (1973), 375–423. MR0320180 (47:8719)

[11] É. Ghys, *Groupes aléatoires*, Astérisque **294** (2004), 173–204. MR2111644 (2005j:20049)

[12] M. Gromov, *Hyperbolic groups*, in Essays in Group Theory, ed. S. M. Gersten, Springer, New York, 1987, 75-265. MR919829 (89e:20070)

[13] M. Gromov, *Asymptotic invariants of infinite groups*, Geometric group theory, Vol. 2 (Sussex, 1991), 1–295, London Math. Soc. Lect. Note Ser. **182**, Cambridge Univ. Press, Cambridge, 1993. MR1253544 (95m:20041)

[14] M. Gromov, *Random walk in random groups*, Geom. Funct. Anal. **13** (2003), 73–146. MR1978492 (2004j:20088a)

[15] M. Gromov and R. Schoen, *Harmonic maps into singular spaces and p-adic superrigidity for lattices in groups of rank one*, Publ. Math. IHES **76** (1992), 165–246. MR1215595 (94e:58032)

[16] A. Guichardet, *Étude de la 1-cohomologie et de la topologie du dual pour les groupes de Lie à radical abélian*, Math. Ann. **228** (1977), 215-232. MR0442151 (56:539)

[17] P. de la Harpe and A. Valette, La Propriété (T) de Kazhdan pour les groupes localement compacts, Astérisque **175**, Soc. Math. France, 1989. MR1023471 (90m:22001)

[18] H. Izeki, T. Kondo and S. Nayatani, *Fixed-point property of random groups*, preprint.

[19] H. Izeki and S. Nayatani, *Combinatorial harmonic maps and superrigidity* (Japanese), Hyperbolic spaces and discrete groups II (Kyoto, 2001), RIMS Kokyuroku **1270** (2002), 182–194. MR1953887

[20] H. Izeki and S. Nayatani, *Combinatorial harmonic maps and superrigidity. The case of a singular target* (Japanese), Perspectives of hyperbolic spaces (Kyoto, 2002), RIMS Kokyuroku **1329** (2003), 1–7.

[21] H. Izeki and S. Nayatani, *Combinatorial harmonic maps and discrete-group actions on Hadamard spaces*, Geom. Dedicata **114** (2005), 147–188. MR2174098 (2006k:58024)

[22] J. Jost, *Convex functionals and generalized harmonic maps into spaces of nonpositive curvature*, Comment. Math. Helv. **70** (1995), 659–673. MR1360608 (96j:58043)

[23] J. Jost, Nonpositive curvature: Geometric and analytic aspects, Lectures in Mathematics ETH Zürich, Birkhäuser, Basel, 1997. MR1451625 (98g:53070)

[24] J. Jost and S.-T. Yau, *Harmonic maps and group representations*, in Differential Geometry and Minimal Submanifolds (B. Lawson and K. Tenenblat, eds.), Longman Scientific, 1991, 241–259. MR1173045 (93g:58036)

[25] J. Jost and S.-T. Yau, *Harmonic maps and superrigidity*, Differential geometry: Partial differential equations on manifolds, Proc. Symp. Pure Math. **54-I**, Amer. Math. Soc., Providence, RI, 1993, 245–280. MR1216587 (94m:58060)

[26] M. Kapovich, Hyperbolic Manifolds and Discrete Groups, Progress in Mathematics, 183. Birkhäuser Boston, Inc., Boston, MA, 2001. MR1792613 (2002m:57018)

[27] D. A. Kazhdan, *Connection of the dual space of a group with the structure of its closed subgroups*, Funct. Anal. Appl., **1**, 63–65.

[28] M. Kotani and T. Sunada, *Standard realizations of crystal lattices via harmonic maps*, Trans. Amer. Math. Soc. **353** (2000), 1–20. MR1783793 (2001j:58028)

[29] F. Labourie, *Existence d'applications harmoniques tordues à valeurs dans les variétés à courbure négative*, Proc. Amer. Math. Soc. **111** (1991), 877–882. MR1049845 (91f:58023)

[30] É. Lebeau, *Applications harmoniques entre graphes finis et un théorème de superrigidité*, Ann. Inst. Fourier **46** (1996), 1183–1203. MR1427122 (97m:58054)

[31] G. Margulis, *Discrete groups of motions of manifolds of nonpositive curvature*, Amer. Math. Soc. Transl. **190** (1977), 33–45. MR0492072 (58:11226)

[32] G. Margulis, Discrete Subgroups of Semisimple Lie Groups, Ergeb. der Math. 17, Springer-Verlag, Berlin, Heidelberg, 1990. MR1090825 (92h:22021)

[33] Y. Matsushima, On the first Betti number of compact quotient spaces of higher-dimensional symmetric spaces, Ann. of Math. (2) **75** (1962), 312–330. MR0158406 (28:1629)

[34] U. F. Mayer, *Gradient flows on nonpositively curved metric spaces and harmonic maps*, Comm. Anal. and Geom. **6** (1998), 199–253. MR1651416 (99m:58067)

[35] N. Mok, Y.-T. Siu and S.-K. Yeung, *Geometric superrigidity*, Invent. Math. **113** (1993), 57–83. MR1223224 (94h:53079)

[36] A. Navas, *Actions de groupes de Kazhdan sur le cercle*, Ann. Scient. Éc. Norm. Sup., **35** (2002), 749–758. MR1951442 (2003j:58013)

[37] Shin Nayatani, *Harmonic maps and rigidity* (Japanese), The Frontier of Differential Geometry — Surveys in Geometry, special edition, H. Nakajima, ed., Baifukan, 2005, 188-221.

[38] Y. Ollivier, A January 2005 invitation to random groups, manuscript (2005).

[39] A. Y. Ol'shanskiĭ, *Almost every group is hyperbolic*, Int. J. Algebra Comput. **2** (1992), 1–17. MR1167524 (93j:20068)

[40] P. Pansu, *Formules de Matsushima, de Garland et propriété* (T) *pour des groupes agissant sur des espaces symétriques ou des immeubles*, Bull. Soc. Math. France **126** (1998), 107–139. MR1651383 (2000d:53067)

[41] P. Pansu, *Groupes aléatoires*, Journée annuelle de la Soc. Math. France (2003), 37–49. MR2202284 (2006k:20087)

[42] M. Raghunathan, Discrete subgroups of Lie groups, Springer, New York, 1972. MR0507234 (58:22394a)

[43] M. A. Ronan, Lectures on Buildings, Academic Press, San Diego, 1989. MR1005533 (90j:20001)

[44] Y.-M. Siu, *The complex-analyticity of harmonic maps and strong rigidity of compact Kähler manifolds*, Ann. of Math. (2) **112** (1980), 73-112. MR584075 (81j:53061)

[45] L. A. Takhtajan and L.-P. Teo, *Weil-Petersson metric on the universal Teichmüller space I: curvature properties and Chern forms*, arXiv:Math.CV/0312172. Mem. Amer. Math. Soc. **183** (2006). MR2251887 (2007e:32011)

[46] E. B. Vinberg, *Rings of definition of dense subgroups of semisimple linear groups*, (Russian) Izv. Akad. Nauk SSSR Ser. Mat. **35** (1971), 45–55. MR0279206 (43:4929)

[47] M.-T. Wang, *A fixed point theorem of discrete group actions on Riemannian manifolds*, J. Diff. Geom. **50** (1998), 249–267. MR1684980 (2000e:53051)

[48] M.-T. Wang, *Generalized harmonic maps and representations of discrete groups*, Comm. Anal. and Geom. **8** (2000), 545–563. MR1775138 (2001m:58039)

[49] Y. Watatani, *Property* (T) *of Kazhdan implies property* (FA) *of Serre*, Math. Japon. **27** (1981), 97-103. MR649023 (83i:22006)

[50] R. Zimmer, Ergodic Theory and Semisimple Groups, Monographs in Mathematics 81, Birkhäuser, Boston, Basel, Stuttgart, 1984. MR776417 (86j:22014)

[51] A. Żuk, *La propriété* (T) *de Kazhdan pour les groupes agissant sur les polyèdres*, C. R. Acad. Sci. Paris **323** (1996), 453–458. MR1408975 (97i:22001)

[52] A. Żuk, *Property* (T) *and Kazhdan constants for discrete groups*, Geom. Funct. Anal. **13** (2003), 643–670. MR1995802 (2004m:20079)

DEPARTMEENT OF MATHEMATICS, KEIO UNIVERSITY, YOKOHAMA 223-8522, JAPAN
E-mail address: izeki@math.keio.ac.jp

GRADUATE SCHOOL OF MATHEMATICS, NAGOYA UNIVERSITY, CHIKUSA-KU, NAGOYA 464-8602, JAPAN
E-mail address: nayatani@math.nagoya-u.ac.jp

Translated by HIROYASU IZEKI AND SHIN NAYATANI

Amer. Math. Soc. Transl.
(2) Vol. **230**, 2010

Rational Semigroups, Random Complex Dynamics and Singular Functions on the Complex Plane

Hiroki Sumi

ABSTRACT. We summarize recent developments in the study of the dynamics of semigroups of holomorphic maps and random complex dynamics. Both fields are related to each other very deeply. In the study of these fields, complex analogues of singular functions and complex analogues of the Takagi function naturally appear.

1. Introduction

Let M be a set and $f : M \to M$ be a map. For a given "initial value" $x_0 \in M$, we consider the sequence $\{x_n\}_{n\in\mathbb{N}}$ determined by the difference equation $x_{n+1} = f(x_n)$, and the action of a family $\{f^n\}_{n\in\mathbb{N}}$ (where f^n denotes the n-th iterates of f) on M. Such a system is called a **(discrete) dynamical system.** If M is a complex manifold and $f : M \to M$ is a holomorphic map, then the study of such systems as above is referred to as **complex dynamics.** In mathematics, we deal with (discrete) dynamical systems and complex dynamics for many reasons. We also deal with these systems when we study mathematical models in various fields other than pure mathematics. For example, let us consider the population of a certain kind of insects which spawn only once per generation. Let x_n $(0 \le x_n \le 1)$ be the percentage of the limiting population present in generation n. One mathematical model (see [**36**]) tells us that the sequence $\{x_n\}$ satisfies

(1)
$$x_{n+1} = ax_n(1 - x_n), \text{ where } a \text{ is a positive constant which does not depend on } n.$$

In this model, we consider the dynamical system induced by the quadratic polynomial $f(x) = ax(1-x)$ on the interval $M = [0, 1]$. In order to perform a deeper mathematical analysis of this model, we allow the initial values x_0 to be in the complex plane $\mathbb{C}$, and furthermore, to be in the Riemann sphere $\hat{\mathbb{C}} := \mathbb{C} \cup \{\infty\} \cong \mathbb{CP}^1 \cong S^2$. As illustrated by this example, it is important to investigate the dynamical systems

This article originally appeared in Japanese in Sūgaku **61** (2) (2009), 133–161.

2000 *Mathematics Subject Classification.* Primary 37F10, 30D05.

Key words and phrases. Rational semigroups, polynomial semigroups, random complex dynamics, random iteration, Markov process, Julia sets, fractal geometry, (backward) iterated function systems, interaction cohomology, complex singular functions, complex Takagi function, devil's coliseum.

This research was partially supported by JSPS KAKENHI 21540216.

of polynomial maps on the Riemann sphere. It is, however, very difficult to analyze the dynamical systems of even quadratic polynomials. Many deep results have been proven by many mathematicians, and sometimes it was very important to apply or develop Teichmüller theory for the study of the dynamics of polynomial (or rational) maps on the Riemann sphere. (For an introduction to complex dynamics, see [**38**] and then [**37**], etc.)

In the above model, some species may have several strategies (maps) $f_1, \ldots, f_m$ to survive in nature, since they must sometimes adapt themselves to changes in climate or environment. In this modified model, we consider a random dynamical system such that at every step we choose a map f_j which maps x_n to x_{n+1} according to a probability distribution.

Thus, it is very natural to consider random dynamics of a family of polynomial (or rational) maps. In this article, we generalize the usual complex dynamics. We consider the action (or dynamics) of a semigroup (with the semigroup operation being functional composition) generated by a family of holomorphic maps on a complex manifold M, and we also consider the corresponding random complex dynamics. In these fields, we will see many interesting new phenomena, which do not hold in the usual iteration dynamics of a single holomorphic map. We systematically investigate these phenomena and their mechanisms.

We now let M be a Riemann surface (one-dimensional connected complex manifold). If M has a complete hyperbolic metric, i.e., if there is a holomorphic universal covering map from the unit disc onto M, then any holomorphic map f on M is not expanding with respect to the hyperbolic metric on M, and so the "Julia set" of f, which is defined in the next section, should be empty. Thus we are not so interested in this case, and so one mainly deals with the case $M = \hat{\mathbb{C}}$ or $M = \mathbb{C}$. In this article, we deal with the case $M = \hat{\mathbb{C}}$.

We now consider the random dynamics of polynomials on $\hat{\mathbb{C}}$. For each initial value $z \in \hat{\mathbb{C}}$, let $T_\infty(z)$ be the probability of the randomly generated orbits tending to $\infty \in \hat{\mathbb{C}}$ starting with the initial value z. Then we have a function $T_\infty : \hat{\mathbb{C}} \to [0,1]$. In this article, we show that under certain conditions, the function $T_\infty : \hat{\mathbb{C}} \to [0,1]$ can be regarded as a complex analogue of **the devil's staircase** (or **the Cantor function**) ([**78**]) and **Lebesgue's singular functions**, i.e., functions which are defined on $\mathbb{R}$ and continuous on $\mathbb{R}$ but have certain singular properties. Moreover, we investigate the (non)differentiability of T_∞ by using ergodic theory, potential theory, complex analysis, and so on (sections 9, 10, Figures 5, 6). Moreover, we show that associated with the above function of probability of tending to ∞, a complex analogue of the **Takagi function** ([**77**]) appears (section 10, Figure 7).

Let us recall the definition of the devil's staircase (the Cantor function), Lebesgue's singular functions, and the Takagi function. Generally, the devil's staircase, Lebesgue's singular functions, and the Takagi function are defined as bounded functions on $[0,1]$ which satisfy some kinds of functional equations and boundary conditions ([**20, 78, 1**]). More precisely (the following definitions look a little bit different from those in [**20, 78, 1**], but it turns out that they are equivalent to those in [**20, 78, 1**]), the devil's staircase is defined as the restriction $\varphi|_{[0,1]}$, where $\varphi : \mathbb{R} \to \mathbb{R}$ is the unique bounded function which satisfies

$$(2) \qquad \frac{1}{2}\varphi(3x) + \frac{1}{2}\varphi(3x-2) = \varphi(x) \ (\forall x \in \mathbb{R}), \ \varphi|_{(-\infty,0]} \equiv 0, \ \varphi|_{[1,+\infty)} \equiv 1,$$

and for each $0 < a < 1$ with $a \neq 1/2$, Lebesgue's singular function L_a is defined as the restriction $\psi_a|_{[0,1]}$, where $\psi_a : \mathbb{R} \to \mathbb{R}$ is the unique bounded function which satisfies

(3) $a\psi_a(2x) + (1-a)\psi_a(2x-1) = \psi_a(x)\ (\forall x \in \mathbb{R}),\ \psi_a|_{(-\infty,0]} \equiv 0,\ \psi_a|_{[1,+\infty)} \equiv 1,$

and the Takagi function is defined as the restriction $\phi/2|_{[0,1]}$, where $\phi : \mathbb{R} \to \mathbb{R}$ is the unique bounded function which satisfies

(4)

$$\frac{1}{2}\phi(2x)+\frac{1}{2}\phi(2x-1)+\psi_{1/2}(2x)-\psi_{1/2}(2x-1) = \phi(x)\ (\forall x \in \mathbb{R}),\ \phi|_{(-\infty,0]\cup[1,+\infty)} \equiv 0.$$

We now give a (relatively new, uncommon) interpretation for these functions in terms of random dynamical systems on $\mathbb{R}$. Let $f_1(x) := 3x,\ f_2(x) := 3(x-1)+1\ (x \in \mathbb{R})$. We consider the random dynamics on $\mathbb{R}$ such that at every step we choose f_1 with probability $1/2$ and f_2 with probability $1/2$. Let $\tilde{\mathbb{R}} := \mathbb{R} \cup \{\pm\infty\}$ be the two-point compactification of $\mathbb{R}$. For each $x \in \mathbb{R}$, let $T_{+\infty}(x)$ be the probability of tending to $+\infty$ starting with the initial value $x \in \mathbb{R}$. We easily see that the function $T_{+\infty}$ is bounded on $\mathbb{R}$ and satisfies (2). Thus the restriction $T_{+\infty}|_{[0,1]} : [0,1] \to [0,1]$ is equal to the devil's staircase (or the Cantor function) (Figure 1). The devil's staircase satisfies the following properties:

- It is continuous on $[0,1]$.
- It varies precisely on the Cantor middle third set C (a kind of thin fractal set), i.e., $T'_{+\infty}(x) = 0$ for $x \in \mathbb{R} \setminus C$ and $T_{+\infty}|_U$ is not constant for each open subset U of $\mathbb{R}$ with $\mathbb{R} \cap C \neq \emptyset$.
- It is monotone.

Similarly, let $g_1(x) := 2x,\ g_2(x) := 2(x-1)+1\ (x \in \mathbb{R})$. For each $0 < a < 1$, we consider the random dynamics on $\mathbb{R}$ such that at every step we choose g_1 with probability a and g_2 with probability $1-a$. Let $T_{+\infty,a}(x)$ be the probability of tending to $+\infty$. We set $L_a := T_{+\infty,a}|_{[0,1]}$. We easily see that the function $T_{+\infty,a}$ is bounded on $\mathbb{R}$ and satisfies (3). Thus the function $L_a : [0,1] \to [0,1]$ is equal to Lebesgue's singular function with parameter a (Figure 1). The function L_a is continuous and monotone on $[0,1]$, and if $a \neq 1/2$, L_a has the following singular property: for almost every point $x \in [0,1]$ with respect to the one-dimensional Lebesgue measure, L_a is differentiable at x and the derivative is zero. Moreover, Sekiguchi and Shiota showed that for each fixed $x \in [0,1]$, the function $a \mapsto L_a(x)$ is real-analytic on $(0,1)$ ([**42**]), and Hata and Yamaguti showed that the function $x \mapsto (1/2)\cdot(\partial L_a(x)/\partial a)|_{a=1/2}$ on $[0,1]$ is equal to the Takagi function ([**28**], Figure 1).

Thus, the devil's staircase and Lebesgue's singular functions can be defined in terms of random dynamics on $\mathbb{R}$; that is, they can be defined as the functions of probability of tending to $+\infty$. Moreover, the Takagi function can be defined as the partial derivative with respect to the probability parameter.

Note that the singular functions described above and the generalizations of them have been investigated for many years by so many researchers, and still this topic attracts so many mathematicians (see [**77, 20, 28, 42, 78, 1**], etc.). We remark that in most of the literature, the theory of random dynamical systems has not been used directly to investigate these singular functions on the interval, although some researchers have used it implicitly. However, as the author points out, it is very natural and straightforward to use the theory of random dynamical systems in the study of these singular functions.

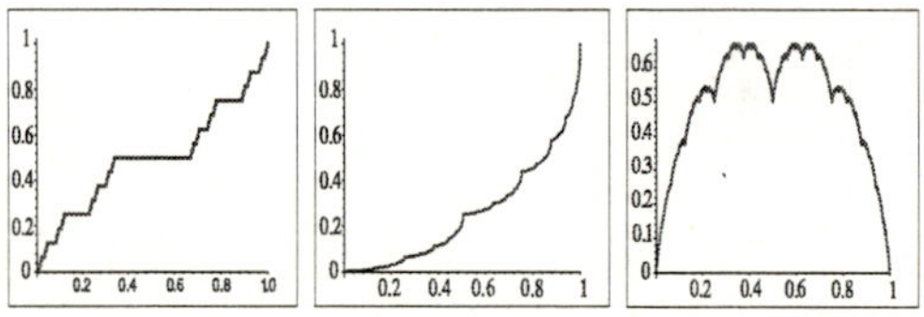

FIGURE 1. (From left to right) the devil's staircase, Lebesgue's singular function, the Takagi function

One of the main purposes of this article is to state that we can consider a complex analogue of the above story by using random complex dynamics on the complex plane (or the Riemann sphere). A complex analogue of the singular functions is defined on $\mathbb{C}$, is continuous on $\mathbb{C}$, and varies only on the "Julia set" of a semigroup generated by several polynomial (or rational) maps.

The strategy to investigate random complex dynamics and the related dynamics of a semigroup generated by a family of holomorphic maps, which the author thinks best, is to investigate "simultaneously" all of the following items:

- dynamics (actions) of semigroups of holomorphic maps,
- spaces of semigroups of holomorphic maps which have a kind of nice property,
- Iterated Function Systems and their generalizations,
- (backward) self-similar sets and their generalizations, fractal geometry,
- representing the interaction of many kinds of maps in a system which forms a (backward) self-similar set by using homological algebra,
- random complex dynamics (a kind of Markov process),
- dynamics of skew products associated with generator systems of semigroups of holomorphic maps, fiberwise (sequencewise) dynamics, and
- dynamics of fibered rational maps (which is defined later).

By using this strategy, each item above will simultaneously take a step forward.

In the following sections, we will see that these many items are related to each other very deeply, and that they join to make a very strong stream in a new field of mathematics. In the research of the above items, we sometimes generalize the results of "usual complex dynamics" (iteration dynamics of a single rational map), and we sometimes find new phenomena.

2. Rational semigroups

A **rational semigroup** is a semigroup generated by a family of nonconstant rational maps on $\hat{\mathbb{C}}$, where the semigroup operation is functional composition. A **polynomial semigroup** is a semigroup generated by a family of nonconstant polynomial maps on $\hat{\mathbb{C}}$. The first study of the dynamics of rational semigroups was initiated by A. Hinkkanen and G. J. Martin ([**29**]), who were interested in the role of the dynamics of polynomial semigroups while studying various one-complex-dimensional moduli spaces for discrete groups, and by F. Ren's group ([**79, 24**]), who studied such semigroups from the perspective of random dynamical systems. Note that the author of this article noticed the idea of the dynamics of rational semigroups and started his research on this subject around 1994, independently from these two groups.

In the following, we consider the spherical metric on $\hat{\mathbb{C}}$.

DEFINITION 2.1. Let G be a rational semigroup. We set

$$F(G) := \{ z \in \hat{\mathbb{C}} \mid \exists \text{ neighborhood } U \text{ of } z \text{ such that } \{g : U \to \hat{\mathbb{C}}\}_{g \in G} \text{ is equicontinuous on } U\}.$$

This is called the **Fatou set** of G. Moreover, we set $J(G) := \hat{\mathbb{C}} \setminus F(G)$, and this is called the **Julia set** of G.

REMARK 2.2. If G is a Kleinian group (discrete group of linear fractional transformations on $\hat{\mathbb{C}}$), then the Julia set $J(G)$ is equal to the **limit set** of G.

DEFINITION 2.3. If a semigroup G is generated by $\{h_1, h_2, \ldots\}$, then we write $G = \langle h_1, h_2, \ldots \rangle$. For a rational map g on $\hat{\mathbb{C}}$, we set $F(g) := F(\langle g \rangle)$ and $J(g) := J(\langle g \rangle)$.

DEFINITION 2.4. For a rational semigroup G, we set $E(G) := \{z \in \hat{\mathbb{C}} \mid \sharp(\bigcup_{g \in G} g^{-1}(\{z\})) < \infty\}$, where $\sharp$ denotes the cardinality. This $E(G)$ is called the **exceptional set** of G.

We now present some fundamental properties of the Julia set and the Fatou set of a rational semigroup G.

LEMMA 2.5 ([**29, 79, 24**]). *Let G be a rational semigroup. Then we have the following.*

(1) *$F(G)$ is open and $J(G)$ is compact.*
(2) *For each $g \in G$, $g(F(G)) \subset F(G)$ and $g^{-1}(J(G)) \subset J(G)$. However, we do not have the equality in general (e.g., if $G = \langle z^2, 2z \rangle$, then $F(G) = \{z \in \hat{\mathbb{C}} \mid |z| > 1\}$ and $J(G) = \{z \in \mathbb{C} \mid |z| \leq 1\}$).*
(3) *If $\sharp(J(G)) \geq 3$, then $J(G)$ is a nonempty compact set which does not have any isolated points. Thus it is a perfect set.*
(4) *If $\sharp(J(G)) \geq 3$, then $\sharp(E(G)) \leq 2$.*
(5) *If $z \in \hat{\mathbb{C}} \setminus E(G)$, then $J(G) \subset \overline{\bigcup_{g \in G} g^{-1}(\{z\})}$. Thus if $z \in J(G) \setminus E(G)$, then $\overline{\bigcup_{g \in G} g^{-1}(\{z\})} = J(G)$.*
(6) *If $\sharp(J(G)) \geq 3$, then $J(G)$ is the smallest set with respect to inclusion in the space of all compact subsets K of $\hat{\mathbb{C}}$ such that $\sharp K \geq 3$ and $g^{-1}(K) \subset K$ for each $g \in G$.*
(7) *If G is a group or a commutative semigroup, then for each $g \in G$, $F(G)$ and $J(G)$ are completely invariant under g (i.e., $g(F(G)) = g^{-1}(F(G)) = F(G), g(J(G)) = g^{-1}(J(G)) = J(G)$).*

REMARK 2.6. For a rational semigroup G, if H is a subsemigroup of G, then $J(H) \subset J(G)$. Also, if $g : \hat{\mathbb{C}} \to \hat{\mathbb{C}}$ is a rational map with $\deg(g) \geq 2$, then $\sharp(J(g)) \geq 3$ ([**38**]). Thus, if a rational semigroup G contains a rational map g with $\deg(g) \geq 2$, then $\sharp(J(G)) \geq 3$.

As pointed out in Lemma 2.5, if G is neither a group nor a commutative semigroup, then it may not hold that "for each $g \in G$, $g(F(G)) = F(G)$ and $g(J(G)) = J(G)$." However, $J(G)$ has the following nice property, so-called **backward self-similarity**, which is stated as follows.

LEMMA 2.7 ([**55**]). *If a rational semigroup G is finitely generated and $G = \langle h_1, \ldots, h_m \rangle$, then*

$$J(G) = h_1^{-1}(J(G)) \cup \cdots \cup h_m^{-1}(J(G)).$$

By Lemma 2.7, a **self-similar set** ([**21, 78**]) constructed by finitely many contracting similitudes on $\mathbb{C}$ can be regarded as the Julia set of a rational semigroup. Moreover, since the Julia sets of rational semigroups have backward self-similarity, the research of Julia sets of rational semigroups is deeply related to fractal geometry (for an introduction to fractal geometry, see [**21, 78**]).

EXAMPLE 2.8. Let $p_1, p_2, p_3 \in \mathbb{C}$ make an equilateral triangle. Let $h_j(z) = 2(z - p_j) + p_j, j = 1, 2, 3$ and $G = \langle h_1, h_2, h_3 \rangle$. Then $J(G)$ is equal to the Sierpiński gasket ([**78**]; see Figure 2).

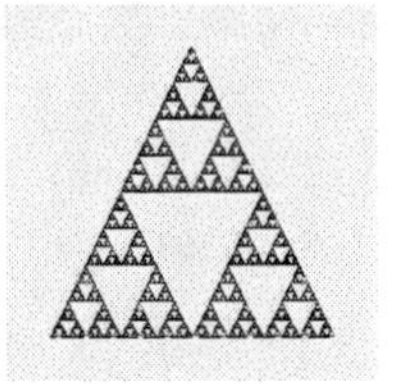

FIGURE 2. Sierpiński gasket

The following theorem is very important and fundamental, and there are many applications of it. This was shown by using Ahlfors's five island theorem in [**29, 79, 24**]. For a simple proof, see [**49, 50**].

THEOREM 2.9 ([**29, 79, 24, 49, 50**]). *Let G be a (possibly infinitely generated) rational semigroup. Suppose* $\sharp(J(G)) \geq 3$. *Then*

$$J(G) = \overline{\{z \in \hat{\mathbb{C}} \mid \exists g \in G, g(z) = z, |m(g, z)| > 1\}} = \overline{\bigcup_{g \in G} J(g)},$$

where the closure is taken in $\hat{\mathbb{C}}$, *and* $m(g, z)$ *denotes the multiplier of* g *at* z *(for the definition of multiplier, see* [**38**]*).*

Regarding the iteration dynamics of a single rational map, by applying and developing Teichmüller theory, D. Sullivan showed a quite remarkable result which states that there is **no wandering domain** ([**54**]). We consider the same problem for the dynamics of general rational semigroups.

DEFINITION 2.10. Let G be a rational semigroup. Let U be a connected component of $F(G)$. For each $g \in G$, we denote by U_g the connected component of $F(G)$ containing $g(U)$. We say that U is a **wandering domain** if $\sharp\{U_g \mid g \in G\} = \infty$.

REMARK 2.11. In [**29**], an infinitely generated polynomial semigroup having a wandering domain was found. Moreover, in [**29**], it was shown that if a rational semigroup is "nearly abelian", then there is no wandering domain. In [**26**], T. Harada gave a positive answer to a conjecture in [**29**] on nearly abelian polynomial semigroups.

The following conjecture has not been solved.

CONJECTURE 2.12 ([**29**]). *Let G be a finitely generated rational semigroup containing an element g with* $\deg(g) \geq 2$. *Then G has no wandering domain.*

In the study of the dynamics of rational semigroups, even if we try to use Teichmüller theory, we sometimes are not able to obtain enough information regarding the dynamics on the Fatou set. This is because the Fatou set may not be completely invariant under each element of the semigroup. In order to see this, for a rational semigroup G, let $CI(G)$ be the smallest set in the space of all compact subsets K of $\hat{\mathbb{C}}$ with $\sharp K \geq 3$ which is completely invariant under each element of G. This $CI(G)$ is called the **completely invariant Julia set** of G ([**46, 47**]). In [**46**], R. Stankewitz showed that if f, g are two polynomials of degree two or more such that $J(f) \neq J(g)$, then $CI(\langle f, g\rangle) = \hat{\mathbb{C}}$. Based on this result, the author of this article gave the following Conjecture 2.13. A measurable $(-1,1)$-form $\mu = \mu(z)(d\overline{z}/dz)$ on a Riemann surface S is called a Beltrami differential on S. Let $\|\mu\|_\infty$ be the L^∞-norm of μ. In the Teichmüller theory, we deal with a space of (equivalence classes) of Beltrami differentials.

CONJECTURE 2.13 ([**65**]). *Let f, g be two polynomials such that* $\deg(f), \deg(g) \geq 2$ *and $J(f) \neq J(g)$. If a Beltrami differential μ on $\hat{\mathbb{C}}$ with $\|\mu\|_\infty < 1$ is invariant under both f and g, then $\mu = 0$. (Note: By* [**27, 46**], *the space of such Beltrami differentials μ is finite dimensional.)*

We remark that it is important to consider the Teichmüller space of a rational semigroup (Harada, [**27**]) in order to consider Conjecture 2.13.

In section 5, we show that if we assume "(semi-)hyperbolicity", then there is no wandering domain. The dynamics of rational semigroups are related to the usual iteration dynamics of a single polynomial map through "generalized polynomial-like maps" (K. Katagata, [**34**]). In [**51**], several counterexamples were given to some conjectures in [**29**], and modified conjectures were given there. For more papers on rational semigroups, see the above papers, [**30, 31, 6, 7, 41, 76, 40**], and the papers of the author of this article. Moreover, R. Stankewitz and his students (W. Conatser, T. Butz, Y. Li, K. Hart) made a very nice computer program to draw Julia sets and attractors of rational semigroups (see http://www.bsu.edu/web/rstankewitz/JuliaHelp2.0/Julia.html).

3. Dynamics of fibered rational maps

In this section, in order to investigate random complex dynamics and the dynamics of rational semigroups, we introduce the dynamics of fibered rational maps. The first study of the dynamics of fibered rational maps was given by M. Jonsson in ([**33**]). Note that for the application to random complex dynamics and the dynamics of rational semigroups, it is enough to consider the dynamics of fibered rational maps on trivial bundles. However, in this article, we consider the dynamics of fibered rational maps on general bundles, since such dynamics appear in higher-dimensional complex dynamics and there are many applications.

DEFINITION 3.1 ([**33**]). In this article, we say that a triplet (π, Y, X) is a $\hat{\mathbb{C}}$-bundle if all of the following hold.

(1) Y and X are compact metric spaces.
(2) $\pi : Y \to X$ is continuous and surjective.
(3) There exists an open covering $\{U_i\}$ of X such that for each i, there exists a homeomorphism $\Phi_i : U_i \times \hat{\mathbb{C}} \to \pi^{-1}(U_i)$ satisfying the following.

- $\Phi_i(\{x\}\times\hat{\mathbb{C}}) = \pi^{-1}(\{x\})$ and for each $x \in U_i \cap U_j$, under the canonical identification $\{x\} \times \hat{\mathbb{C}} \cong \hat{\mathbb{C}}$, $\Phi_j^{-1} \circ \Phi_i : \{x\} \times \hat{\mathbb{C}} \to \{x\} \times \hat{\mathbb{C}}$ is a holomorphic automorphism on $\hat{\mathbb{C}}$.

REMARK 3.2. Let (π, Y, X) be a $\hat{\mathbb{C}}$-bundle. For each $x \in X$, we set $Y_x := \pi^{-1}(\{x\})$. From the definition of a $\hat{\mathbb{C}}$-bundle, each fiber Y_x has a complex structure. Moreover, by using $\{\Phi_i\}$, for each $x_0 \in X$, we can consider a continuous family $\{i_x : \hat{\mathbb{C}} \to Y_x\}$ of homeomorphisms, where x belongs to a small neighborhood of x_0. Such a family $\{i_x\}$ is called a local parameterization. Since we are assuming that X is compact, we may assume that there exists a compact family M_0 of the space $\text{Aut}(\hat{\mathbb{C}})$ of all holomorphic automorphisms on $\hat{\mathbb{C}}$ such that for any local parameterizations $\{i_x\}, \{j_x\}$, $j_x^{-1} \circ i_x \in M_0$.

In this article, for a $\hat{\mathbb{C}}$-bundle (π, Y, X), we always assume the following.

- For each $x \in X$, there exists a smooth positive $(1,1)$-form ω_x on Y_x such that the distance on Y_x which is induced by ω_x is equal to the relative distance induced by the distance on Y. Moreover, $x \mapsto \omega_x$ is continuous; i.e., if $\{i_x\}$ is a local parameterization, then $x \mapsto i_x^*\omega_x$ is continuous with respect to the C^∞ topology on the space of all smooth $(1,1)$-forms on $\hat{\mathbb{C}}$.

DEFINITION 3.3. Let $(\pi, Y = X \times \hat{\mathbb{C}}, X)$ be a trivial $\hat{\mathbb{C}}$-bundle. Let $\pi_{\hat{\mathbb{C}}} : Y \to \hat{\mathbb{C}}$ be the canonical projection. In this case, we assume that under the identification $\pi_{\hat{\mathbb{C}}} : Y_x \cong \hat{\mathbb{C}}$, ω_x is equal to the spherical metric (Fubini-Study metric) on $\hat{\mathbb{C}}$.

DEFINITION 3.4 ([**33**]). Let (π, Y, X) be a $\hat{\mathbb{C}}$-bundle. Let $f : Y \to Y, g : X \to X$ be continuous maps. We say that $f : Y \to Y$ is a **fibered rational map** over $g : X \to X$ if $\pi \circ f = g \circ \pi$ and for each $x \in X$, $f|_{Y_x} : Y_x \cong \hat{\mathbb{C}} \to Y_{g(x)} \cong \hat{\mathbb{C}}$ is a nonconstant rational map.

DEFINITION 3.5. Under the above definition, if (π, Y, X) is a trivial $\hat{\mathbb{C}}$-bundle, $f : Y \to Y$ is called a **rational skew product**.

DEFINITION 3.6. Let (π, Y, X) be a $\hat{\mathbb{C}}$-bundle and let $f : Y \to Y$ be a fibered rational map over $g : X \to X$. For each $x \in X$ and $n \in \mathbb{N}$, we set $f_x^n = f^n|_{Y_x} : Y_x \to Y_{g^n(x)}$. Moreover, we set $f_x := f_x^1$. Furthermore, we set $d_n(x) := \deg(f_x^n)$, $d(x) := \deg(f_x)$.

REMARK 3.7. From the definition of fibered rational maps, $x \mapsto d(x) \in \mathbb{N}$ is continuous.

DEFINITION 3.8. Let $(\pi, Y = X \times \hat{\mathbb{C}}, X)$ be a trivial $\hat{\mathbb{C}}$-bundle and let $f : Y \to Y$ be a rational skew product over $g : X \to X$. For each $x \in X$ and each $n \in \mathbb{N}$, let $f_{x,n} : \hat{\mathbb{C}} \to \hat{\mathbb{C}}$ be the rational map defined by $f_{x,n}(y) := \pi_{\hat{\mathbb{C}}}(f^n(x,y))$. If $f_{x,1} : \hat{\mathbb{C}} \to \hat{\mathbb{C}}$ is a polynomial map for each $x \in X$, then $f : Y \to Y$ is called a **polynomial skew product**.

DEFINITION 3.9. Let Rat be the space of all nonconstant rational maps on $\hat{\mathbb{C}}$ endowed with the topology induced by the uniform convergence on $\hat{\mathbb{C}}$. Let $\mathcal{Y}$ be the space of all polynomial maps g with $\deg(g) \geq 2$, endowed with the relative topology from Rat. (Note: Rat has countably many connected components. Each connected component is an open subset of Rat and is a finite-dimensional complex manifold. A similar thing holds for $\mathcal{Y}$. Moreover, in $\mathcal{Y}$, $h_n \to h$ if and only if

$\deg(h_n) = \deg(h)$ for each sufficiently large $n \in \mathbb{N}$, and each coefficient of h_n tends to the corresponding coefficient of h as $n \to \infty$.)

DEFINITION 3.10. Let Γ be a nonempty compact subset of Rat. We set $\Gamma^{\mathbb{N}} := \{\gamma = (\gamma_1, \gamma_2, \ldots) \mid \gamma_j \in \Gamma\ (\forall j \in \mathbb{N})\}$. We define $\text{Rat}^{\mathbb{N}}$ similarly as above, and we endow it with the product topology. We endow $\Gamma^{\mathbb{N}}$ with the relative topology from $\text{Rat}^{\mathbb{N}}$. Then $\Gamma^{\mathbb{N}}$ is a compact metric space. Let $\sigma : \Gamma^{\mathbb{N}} \to \Gamma^{\mathbb{N}}$ be the map defined by $\sigma(\gamma_1, \gamma_2, \ldots) := (\gamma_2, \gamma_3, \ldots)$. This is called the **shift map**. Moreover, let $f : \Gamma^{\mathbb{N}} \times \hat{\mathbb{C}} \to \Gamma^{\mathbb{N}} \times \hat{\mathbb{C}}$ be the map defined by $(\gamma, y) \mapsto (\sigma(\gamma), \gamma_1(y))$, where $\gamma = (\gamma_1, \gamma_2, \ldots)$. Then $f : \Gamma^{\mathbb{N}} \times \hat{\mathbb{C}} \to \Gamma^{\mathbb{N}} \times \hat{\mathbb{C}}$ is a rational skew product over $\sigma : \Gamma^{\mathbb{N}} \to \Gamma^{\mathbb{N}}$.

$f : \Gamma^{\mathbb{N}} \times \hat{\mathbb{C}} \to \Gamma^{\mathbb{N}} \times \hat{\mathbb{C}}$ is called the **rational skew product associated with the compact family** Γ.

DEFINITION 3.11. For each $m \in \mathbb{N}$, we endow the finite set $\{1, \ldots, m\}$ with the discrete topology, and we set $\Sigma_m := \{1, \ldots, m\}^{\mathbb{N}}$ endowed with the product topology (Σ_m is a compact metrizable space). Let $\sigma : \Sigma_m \to \Sigma_m$ be the shift map, i.e. $\sigma(w_1, w_2, \ldots) = (w_2, w_3, \ldots)$. For each $f = (f_1, \ldots, f_m) \in (\text{Rat})^m$, let $\tilde{f} : \Sigma_m \times \hat{\mathbb{C}} \to \Sigma_m \times \hat{\mathbb{C}}$ be the map defined by $(w, y) \mapsto (\sigma(w), f_{w_1}(y))$, where $w = (w_1, w_2, \ldots) \in \Sigma_m$. Then $\tilde{f} : \Sigma_m \times \hat{\mathbb{C}} \to \Sigma_m \times \hat{\mathbb{C}}$ is a rational skew product over the shift map $\sigma : \Sigma_m \to \Sigma_m$. This $\tilde{f}$ is called the **rational skew product associated with a multi-map** $f = (f_1, \ldots, f_m) \in (\mathbf{Rat})^m$.

DEFINITION 3.12. Let (π, Y, X) be a $\hat{\mathbb{C}}$-bundle and let $f : Y \to Y$ be a fibered rational map over $g : X \to X$. Let $x \in X$. We denote by $F^x(f)$ the set of all points $y \in Y_x$ that satisfy the following: "the family $\{f_x^n : Y_x \to Y\}_{n \in \mathbb{N}}$ is equicontinuous in a neighborhood of y in Y_x". $F^x(f)$ is called the **fiber Fatou set** of f at x. Moreover, we set $J^x(f) := Y_x \setminus F^x(f)$, and this is called the **fiber Julia set** of f at x. Furthermore, we set $\tilde{J}(f) := \overline{\bigcup_{x \in X} J^x(f)}$, where the closure is taken in Y. We set $\tilde{F}(f) := Y \setminus \tilde{J}(f)$. Finally, for each $x \in X$, we set $\hat{J}^x(f) := \tilde{J}(f) \cap Y_x$.

LEMMA 3.13 ([**33, 61**]). *Let (π, Y, X) be a $\hat{\mathbb{C}}$-bundle and let $f : Y \to Y$ be a fibered rational map over $g : X \to X$. Then we have all of the following.*

(1) *For each $x \in X$, $F^x(f)$ is an open subset of Y_x, and $J^x(f)$ and $\hat{J}^x(f)$ are compact subsets of Y_x.*
(2) *For each $x \in X$, $f_x^{-1}(J^{g(x)}(f)) = J^x(f)$ and $\hat{J}^x(f) \supset J^x(f)$. Thus $f(\tilde{J}(f)) \subset \tilde{J}(f)$. However, $\hat{J}^x(f) = J^x(f)$ does not hold in general.*
(3) *If, in addition to the assumption, $g : X \to X$ is an open map, then $f^{-1}(\tilde{J}(f)) = \tilde{J}(f)$ and $f(\tilde{F}(f)) \subset \tilde{F}(f)$.*
(4) *If, in addition to the assumption, $g : X \to X$ is an open and surjective map, then $f^{-1}(\tilde{J}(f)) = \tilde{J}(f) = f(\tilde{J}(f))$, $f^{-1}(\tilde{F}(f)) = \tilde{F}(f) = f(\tilde{F}(f))$, and for each $x \in X$, $f_x^{-1}(\hat{J}^{g(x)}(f)) = \hat{J}^x(f)$.*

DEFINITION 3.14. Let $(\pi, Y = X \times \hat{\mathbb{C}}, X)$ be a trivial $\hat{\mathbb{C}}$-bundle and let $f : Y \to Y$ be a rational skew product over $g : X \to X$. Then for each $x \in X$, we set $F_x(f) := \pi_{\hat{\mathbb{C}}}(F^x(f))$, $J_x(f) := \pi_{\hat{\mathbb{C}}}(J^x(f))$, and $\hat{J}_x(f) := \pi_{\hat{\mathbb{C}}}(\hat{J}^x(f))$. If f is the rational skew product associated with a compact set $\Gamma \subset$ Rat, then $J_x(f)$ is called the **Julia set** of x.

LEMMA 3.15 ([**70**]). *Let Γ be a nonempty compact subset of* Rat *and let $f : \Gamma^{\mathbb{N}} \times \hat{\mathbb{C}} \to \Gamma^{\mathbb{N}} \times \hat{\mathbb{C}}$ be the rational skew product associated with Γ. Let G be the rational* *semigroup generated by* Γ. *Suppose* $\sharp(J(G)) \geq 3$. *Then* (1) $\pi_{\hat{\mathbb{C}}}(\tilde{J}(f)) = J(G)$ *and* (2) *for each* $\gamma = (\gamma_1, \gamma_2, \gamma_3, \ldots) \in \Gamma^{\mathbb{N}}$, $\hat{J}_\gamma(f) = \bigcap_{j=1}^{\infty} \gamma_1^{-1} \cdots \gamma_j^{-1}(J(G))$.

REMARK 3.16. The reason we take the closure of $\bigcup_{x\in X} J^x(f)$ in Y in the definition of $\tilde{J}(f)$ is that the subset $\bigcup_{x\in X} J^x(f)$ is not a compact subset of Y in general. For example, let $z_1 \in \mathbb{C}$ be a point and let h_1 be a polynomial with $\deg(h_1) \geq 2$ such that h_1 has a Siegel disk with center z_1, and let h_2 be a polynomial with $\deg(h_2) \geq 2$ such that z_1 is a repelling fixed point of h_2. Let $\Gamma = \{h_1, h_2\}$. Let $f : \Gamma^{\mathbb{N}} \times \hat{\mathbb{C}} \to \Gamma^{\mathbb{N}} \times \hat{\mathbb{C}}$ be the rational skew product associated with Γ. We set $x = (h_1, h_1, h_1, \ldots) \in \Gamma^{\mathbb{N}}$. Then, it is easy to see that $(x, z_1) \in \hat{J}^x(f) \setminus J^x(f)$. In particular, $\bigcup_{x\in X} J^x(f)$ is not compact. This fact ($\bigcup_{x\in X} J^x(f)$ may not be compact) causes many difficulties in the study of the dynamics of fibered rational maps. Thus it causes many difficulties in the study of random complex dynamics and the dynamics of rational semigroups.

Let (π, Y, X) be a $\hat{\mathbb{C}}$-bundle and let $f : Y \to Y$ be a fibered rational map over $g : X \to X$. We now consider the case $d(x) \geq 2$ for each $x \in X$. Then, by using M. Jonsson's arguments ([**33**]), in which pluripotential theory was used, and by using its slight generalization ([**58**]), we can construct a "canonical" family $\{\mu^x\}_{x\in X}$ of measures on Y such that for each $x \in X$, the support of μ^x is equal to the fiber Julia set $J^x(f)$ as follows: First, let $x \in X$ be a point and for each $n \in \mathbb{N}$, we set $x_n = g^n(x)$. Let i_x (resp. i_{x_1}) be a local parameterization around x (resp. x_1) and let $Q_x := i_{x_1}^{-1} \circ f_x \circ i_x$. We lift Q_x to a homogeneous polynomial map R_x on $\mathbb{C}^2 \setminus \{0\}$ of degree $d(x)$ such that $\sup\{|R_x(z,w)| : |(z,w)| = 1\} = 1$. Similarly, for each point $x_1, x_2, \ldots$, we take a local parameterization $i_{x_1}, i_{x_2}, \ldots$, and we take a homogeneous polynomial map $R_{x_1}, R_{x_2}, \ldots$. Then let $R_x^n = R_{x_{n-1}} \circ \cdots \circ R_x$. For each n, let $G_{x_n,0}$ be a smooth plurisubharmonic function on $\mathbb{C}^2 \setminus \{0\}$ such that it is a potential of ω_{x_n}; i.e., for any local holomorphic section $s : V \subset \hat{\mathbb{C}} = \mathbb{CP}^1 \to \mathbb{C}^2 \setminus \{0\}$ of the canonical projection $\mathbb{C}^2 \setminus \{0\} \to \mathbb{CP}^1 = \hat{\mathbb{C}}$, we have $\omega_{x_n} = dd^c(G_{x_n,0} \circ s \circ i_{x_n}^{-1})$, $G_{x_n,0}(z,w) \leq \log|(z,w)| + O(1)$ as $|(z,w)| \to \infty$, and $G_{x_n,0}(\lambda z, \lambda w) = G_{x_n,0}(z,w) + \log|\lambda|$ ($\forall \lambda \in \mathbb{C} \setminus \{0\}$). Let $G_{x,n} := (1/d_n(x)) \cdot G_{x_n,0} \circ R_x^n$. This is a plurisubharmonic function on $\mathbb{C}^2 \setminus \{0\}$. Letting $n \to \infty$, $G_{x,n}$ tends to a continuous plurisubharmonic function G_x on $\mathbb{C}^2 \setminus \{0\}$, and we can see that G_x does not depend on the choices of the local parameterizations i_{x_n} around x_n and the choices of $G_{x_n,0}$ and R_{x_n}. Moreover, we can see that $x \mapsto G_x$ is continuous. Furthermore, we can see that there exists a Borel probability measure μ^x on Y_x such that for any local holomorphic section $s : V \subset \hat{\mathbb{C}} \to \mathbb{C}^2 \setminus \{0\}$, $\mu^x = dd^c(G_x \circ s \circ i_x^{-1})$. By using this family $\{\mu^x\}_{x\in X}$ of measures, we obtain the following proposition. Note that this proposition and Theorems 3.19, 3.22 (which will be stated later) generalize some results of H. Brolin, M. Lyubich, Freire-Lopes-Mañé, Fornaess-Sibony in the usual iteration dynamics of a single rational map.

PROPOSITION 3.17 ([**33, 58, 61**]). *Let* (π, Y, X) *be a* $\hat{\mathbb{C}}$*-bundle and let* $f : Y \to Y$ *be a fibered rational map over* $g : X \to X$. *Suppose that for each* $x \in X$, $d(x) \geq 2$. *Then, regarding the family* $\{\mu^x\}_{x\in X}$ *of measures constructed above and fiber Julia sets* $J^x(f)$, *we have all of the following.*

(1) *For each* $x \in X$, $\operatorname{supp} \mu^x = J^x(f)$.

(2) *We regard μ^x as a Borel probability measure on Y for each $x \in X$. Then, $x \mapsto \mu^x$ is continuous with respect to the weak* topology of the space of all Borel probability measures on Y.*
(3) *For each $x \in X$, $J^x(f)$ is a nonempty compact set that does not have any isolated point. In particular, $J^x(f)$ has uncountably many points.*
(4) *Regarding $J^x(f)$ as a compact subset of Y, $x \mapsto J^x(f)$ is lower semicontinuous; i.e., for each $x \in X, y \in J^x(f)$ and for each sequence $\{x^n\}_{n\in\mathbb{N}}$ in X tending to x, there exists a sequence $\{y^n\}_{n\in\mathbb{N}}$ in Y such that $y^n \in J^{x^n}(f)$ $(\forall n \in \mathbb{N})$ and $y^n \to y$ $(n \to \infty)$.*

REMARK 3.18. As mentioned above, if "$d(x) \geq 2$ for each $x \in X$", then $x \mapsto J^x(f)$ is lower semicontinuous. We remark that even if we assume the above condition "$\cdots$", then $x \mapsto J^x(f)$ is not continuous with respect to the Hausdorff topology in the space of all nonempty compact subsets of Y in general. In fact, even under the above assumption "$\cdots$", if $\bigcup_{x\in X} J^x(f)$ is not a compact subset of Y (e.g., the example in Remark 3.16), then $x \mapsto J^x(f)$ is not continuous. As mentioned in Remark 3.16, this causes a difficulty in the study of the dynamics of fibered rational maps, random complex dynamics, and the dynamics of rational semigroups, which we do not have in the usual iteration dynamics of a single rational map.

We now consider the dynamics of a continuous map f on a general compact metric space Y. There is a well-known fact, the so-called **variational principle**, which states that the "topological entropy" of f is equal to the supremum of "metric entropies" of f-invariant Borel probability measures on Y. In the theory of dynamical systems, it is a very important and interesting problem to determine the topological entropy and to investigate the existence and uniqueness of the invariant measure which maximizes the metric entropy. Regarding this problem, we have the following results. First, we introduce some notation. For a continuous map f on a compact metric space Y, we denote by $h(f)$ the **topological entropy** of $f : Y \to Y$. For an f-invariant Borel probability measure ν, we denote by $h_\nu(f)$ the **metric entropy** of the couple (f, ν). In the above setting, if there is a compact metric space X, a continuous map g on X, and a continuous map $\pi : Y \to X$ such that $g \circ \pi = \pi \circ f$, then we denote by $h_\nu(f|g)$ the **relative metric entropy** of (f, ν) with respect to $g : X \to X$. Moreover, if (π, Y, X) is a $\hat{\mathbb{C}}$-bundle and if $f : Y \to Y$ is a fibered rational map over $g : X \to X$ such that $d(x) \geq 2$ for each $x \in X$, for any g-invariant Borel probability measure μ' on X, we associate a Borel probability measure μ on Y defined as:

$$\langle \mu, \varphi \rangle := \int_X \left(\int_{Y_x} \varphi \, d\mu^x \right) d\mu', \tag{5}$$

for any continuous function φ on Y, where $\{\mu^x\}_{x\in X}$ is the family of measures on Y constructed just before Proposition 3.17. Then, we can see that μ is f-invariant and $\pi_*(\mu) = \mu'$. Under this notation, M. Jonsson showed the following.

THEOREM 3.19 ([**33**]). *Let (π, Y, X) be a $\hat{\mathbb{C}}$-bundle and let $f : Y \to Y$ be a fibered rational map over $g : X \to X$. Suppose that there exists a positive integer d with $d \geq 2$ such that for each $x \in X, d(x) = d$. Then we have all of the following.*
(1) $h(f) = h(g) + \log d$.
(2) *For any f-invariant Borel probability measure ν on Y, $h_\nu(f|g) \leq \log d$.*

(3) *Let μ' be a g-invariant Borel probability measure on X and let μ be the measure on Y in (5). Then $h_\mu(f|g) = \log d$ and $h_\mu(f) = h_{\mu'}(g) + \log d$. Moreover,*
(a) if μ' is ergodic with respect to $g : X \to X$, then μ is ergodic with respect to $f : Y \to Y$,
(b) if μ' is mixing with respect to $g : X \to X$, then μ is mixing with respect to $f : Y \to Y$,
(c) if μ' is a maximum entropy measure for $g : X \to X$ (i.e., the metric entropy of μ' is a maximum one and is equal to the topological entropy), then μ is a maximum entropy measure for $f : Y \to Y$.

(4) *Let μ' be a g-invariant Borel probability measure on X and let μ be the measure on Y in (5). Let ν be an f-invariant Borel probability measure on Y such that $\pi_*(\nu) = \mu'$. If $h_\nu(f|g) = \log d$, then $\nu = \mu$.*

(5) *Let μ' be a g-invariant Borel probability measure on X and let μ be the measure on Y in (5).*
If μ' is the unique maximum entropy measure for $g : X \to X$, then μ is the unique maximum entropy measure for $f : Y \to Y$.

If we apply Theorem 3.19 to the random complex dynamics and the dynamics of rational semigroups, then we want to remove the condition "there exists a positive integer d with $d \geq 2$ such that for each $x \in X, d(x) = d$". We present a result without the above condition.

LEMMA 3.20 ([**65**]). *Let (π, Y, X) be a $\hat{\mathbb{C}}$-bundle and let $f : Y \to Y$ be a fibered rational map over $g : X \to X$. Then, we have all of the following.*

(1) *For any f-invariant Borel probability measure μ on Y, $h_\mu(f|g) \leq \int_X \log d(x)\ d(\pi_*(\mu))(x)$.*

(2) *Suppose that for each $x \in X$, $d(x) \geq 2$. Let μ' be a g-invariant Borel probability measure on X and let μ be the measure on Y in (5). Then statements (a) and (b) in Theorem 3.19(3) hold. Moreover, we have the following equation:*

$$h_\mu(f|g) = \sup_\nu h_\nu(f|g) = \int_X \log d(x)\ d\mu'(x), \tag{6}$$

where the supremum is taken in the space of all f-invariant Borel probability measures ν on Y with $\pi_*(\nu) = \mu'$.

The author has a conjecture that under the assumptions of Lemma 3.20, there exists a unique ν of which $h_\nu(f|g)$ is equal to the supremum in (6). However, it is still open (we cannot generalize the arguments in [**33**]). We give the precise form of this conjecture as follows.

CONJECTURE 3.21. *Let (π, Y, X) be a $\hat{\mathbb{C}}$-bundle and let $f : Y \to Y$ be a fibered rational map over $g : X \to X$. Suppose that for each $x \in X$, $d(x) \geq 2$. Let μ' be a g-invariant Borel probability measure on X and let μ be the measure on Y in (5). Then there exists a unique ν of which $h_\nu(f|g)$ is equal to the supremum in (6). Moreover, the measure ν is equal to μ.*

We remark that if Γ is a finite subset of Rat and if $f : \Gamma^{\mathbb{N}} \times \hat{\mathbb{C}} \to \Gamma^{\mathbb{N}} \times \hat{\mathbb{C}}$ is a rational skew product associated with Γ, then Conjecture 3.21 is true under a mild assumption. More precisely, we have the following results.

We also remark that the probability measure ρ^a in the following theorem appears again in the results of (non)differentiability of the function T_∞ (a complex analogue of the devil's staircase or Lebesgue's singular functions) of probability of tending to ∞ in the random dynamics of polynomials in section 10. To state the theorem, we introduce some notation. First, let $m \in \mathbb{N}$ and let $f = (f_1, \dots, f_m) \in (\text{Rat})^m$ be a multi-map. We consider the skew product map $\tilde{f} : \Sigma_m \times \hat{\mathbb{C}} \to \Sigma_m \times \hat{\mathbb{C}}$ associated with f. An element $a = (a_1, a_2, \dots, a_m) \subset \mathbb{R}^m$ with $\sum_{j=1}^m a_j = 1, a_j > 0$ $(\forall j)$ is called a "weight", and we denote by $\mathcal{W}$ the set of all weights. Let K be a nonempty compact subset of $\Sigma_m \times \hat{\mathbb{C}}$ such that $\tilde{f}^{-1}(K) \subset K$. Let $C(K)$ be the space of all continuous complex-valued functions on K endowed with the supremum norm $\|\cdot\|_K$. For each $a \in \mathcal{W}$, let $B_a : C(K) \to C(K)$ be the operator on $C(K)$ defined as $B_a\varphi(z) = \sum_{\zeta \in \tilde{f}^{-1}(z)} \varphi(\zeta)\psi_a(\zeta)$, where $\psi_a : \Sigma_m \times \hat{\mathbb{C}} \to \mathbb{R}$ is the function defined as $\psi_a(\zeta) = a_{w_1}/\deg(f_{w_1})$ if $\pi(\zeta) = (w_1, w_2, \dots) \in \Sigma_m$. Under this notation, we present the following result.

THEOREM 3.22 ([**57**]). *Let $m \in \mathbb{N}$ and let $f = (f_1, \dots, f_m) \in (\text{Rat})^m$ be a multi-map ($\deg(f_j)$ might be mutually distinct, and there might be a k with $\deg(f_k) = 1$). Let $\tilde{f} : \Sigma_m \times \hat{\mathbb{C}} \to \Sigma_m \times \hat{\mathbb{C}}$ be the rational skew product associated with f. Let $G = \langle f_1, \dots, f_m \rangle$. Suppose that G contains an element of degree two or more, and suppose $E(G) \subset F(G)$. Moreover, setting $H := \{h^{-1} \mid h \in \text{Aut}(\hat{\mathbb{C}}) \cap G\}$, we suppose that $F(H) \supset J(G)$ (if $H = \emptyset$, then we set $F(H) = \hat{\mathbb{C}}$). Then, we have all of the following.*

(1) *For each $a \in \mathcal{W}$, there exists a unique Borel probability measure ρ^a on $\Sigma_m \times \hat{\mathbb{C}}$ which satisfies the following: if K is a nonempty compact subset of $\Sigma_m \times (\hat{\mathbb{C}} \setminus E(G))$ with $\tilde{f}^{-1}(K) \subset K$, then for any $\varphi \in C(K)$, $\|B_a^n\varphi - (\int_K \varphi \, d\rho^a) \cdot \mathbf{1}\|_K \to 0$ $(n \to \infty)$, where $\mathbf{1}$ denotes the constant function taking its value 1.*
(2) *Regarding B_a as an operator on $C(\Sigma_m \times \hat{\mathbb{C}})$, we have $B_a^*\rho^a = \rho^a$. Moreover, ρ^a is $\tilde{f}$-invariant. Furthermore, $\pi_*(\rho^a)$ is equal to the Bernoulli measure τ_a on Σ_m with respect to the weight a.*
(3) *The support of ρ^a is equal to $\tilde{J}(\tilde{f})$.*
(4) *$(\tilde{f}, \rho^a)$ is exact. In particular, ρ^a is ergodic with respect to $\tilde{f}$.*
(5) *Let $\tilde{a} = \left(\deg(f_1)/\sum_{j=1}^m \deg(f_j), \dots, \deg(f_m)/\sum_{j=1}^m \deg(f_j)\right) \in \mathcal{W}$. Then $h(\tilde{f}) = h_{\rho^{\tilde{a}}}(\tilde{f}) = \log(\sum_{j=1}^m \deg(f_j))$, and $\rho^{\tilde{a}}$ is the unique maximum entropy measure for $\tilde{f} : \Sigma_m \times \hat{\mathbb{C}} \to \Sigma_m \times \hat{\mathbb{C}}$. In particular, the image of the maximum entropy measure $\rho^{\tilde{a}}$ for $\tilde{f}$ under π_* is equal to the Bernoulli measure on Σ_m with respect to the weight $\tilde{a}$. (Thus, if the numbers $\deg(f_j)$ are not the same, then statement (c) in Theorem 3.19(3) does not hold for the map $\tilde{f} : \Sigma_m \times \hat{\mathbb{C}} \to \Sigma_m \times \hat{\mathbb{C}}$ in this theorem.)*
(6) *For each $a \in \mathcal{W}$, we have $h_{\rho^a}(\tilde{f}|\sigma) = \sup_\rho h_\rho(\tilde{f}|\sigma) = \sum_{j=1}^m a_j \log(\deg(f_j))$, where the supremum is taken in the space of all $\tilde{f}$-invariant Borel probability measures ρ on $\Sigma_m \times \hat{\mathbb{C}}$ with $\pi_*(\rho) = \tau_a$.*
(7) *Let $a \in \mathcal{W}$. If an ergodic $\tilde{f}$-invariant Borel probability measure ρ on $\Sigma_m \times \hat{\mathbb{C}}$ satisfies $\pi_*(\rho) = \tau_a$ and $h_\rho(\tilde{f}|\sigma) = \sum_{j=1}^m a_j \log(\deg(f_j))$, then $\rho = \rho^a$.*

REMARK 3.23. Under the assumptions of Theorem 3.22, if we assume further that $\deg(f_j) \geq 2$ for each $j = 1, \dots, m$, then $E(G) \subset F(G)$ and we have all of the following.

(1) We can remove "ergodic" in Theorem 3.22(7). (Note: This matter was not mentioned in [**57**], but if we combine the arguments in [**33**, Lemma 6.4] and statement (7) above, then we can show it.)
(2) Let τ_a be the Bernoulli measure on Σ_m with respect to the weight a. Setting $\mu' = \tau_a$, let μ be the $\tilde{f}$-invariant Borel probability measure on $Y = \Sigma_m \times \hat{\mathbb{C}}$ as in (5). Then μ is equal to ρ^a. (Note: This matter was not mentioned in [**57**], but we can easily show it by using statement (7) above and Lemma 3.20.)

REMARK 3.24. Theorem 3.19 and Theorem 3.22 were obtained at almost the same time, but the methods of the proofs are different. For other research on the dynamics of fibered rational maps, see the work [**32**] of M. Jonsson and O. Sester's works [**43, 44**], which the author of this article thinks are very important.

4. Uniform perfectness and interior points of Julia sets

In complex analysis and potential theory, we sometimes consider the property of so-called "uniform perfectness" of compact sets. In this section, we present some results on the uniform perfectness of fiber Julia sets of fibered rational maps or Julia sets of rational semigroups. Applying these results, we find a new phenomenon that the Julia set $J(G)$ of a finitely generated rational semigroup may have nonempty interior even if $J(G) \neq \hat{\mathbb{C}}$. Note that this phenomenon does not hold in the theory of Kleinian groups or the iteration dynamics of a single rational map.

First, we define uniform perfectness.

DEFINITION 4.1 ([**35**]). Let A be a domain in $\hat{\mathbb{C}}$. If $\hat{\mathbb{C}} \setminus A$ consists of two connected components, then we say that A is a **doubly connected domain**. If A is a doubly connected domain and $\sharp(\hat{\mathbb{C}} \setminus A) \geq 3$, then A is holomorphically isomorphic to some annulus $\{z \in \mathbb{C} \mid 1 < |z| < r\}$ $(1 < r \leq \infty)$, and we set $M(A) := (\frac{1}{2\pi}) \cdot \log r$. If $\sharp(\hat{\mathbb{C}} \setminus A) = 2$, we set $M(A) := \infty$. $M(A)$ is called the **modulus** of A. Moreover, for a doubly connected domain A and a subset K of $\hat{\mathbb{C}}$, we say that A **separates** K if each connected component of $\hat{\mathbb{C}} \setminus A$ intersects K.

DEFINITION 4.2 ([**53**]). Let K be a compact subset of $\hat{\mathbb{C}}$ with $\sharp K \geq 2$. Let $C > 0$. We say that K is **C- uniformly perfect** if for any doubly connected domain A separating K, we have $M(A) \leq C$. Moreover, if K is C-uniformly perfect with some $C > 0$, then K is said to be **uniformly perfect**.

In [**30**], Hinkkanen and Martin showed that there is a connection between the existence of interior points in the Julia set and the uniform perfectness of the Julia set as follows.

NOTATION. For any subset J of a topological space Y, we denote by $\mathrm{int}(J)$ the set of all interior points of J.

THEOREM 4.3. *Let G be a rational semigroup. Let $a \in J(G)$ and suppose that there exists $g \in G$ such that $g(a) = a$ and $g'(a) = 0$. If $J(G)$ is uniformly perfect, then $a \in \mathrm{int}(J(G))$.*

In fact, as Stankewitz showed in the following theorem, the Julia sets of many rational semigroups are uniformly perfect.

THEOREM 4.4 ([**48**]). *Let Γ be a nonempty relatively compact subset of Rat. Let G be a rational semigroup generated by Γ. Then $J(G)$ is uniformly perfect. In particular, the Julia set of any finitely generated rational semigroup is uniformly perfect.*

REMARK 4.5. Combining Theorem 4.3 and Theorem 4.4, we can construct many examples of rational semigroups G such that $J(G) \neq \hat{\mathbb{C}}$ and $\text{int}(J(G)) \neq \emptyset$. For example, suppose h_1, h_2 are two polynomials with $\deg(h_i) \geq 2$, each fixing $a \in \mathbb{C}$. Suppose also that $h_1'(a) = 0$ and $|h_2'(a)| > 1$. Let $G = \langle h_1, h_2 \rangle$. Then $J(G) \neq \hat{\mathbb{C}}$ and $a \in \text{int}(J(G))$ ([**30**]). Note that in [**30**], it was shown that there exists an infinitely generated rational semigroup G such that $J(G)$ is not uniformly perfect.

Regarding the uniform perfectness and the existence of interior points of Julia sets, by using Proposition 3.17 (an application of potential theory) in section 3, we can show the following.

THEOREM 4.6 ([**61**]). *Let (π, Y, X) be a $\hat{\mathbb{C}}$-bundle and let $f : Y \to Y$ be a fibered rational map over $g : X \to X$. Suppose that for each $x \in X$, $d(x) \geq 2$. Then we have all of the following.*

(1) *There exists a constant $C_1 > 0$ such that for each $x \in X$, $J^x(f)$ and $\hat{J}^x(f)$ are C_1-uniformly perfect.*
(2) *There exists a constant $C_2 > 0$ such that for each $x \in X$, we have* $\text{diam} J_x > C_2$, *where* diam *denotes the diameter in the metric space Y.*

Applying Theorem 4.6(1)(2) to the rational skew product associated with a compact subset Γ of Rat such that $\deg(g) \geq 2$ for each $g \in \Gamma$, and then combining it with Theorem 2.9, we can show the following theorem.

THEOREM 4.7 ([**61**]). *Let Γ be a nonempty compact subset of Rat such that $\deg(g) \geq 2$ for each $g \in \Gamma$. Let G be a rational semigroup generated by Γ. Then there exists a constant $C > 0$ such that for each subsemigroup H of G, $J(H)$ is C-uniformly perfect.*

5. Hyperbolicity and semihyperbolicity

When we investigate the dynamics of smooth maps on a smooth manifold, we often assume a nice property, the so-called "hyperbolicity" on the dynamics. It is very interesting and important to consider when the hyperbolicity holds. In this section, regarding the dynamics of rational semigroups and fibered rational maps, we define the hyperbolicity and its weakened condition "semihyperbolicity", and we investigate the properties of them. In particular, we present a result that for a semihyperbolic fibered rational map $f : Y \to Y$, the set $\bigcup_{x \in X} J^x(f)$ is compact in Y. (Note: If we do not assume semihyperbolicity, this does not hold in general.) This has many applications. We can apply this result to various objects in the study of the dynamics of rational semigroups, for example, the estimate of the Hausdorff dimension of the Julia sets of rational semigroups.

DEFINITION 5.1. For a rational semigroup G, we set

$$P(G) := \overline{\bigcup_{g \in G} \{\text{critical values of } g : \hat{\mathbb{C}} \to \hat{\mathbb{C}}\}} \ (\subset \hat{\mathbb{C}})$$

and we call $P(G)$ the **postcritical set** of G.

REMARK 5.2. If G is generated by a nonempty subset Γ of Rat, then

$$P(G) = \overline{\bigcup_{g \in G \cup \{Id\}} g(\bigcup_{h \in \Gamma} \{\text{critical values of } h : \hat{\mathbb{C}} \to \hat{\mathbb{C}}\})}.$$

From this formula, we can obtain much information on $P(G)$. We have that for any $g \in G$, $g(P(G)) \subset P(G)$.

DEFINITION 5.3. Let G be a rational semigroup.

(1) We say that G is **hyperbolic** if $P(G) \subset F(G)$.
(2) We say that G is **subhyperbolic** if $P(G) \cap F(G)$ is compact and $\sharp(P(G) \cap J(G)) < \infty$.

DEFINITION 5.4. Let G be a rational semigroup and let $N \in \mathbb{N}$.

(1) We denote by $SH_N(G)$ the set of points $z \in \hat{\mathbb{C}}$ satisfying the following: " there exists a neighborhood U of z such that for each $g \in G$ and for any connected component V of $g^{-1}(U)$, $\deg(g : V \to U) \leq N$. Here we denote by $\deg(\cdot)$ the degree of a finite branched covering of $g : V \to U$".
(2) We set $UH(G) := \hat{\mathbb{C}} \setminus \bigcup_{N \in \mathbb{N}} SH_N(G)$. We say that G is **semihyperbolic** if $UH(G) \subset F(G)$ (i.e., there exists an $N \in \mathbb{N}$ with $J(G) \subset SH_N(G)$).

REMARK 5.5. If a rational semigroup G is hyperbolic, then G is subhyperbolic and semihyperbolic.

We next define the hyperbolicity and semihyperbolicity for fibered rational maps.

NOTATION. For each $x \in \hat{\mathbb{C}}, r > 0$, we set $B(x, r) := \{z \in \hat{\mathbb{C}} \mid d(x, z) < r\}$, where d denotes the spherical distance.

DEFINITION 5.6. Let (π, Y, X) be a $\hat{\mathbb{C}}$-bundle and let $f : Y \to Y$ be a fibered rational map over $g : X \to X$.

(1) We set $C(f) := \bigcup_{x \in X} \{y \in Y_x \mid y \text{ is a critical point of } f_x\} (\subset Y)$ and this is called the **fiber critical set** of f.
(2) We set $P(f) := \overline{\bigcup_{n \in \mathbb{N}} f^n(C(f))}$ $(\subset Y)$ and this is called the **fiber postcritical set** of f.
(3) We say that f is **hyperbolic (along fibers)** if $P(f) \subset \tilde{F}(f)$.
(4) For each $z \in Y$ and $n \in \mathbb{N}$, we denote by $(f^n)'(z)$ the derivative of $f^n_{\pi(z)} : Y_{\pi(z)} \to Y_{g^n(\pi(z))}$ at z. Moreover, we denote by $\|(f^n)'(z)\|$ the norm of the derivative $(f^n)'(z)$ measured from the Riemannian metric corresponding to $\omega_{\pi(z)}$ on $Y_{\pi(z)}$ to the Riemannian metric corresponding to $\omega_{g^n(\pi(z))}$ on $Y_{g^n(\pi(z))}$.
(5) We say that f is **expanding (along fibers)** if $\tilde{J}(f) \neq \emptyset$ and there exists a constant $C > 0$ and a constant $\lambda > 1$ such that $\inf_{z \in \tilde{J}(f)} \|(f^n)'(z)\| \geq C\lambda^n$.

(6) For any $N \in \mathbb{N}$, we denote by $SH_N(f)$ the set of points $z \in Y$ satisfying that there exists a positive number δ, a neighborhood U of $\pi(z)$ in X, and a local parameterization $\{i_x\}_{x \in U}$ in U such that for each $x \in U$, each $n \in \mathbb{N}$, each $x^n \in g^{-n}(\{x\})$ and each connected component V of $f_{x^n}^{-1}(i_x(B(i_{\pi(z)}^{-1}(z), \delta)))$,
$$\deg(f_{x^n} : V \to i_x(B(i_{\pi(z)}^{-1}(z), \delta))) \leq N.$$

(7) We set $UH(f) := Y \setminus \bigcup_{N \in \mathbb{N}} SH_N(f)$. We say that f is **semihyperbolic (along fibers)** if $UH(f) \subset \tilde{F}(f)$ (i.e., if there exists an $N \in \mathbb{N}$ with $\tilde{J}(f) \subset SH_N(f)$).

REMARK 5.7. If f is expanding, then f is hyperbolic. If f is hyperbolic, then f is semihyperbolic.

REMARK 5.8. Let Γ be a nonempty compact subset of Rat and let $f : \Gamma^{\mathbb{N}} \times \hat{\mathbb{C}} \to \Gamma^{\mathbb{N}} \times \hat{\mathbb{C}}$ be the rational skew product associated with Γ. Let G be the rational semigroup generated by Γ. Then G is hyperbolic if and only if f is hyperbolic. Moreover, G is semihyperbolic if and only if f is semihyperbolic.

We present some results on the dynamics of semihyperbolic rational semigroups.

THEOREM 5.9 ([**58**]). *Let G be a (possibly noncompactly generated) semihyperbolic rational semigroup containing an element of degree two or more. Then, for each $z \in F(G)$, $\overline{\{g(z) \mid g \in G\}} \subset F(G)$. In particular, there exists no wandering domain (since $\overline{\{g(z) \mid g \in G\}}$ is a compact subset of $F(G)$ by the above result).*

DEFINITION 5.10. Let G be a rational semigroup. We set
$$A(G) := \overline{\bigcup_{h \in G} h(\{z \in \hat{\mathbb{C}} \mid \exists g \in G, g(z) = z \text{ and } |m(g,z)| < 1\})} \ (\subset \hat{\mathbb{C}}).$$

THEOREM 5.11 ([**58**]). *Let $G = \langle h_1, \ldots, h_m \rangle$ be a finitely generated semihyperbolic rational semigroup containing an element of degree two or more. Suppose that either $\mathrm{Aut}(\hat{\mathbb{C}}) \cap G = \emptyset$ or each element of $\mathrm{Aut}(\hat{\mathbb{C}}) \cap G$ is loxodromic. (Note: An element g of $\mathrm{Aut}(\hat{\mathbb{C}})$ is said to be loxodromic if there exist exactly two fixed points z_1, z_2 of g and $|m(g, z_i)| \neq 1$ for each i.) Moreover, suppose $F(G) \neq \emptyset$. Then, $\emptyset \neq A(G) \subset F(G)$, and for each nonempty compact subset L of $F(G)$,*
$$\sup\{d(h_{i_n} \circ \cdots \circ h_{i_1}(z), A(G)) \mid z \in L, (i_1, \ldots, i_n) \in \{1, \ldots, m\}^n\} \to 0 \ (n \to \infty),$$
where d denotes the spherical distance. Furthermore, if $\{g_n\}_{n \in \mathbb{N}}$ is a sequence in G consisting of mutually distinct elements and if it converges to a $\varphi : V \to \hat{\mathbb{C}}$ on a nonempty subdomain V of $F(G)$, then there exists a point $\zeta \in A(G)$ such that $\varphi \equiv \zeta$.

In [**58**], a result on finitely generated subhyperbolic rational semigroups which is similar to Theorem 5.11 was shown.

In the latter half of this section, we discuss the problem "when are the semigroups semihyperbolic ?". Before that, we present some results on the dynamics of semihyperbolic fibered rational maps. We introduce the following technical condition.

DEFINITION 5.12. Let (π, Y, X) be a $\hat{\mathbb{C}}$-bundle and let $x \in X, r > 0, z \in Y_x$. Then we set $\tilde{B}(z, r) := \{y \in Y_x \mid d_x(y, z) < r\}$, where d_x denotes the relative distance on Y_x induced from the metric space Y.

DEFINITION 5.13. Let (π, Y, X) be a $\hat{\mathbb{C}}$-bundle and let $f : Y \to Y$ be a fibered rational map over $g : X :\to X$. We say that f satisfies **condition** $(*)$ if for each $x_0 \in X$, there exists a neighborhood U and a family $\{D_x\}_{x \in U}$ $(D_x \subset Y_x)$ of topological disks such that all of the following hold.

(1) For each $x \in U$, there exists a $z_x \in Y_x$ and an $r_x > 0$ such that $D_x = \tilde{B}(z_x, r_x)$.
(2) $\overline{\bigcup_{x \in U} \bigcup_{n \geq 0} f_x^n(D_x)} \subset \tilde{F}(f)$.
(3) For each $x \in U$, $\mathrm{diam}(f_x^n(D_x)) \to 0$ $(n \to \infty)$.
(4) $x \mapsto z_x \in Y$ and $x \mapsto r_x \in \mathbb{R}$ are continuous on U.

REMARK 5.14.

(1) Let $(\pi, Y = X \times \hat{\mathbb{C}}, X)$ be a trivial $\hat{\mathbb{C}}$-bundle and let $f : Y \to Y$ be a polynomial skew product over $g : X \to X$. Suppose that for each $x \in X$, $d(x) \geq 2$. Then f satisfies condition $(*)$ (take a small disk neighborhood D of $\infty \in \hat{\mathbb{C}}$ and let $D_x = \{x\} \times D$ for each $x \in X$).
(2) Let $G = \langle h_1, \ldots, h_m \rangle$ be a finitely generated semihyperbolic rational semigroup containing an element of degree two or more. Moreover, suppose that either $\mathrm{Aut}(\hat{\mathbb{C}}) \cap G = \emptyset$ or each element of $\mathrm{Aut}(\hat{\mathbb{C}})$ is loxodromic. Furthermore, suppose $F(G) \neq \emptyset$. Then, the rational skew product associated with the multi-map $(h_1, \ldots, h_m)$ satisfies condition $(*)$. (Use Theorem 5.11.)

The following theorem is an important result regarding the case that there exists a nonconstant limit function on a fiber Fatou set. There are many applications of the following theorem to the dynamics of semihyperbolic rational semigroups and the dynamics of polynomial semigroups with bounded planar postcritical set which will be introduced in section 7. Moreover, by using the following theorem, we can show the compactness of $\bigcup_{x \in X} J^x(f)$ for semihyperbolic fibered rational maps f.

THEOREM 5.15 ([**58, 61**]). *Let (π, Y, X) be a $\hat{\mathbb{C}}$-bundle and let $f : Y \to Y$ be a fibered rational map over $g : X \to X$. Suppose that f satisfies condition $(*)$. Let $z \in Y$ be a point and suppose that $z \in F^{\pi(z)}(f)$. Suppose also that there exists a strictly increasing sequence $\{n_j\}$ of positive integers such that $f^{n_j}(z)$ tends to a point $z_0 \in Y$ as $j \to \infty$. Let $\{i_x\}$ be a local parameterization around $\pi(z)$ and $\pi(z_0)$ and let U be a connected open neighborhood of $i_{\pi(z)}^{-1}(z)$ in $\hat{\mathbb{C}}$. Moreover, regarding the above $\{n_j\}$, let $R_j := i_{\pi(f^{n_j}(z))}^{-1} \circ f_{\pi(z)}^{n_j} \circ i_{\pi(z)}$ and suppose $\{R_j\}$ tends to a nonconstant map $\psi : U \to \hat{\mathbb{C}}$ uniformly on U as $j \to \infty$. For each (i, j) with $1 \leq i \leq j$, let $S_{i,j} := f_{g^{n_i}(\pi(z))}^{n_j - n_i}$. Moreover, let*

$$V := \{a \in Y_{\pi(z_0)} \mid \exists \epsilon > 0, \limsup_{i \to \infty} \sup_{j > i} \sup_{\xi \in \tilde{B}(a, \epsilon)} d(S_{i,j} \circ i_{g^{n_i}(\pi(z))} \circ i_{\pi(z_0)}^{-1}(\xi), \xi) = 0\},$$

where d denotes the distance on the metric space Y. Then V is a nonempty open subset of $Y_{\pi(z_0)}$, $V \neq Y_{\pi(z_0)}$, and

$$\emptyset \neq \partial V \subset \hat{J}^{\pi(z_0)}(f) \cap UH(f) \subset \hat{J}^{\pi(z_0)}(f) \cap P(f),$$

where the boundary ∂V of V is taken regarding the topology in $Y_{\pi(z_0)}$.

By Theorem 5.15, we obtain the following result on the dynamics of semihyperbolic fibered rational maps.

NOTATION. Let Y_1, Y_2 be two topological spaces and let $f : Y_1 \to Y_2$ be a continuous map. For each subset Z of Y_2, we denote by $c(Z, f)$ the set of all connected components of $f^{-1}(Z)$.

THEOREM 5.16 ([**58, 61**]). *Let (π, Y, X) be a $\hat{\mathbb{C}}$-bundle and let $f : Y \to Y$ be a semihypberbolic fibered rational map over $g : X \to X$. Moreover, suppose that f satisfies condition $(*)$. Then, we have all of the following.*

(1) *$\bigcup_{x\in X} J^x(f)$ is a compact subset of Y, and we have $\tilde{J}(f) = \bigcup_{x\in X} J^x(f)$.*
(2) *There exist some positive constants δ, L, λ $(0 < \lambda < 1)$ such that for each $n \in \mathbb{N}$,*

$$\sup\{\text{diam } U \mid U \in c(\tilde{B}(y,\delta), f^n_{x^n}),\ y \in \tilde{J}(f), x^n \in g^{-n}(\{\pi(y)\})\} \le L\lambda^n.$$

(3) *Under the assumption of our theorem, suppose further that for each $x \in X$, $d(x) \ge 2$. Then, $x \mapsto J^x(f)$ is continuous with respect to the Hausdorff metric in the space of all nonempty compact subsets of Y.*
(4) *Under the assumption of our theorem, suppose further that for each $x \in X$, $d(x) \ge 2$. Then, for each nonempty compact subset K of $\tilde{F}(f)$, $\overline{\bigcup_{n\ge 0} f^n(K)} \subset \tilde{F}(f)$. Moreover, there exist constants $C > 0$ and $0 < \tau < 1$ such that for each $n \in \mathbb{N}$, $\sup_{z\in K} \|(f^n)'(z)\| \le C\tau^n$.*

By using Theorem 5.16, we can show that hyperbolicity and expandingness are equivalent:

THEOREM 5.17 ([**58**]). *Let (π, Y, X) be a $\hat{\mathbb{C}}$-bundle and let $f : Y \to Y$ be a fibered rational map over $g : X \to X$. Suppose that f satisfies condition $(*)$. Then, f is hyperbolic if and only if f is expanding.*

By applying Theorem 5.16(2), we can show that for a semihyperbolic polynomial skew product f with certain conditions, each fiber attracting basin of ∞ is a "John domain" and each fiber Julia set is locally connected. In order to state the details, we introduce some notation.

DEFINITION 5.18 ([**39**]). Let V be a nonempty subdomain of $\hat{\mathbb{C}}$ with $\partial V \subset \mathbb{C}$. Let $c > 0$. We say that V is a **c-John domain** if the following hold. There exists a point $z_0 \in V$ (when $\infty \in V$, $z_0 = \infty$) satisfying that for each $z_1 \in V$, there exists a curve $\xi \subset V$ connecting z_1 to z_0 such that for each $z \in \xi$, $\min\{|z - a| \mid a \in \partial V\} \ge c|z - z_1|$. Moreover, a c-John domain with some $c > 0$ is called a **John domain**.

REMARK 5.19 ([**39**]). If V is a simply connected subdomain of $\hat{\mathbb{C}}$ with $\partial V \subset \mathbb{C}$ and if it is a John domain, then ∂V is locally connected. Moreover, for any simple closed curve $\xi \subset \mathbb{C}$, ξ is a quasicircle ([**35**]) if and only if each of the two connected components of $\hat{\mathbb{C}} \setminus \xi$ is a John domain.

DEFINITION 5.20. Let $(\pi, Y = X \times \hat{\mathbb{C}}, X)$ be a trivial $\hat{\mathbb{C}}$-bundle and let $f : Y \to Y$ be a fibered polynomial skew product over $g : X \to X$. Then, for each $x \in X$, we set $A^x(f) := \{y \in Y_x \mid \pi_{\hat{\mathbb{C}}}(f^n_x(y)) \to \infty\ (n \to \infty)\}$ and $A_x(f) := \pi_{\hat{\mathbb{C}}}(A^x(f))$.

REMARK 5.21. Under the above definition, if $d(x) \ge 2$ for each $x \in X$, then for each $x \in X$, $A_x(f)$ is an open neighborhood of ∞, $A_x(f)$ is equal to the connected component of $F_x(f)$ containing ∞, and $\partial(A_x(f)) = J_x(f)$.

We now present a result which generalizes one direction of a theorem (Carleson-Jones-Yoccoz, [**13**]) on the equivalence of semihyperbolicity and Johnness of the attracting basin of ∞ for the iteration dynamics of a single polynomial map.

THEOREM 5.22 ([**61**]). *Let* $(\pi, Y = X \times \hat{\mathbb{C}}, X)$ *be a trivial* $\hat{\mathbb{C}}$*-bundle and let* $f : Y \to Y$ *be a semihyperbolic polynomial skew product over* $g : X \to X$. *Suppose that for each* $x \in X$, $d(x) \geq 2$. *Suppose also that* $\pi_{\hat{\mathbb{C}}}(P(f)) \setminus \{\infty\}$ *is bounded in* $\mathbb{C}$. *Then there exists a positive constant c such that for each* $x \in X$, $A_x(f)$ *is a c-John domain. In particular, for each* $x \in X$, $J_x(f)$ *is connected and locally connected.*

In [**70**], by applying Theorem 5.22, the author of this article classified (semi)hyperbolic polynomial semigroups G generated by a compact family Γ in $\mathcal{Y}$ with bounded planar postcritical set into three classes, and he showed that in one of the three classes, "for almost every $\gamma \in \Gamma^{\mathbb{N}}$, $J_\gamma(f)$ is a simple closed curve but not a quasicircle, and the connected component of $F_\gamma(f)$ containing ∞ is a John domain" (Figure 3 illustrates one of the examples). Note that this phenomenon "$\cdots$" does not hold in the usual iteration dynamics of a single polynomial map. By using this result, he also showed that a similar phenomenon holds for an Axiom A polynomial skew product on $\mathbb{C}^2$ (see [**70**]). Note also that he discovered that this phenomenon in an Axiom A polynomial skew product was a counterexample to one of the main results in [**18**] (see [**19, 70**] for the details). We remark that in the usual iteration dynamics of a single polynomial map, if the Julia set is a simple closed curve but not a quasicircle, then the connected component of the Fatou set containing ∞ is not a John domain ([**13**]).

In [**58**], a necessary and sufficient condition for a finitely generated rational semigroup to be semihyperbolic was given. By using it (or directly), we can show that for a subhyperbolic rational semigroup G, if there exists no superattracting fixed point of any element of G in $J(G)$, then G is semihyperbolic:

PROPOSITION 5.23 ([**58**]). *Let* $G = \langle h_1, \dots, h_m \rangle$ *be a finitely generated subhyperbolic rational semigroup containing an element of degree two or more. Suppose that either* $\mathrm{Aut}(\hat{\mathbb{C}}) \cap G = \emptyset$ *or each element of* $\mathrm{Aut}(\hat{\mathbb{C}}) \cap G$ *is loxodromic. Suppose also that* $F(G) \neq \emptyset$. *Suppose further that for any* $g \in G$, *there exists no point* $z \in J(G)$ *such that* $g(z) = z$ *and* $m(g, z) = 0$. *Then* G *is semihyperbolic.*

EXAMPLE 5.24. Let $G = \langle z^2 + 2, z^2 - 2 \rangle$. Then $P(G) = \{\infty, -2, 2\}$ and $P(G) \cap J(G) = \{-2, 2\}$. Thus G is not hyperbolic but is subhyperbolic. Moreover, by Proposition 5.23, G is semihyperbolic.

We also introduce the notion of an "expanding rational semigroup".

DEFINITION 5.25. Let G be a rational semigroup generated by a finite set Γ of Rat. We say that G is **expanding** if the rational skew product associated with Γ is expanding.

REMARK 5.26. (1) Under the above definition, the expandingness does not depend on the choice of Γ. (2) If G is expanding, then G is hyperbolic. (3) Conversely, for a finitely generated hyperbolic rational semigroup G containing an element of degree two or more, if $\mathrm{Aut}(\hat{\mathbb{C}}) \cap G = \emptyset$ or each element of $\mathrm{Aut}(\hat{\mathbb{C}}) \cap G$ is loxodromic, then G is expanding (by Remark 5.14(2) and Theorem 5.17).

Regarding hyperbolicity and expandingness, we can show the following ([**55, 74**]).

DEFINITION 5.27. We set $\mathrm{Exp}(m) := \{(h_1, \dots, h_m) \in (\mathrm{Rat})^m \mid \langle h_1, \dots, h_m \rangle \text{ is expanding}\}$.

LEMMA 5.28 ([**55, 74**]). Exp(m) *is an open subset of* $(\mathrm{Rat})^m$.

EXAMPLE 5.29. If $c \in \mathbb{C}, |c| > 2$, then $\langle z^2 + c,\ z^2 - c\rangle$ is hyperbolic (thus expanding).

6. Hausdorff dimension of Julia sets

In this section, we consider the Hausdorff dimension of Julia sets of semihyperbolic rational semigroups and expanding rational semigroups. We introduce some notation.

DEFINITION 6.1. Let G be a rational semigroup which has at most countably many elements. For any $t \geq 0$ and any $z \in \hat{\mathbb{C}}$, we set

$$S_G(z,t) := \sum_{g\in G} \sum_{g(y)=z} \|g'(y)\|^{-t}$$

counting multiplicities, where $\|\cdot\|$ denotes the norm of the derivative with respect to the spherical metric. Moreover, we set $S_G(z) := \inf\{t \geq 0 \mid S_G(z,t) < \infty\}$. (If there exists no t with $S_G(z,t) < \infty$, then we set $S_G(z) = \infty$.) Furthermore, we set $s_0(G) := \inf\{S_G(z) \mid z \in \hat{\mathbb{C}}\}(\in [0,\infty])$. $S_G(z)$ is called the **critical exponent of the Poincaré series of** G **at** z, and $s_0(G)$ is called the **critical exponent of the Poincaré series of** G.

DEFINITION 6.2. For an $m \in \mathbb{N}$, we set $\Sigma_m^* := \bigcup_{j=1}^{\infty}\{1,\dots,m\}^j$ (disjoint union). Moreover, for any $f = (f_1,\dots,f_m) \in (\mathrm{Rat})^m$ and any $w = (w_1,\dots,w_n) \in \Sigma_m^*$, we set $f_w := f_{w_n} \circ \cdots \circ f_{w_1}$.

DEFINITION 6.3. Let $m \in \mathbb{N}$, $f = (f_1,\dots,f_m) \in (\mathrm{Rat})^m$, $t \geq 0$, and $z \in \hat{\mathbb{C}}$. Then we set $Z_f(z,t) := \sum_{w\in\Sigma_m^*}\sum_{f_w(y)=z} \|f_w'(y)\|^{-t}$ counting multiplicities. Moreover, we set $Z_f(z) := \inf\{t \geq 0 \mid Z_f(z,t) < \infty\}$. (If there exists no t with $Z_f(z,t) < \infty$, then we set $Z_f(z) = \infty$.) Furthermore, we set $z_0(f) := \inf\{Z_f(z) \mid z \in \hat{\mathbb{C}}\}(\in [0,\infty])$. $Z_f(z)$ is called the **critical exponent of the Poincaré series of** f **at** z, and $z_0(f)$ is called the critical exponent of the Poincaré series of f.

REMARK 6.4. Let $f = (f_1,\dots,f_m) \in (\mathrm{Rat})^m, t \geq 0$, and $z \in \hat{\mathbb{C}}$. Let $G = \langle f_1,\dots,f_m\rangle$. Then, $S_G(z,t) \leq Z_f(z,t), S_G(z) \leq Z_f(z)$, and $s_0(G) \leq z_0(f)$. Moreover, for almost every $f = (f_1,\dots,f_m) \in (\mathrm{Rat})^m$ with respect to the Lebesgue measure, $G = \langle f_1,\dots,f_m\rangle$ is a free semigroup. Furthermore, for any such f, for any $t \geq 0$, and for any $z \in \hat{\mathbb{C}}$, $S_G(z,t) = Z_f(z,t)$, $S_G(z) = Z_f(z)$, and $s_0(G) = z_0(f)$.

DEFINITION 6.5. Let A be a subset of $\hat{\mathbb{C}}$. We denote by $\dim_H(A), \underline{\dim}_B(A)$, and $\overline{\dim}_B(A)$, the Hausdorff dimension, the lower Box dimension, and the upper Box dimension, respectively, of A with respect to the spherical distance ([**21**]). (Note that $\dim_H(A) \leq \underline{\dim}_B(A) \leq \overline{\dim}_B(A)$.)

By Theorem 5.16, we can show the following.

THEOREM 6.6 ([**58**]). *Let G be a finitely generated semihyperbolic rational semigroup containing an element of degree two or more. Suppose that either* $\mathrm{Aut}(\hat{\mathbb{C}}) \cap G = \emptyset$ *or each element of* $\mathrm{Aut}(\hat{\mathbb{C}}) \cap G$ *is loxodromic. Moreover, suppose that* $F(G) \neq \emptyset$. *Then* $\dim_H(J(G)) \leq s_0(G)$.

DEFINITION 6.7. Let $h = (h_1, \ldots, h_m) \in (\mathrm{Rat})^m$ and let U be a nonempty open subset of $\hat{\mathbb{C}}$. We say that h satisfies the **open set condition** with respect to U if $\bigcup_{j=1}^m h_j^{-1}(U) \subset U$ and for each (i,j) with $i \neq j$, $h_i^{-1}(U) \cap h_j^{-1}(U) = \emptyset$. Moreover, we say that h satisfies the open set condition if h satisfies the open set condition with respect to some U.

The following theorems (6.8, 6.10, 6.11) play very important roles in section 10. Note that these theorems generalize both the study by thermodynamic formalisms in the usual iteration dynamics of a single rational map and the study of the fractal dimensions of self-similar sets in fractal geometry.

THEOREM 6.8 ([**61**]). *Let $h = (h_1, \ldots, h_m) \in (\mathrm{Rat})^m$ and let $G = \langle h_1, \ldots, h_m \rangle$. Suppose that G contains an element of degree two or more and that G is semihyperbolic. Moreover, suppose that h satisfies the open set condition with respect to some U. Then either $J(G) = \overline{U}$ or $\dim_H(J(G)) \leq \overline{\dim}_B(J(G)) < 2$.*

DEFINITION 6.9 ([**60**]). Let $f = (f_1, \ldots, f_m) \in \mathrm{Exp}(m)$. Let $t \in \mathbb{R}$. We denote by $P(t,f)$ the topological pressure for the skew product $\tilde{f} : \tilde{J}(\tilde{f}) \to \tilde{J}(\tilde{f})$ associated with f, with respect to the potential function $\varphi(z) := -t \log \|\tilde{f}'(z)\|$. Moreover, we denote by $v(f)$ the unique zero of $t \mapsto P(t,f)$ (the existence and the uniqueness of the zero was shown in [**60**]).

THEOREM 6.10 ([**60**]). *Let $f = (f_1, \ldots, f_m) \in \mathrm{Exp}(m)$ and let $G = \langle f_1, \ldots, f_m \rangle$. Then, for any $z \in \hat{\mathbb{C}} \setminus (A(G) \cup P(G))$,*

$$\dim_H(J(G)) \leq \overline{\dim}_B(J(G)) \leq s_0(G) \leq S_G(z) \leq v(f) = Z_f(z) = z_0(f). \tag{7}$$

Moreover, letting M be the operator on the space $C(\tilde{J}(\tilde{f}))$ of all continuous functions on $\tilde{J}(\tilde{f})$ defined by $M\varphi(x) = \sum_{\tilde{f}(y)=x} \|\tilde{f}'(y)\|^{-v(f)} \varphi(y)$, there exists a unique Borel probability measure ν on $\tilde{J}(\tilde{f})$ such that $M^*(\nu) = \nu$. Furthermore, letting $\mathbf{1}$ be the constant function taking its value 1, the limit $\alpha := \lim_{n\to\infty} M^n \mathbf{1} \in C(\tilde{J}(\tilde{f}))$ exists, and $\alpha \cdot \nu$ is an ergodic $\tilde{f}$-invariant probability measure which satisfies the following:

$$v(f) = \frac{h_{\alpha\nu}(\tilde{f})}{\int_{\tilde{J}(\tilde{f})} (\log \|\tilde{f}'\|) \cdot \alpha \, d\nu} \leq \frac{\log(\sum_{j=1}^m \deg(f_j))}{\int_{\tilde{J}(\tilde{f})} (\log \|\tilde{f}'\|) \cdot \alpha \, d\nu}. \tag{8}$$

From this theorem, for expanding rational semigroups G, we can give an upper estimate of $\dim_H(J(G))$ by using the norm of the derivatives of generators of G. By using this estimate, we can give many examples of expanding rational semigroups G such that $\dim_H(J(G)) < 2$ (and thus $J(G)$ does not have any interior points). Moreover, regarding the condition for the inequalities in (7) in Theorem 6.10 to be the equalities, we have the following result.

THEOREM 6.11 ([**60**]). *Let $f = (f_1, \ldots, f_m) \in \mathrm{Exp}(m)$ and let $G = \langle f_1, \ldots, f_m \rangle$. Moreover, suppose that f satisfies the open set condition. Then, for any $z \in \hat{\mathbb{C}} \setminus (A(G) \cup P(G))$,*

$$\begin{aligned} \dim_H(J(G)) &= \underline{\dim}_B(J(G)) = \overline{\dim}_B(J(G)) = s_0(G) \\ &= S_G(z) = v(f) = Z_f(z) = z_0(f). \end{aligned} \tag{9}$$

Moreover, denoting the $v(f)$-dimensional Hausdorff measure by $H^{v(f)}$, we have $0 < H^{v(f)}(J(G)) < \infty$. Moreover, regarding the measure ν in Theorem 6.10, $(\pi_{\hat{\mathbb{C}}})_*(\nu) =$

$(H^{v(f)}|_{J(G)})/(H^{v(f)}(J(G)))$. *Furthermore, there exists a positive constant* C *such that for any* $0 < r < \text{diam } \hat{\mathbb{C}}$, $C^{-1} \leq \left(((\pi_{\hat{\mathbb{C}}})_*(\nu))(B(x,r))\right)/r^{v(f)} \leq C$.

It is easily shown that $f \mapsto v(f)$ is continuous on $\text{Exp}(m)$. By using the perturbation theory for linear operators and the fact that $e^{P(t,f)}$ is the maximum isolated eigenvalue of an operator, we can show the following. We first note that since $\text{Exp}(m)$ is an open subset of $(\text{Rat})^m$ (Lemma 5.28), each connected component of $\text{Exp}(m)$ is a finite-dimensional complex manifold.

THEOREM 6.12 ([**74**]). *Let* $m \in \mathbb{N}$. *Then* $f \mapsto v(f)$ *is real-analytic and plurisubharmonic on* $\text{Exp}(m)$.

As mentioned above, $\text{Exp}(m)$ is an open subset of $(\text{Rat})^m$. It is very natural to ask in which place of $\text{Exp}(m)$ the open set condition is satisfied, and in which place of $\text{Exp}(m)$ the open set condition is satisfied and the corresponding Julia set is connected. For these questions, the results in section 10 give an answer. In fact, in section 10, the results of this section are applied many times. More precisely, the results of this section are applied to investigate the complex analogues of singular functions in random complex dynamics.

We also remark that there are many examples of f such that $v(f) > 2$ ([**60, 74**]), and there are many interesting results regarding this case.

7. Polynomial semigroups with bounded planar postcritical set

In this section, we investigate the dynamics of polynomial semigroups with a bounded planar postcritical set. We show some phenomena which cannot hold in the usual iteration dynamics of a single rational map. Note that the results of this section are applied to random complex dynamics.

DEFINITION 7.1. Let G be a polynomial semigroup. We set $P^*(G) := P(G) \setminus \{\infty\}$ and this is called the **planar postcritical set** of G.

DEFINITION 7.2. We denote by $\mathcal{G}$ the set of all polynomial semigroups G satisfying the following. For any $g \in G$, $\deg(g) \geq 2$, and $P^*(G)$ is bounded in $\mathbb{C}$. Moreover, we set $\mathcal{G}_{dis} := \{G \in \mathcal{G} \mid J(G) \text{ is disconnected}\}$.

EXAMPLE 7.3. Let $\Gamma = \{h(z) = cz^a(1-z)^b \mid a, b \in \mathbb{N}, c > 0, c(a/(a+b))^a(b/(a+b))^b \leq 1\}$. Let G be the polynomial semigroup generated by Γ. Then any subsemigroup H of G belongs to $\mathcal{G}$.

REMARK 7.4. For a polynomial g with $\deg(g) \geq 2$, $\langle g \rangle \in \mathcal{G}$ if and only if $J(\langle g \rangle)$ is connected ([**38**]). However, for general polynomial semigroups, there exist many examples of G such that $G \in \mathcal{G}$ and $J(G)$ is disconnected (e.g. $G = \langle z^3, z^2/4 \rangle$).

QUESTION 7.5. What happens if $G \in \mathcal{G}_{dis}$?

DEFINITION 7.6 ([**68, 71, 67**]). Let K_1, K_2 be two bounded connected subsets of $\mathbb{C}$.

- $K_1 <_s K_2$ indicates that K_1 is included in a bounded component of $\mathbb{C} \setminus K_2$.
- $K_1 \leq_s K_2$ indicates that either $K_1 <_s K_2$ or $K_1 = K_2$.

REMARK 7.7. In the space of all nonempty connected compact subsets of $\mathbb{C}$, "$\leq_s$" determines a partial order. This "$\leq_s$" is called the **surrounding order**.

DEFINITION 7.8. For a polynomial semigroup G, we set $\hat{K}(G) := \{z \in \mathbb{C} \mid \{g(z) \mid g \in G\}$ is bounded in $\mathbb{C}\}$.

DEFINITION 7.9. For a topological space X, we denote by $\mathrm{Con}(X)$ the set of all connected components of X.

Under the above notation, we have the following results (obtained by the author of this article).

THEOREM 7.10 ([**68, 71, 67**]). *Let G be a (possibly noncompactly generated) polynomial semigroup in $\mathcal{G}_{dis}$. Then we have all of the following.*

(1) $J(G) \subset \mathbb{C}$ *and* $(\mathrm{Con}(J(G)), \leq_s)$ *is totally ordered. (See Figures* 3, 4.*)*
(2) *There exists a unique maximal element* $J_{\max}(G)$ *and a unique minimal element* $J_{\min}(G)$ *in* $(\mathrm{Con}(J(G)), \leq_s)$.
(3) *Each connected component of* $F(G)$ *is either simply connected or doubly connected.*
(4) *Let* $\mathcal{A} := \{U \in \mathrm{Con}(F(G)) \mid U$ *is doubly connected*$\}$. *Then* $(\mathcal{A}, \leq_s)$ *is totally ordered.*
(5) $\mathrm{int}(\hat{K}(G)) \neq \emptyset$.

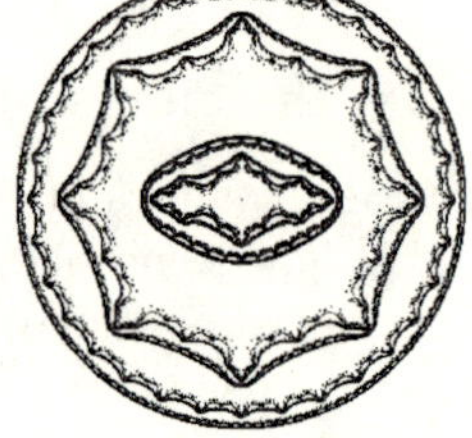

FIGURE 3. The Julia set of $G = \langle h_1, h_2 \rangle$, where $g_1(z) := z^2 - 1, g_2(z) := z^2/4, h_1 := g_1^2, h_2 := g_2^2$, $G \in \mathcal{G}_{dis}$, $\sharp(\mathrm{Con}(J(G))) > \aleph_0$. G is hyperbolic. (h_1, h_2) satisfies the open set condition. Moreover, $\forall J \in \mathrm{Con}(J(G)), \exists! \gamma \in \{h_1, h_2\}^{\mathbb{N}}$ such that $J = J_\gamma(f)$. For almost every $\gamma \in \{h_1, h_2\}^{\mathbb{N}}$ with respect to a Bernoulli measure, $J_\gamma(f)$ is a simple closed curve but not a quasicircle, and $A_\gamma(f)$ is a John domain.

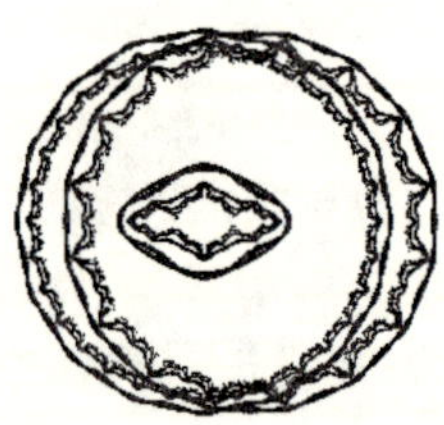

FIGURE 4. The Julia set of a 3-generator hyperbolic polynomial semigroup $G \in \mathcal{G}_{dis}$ with $\sharp(\mathrm{Con}(J(G))) = \aleph_0$.

THEOREM 7.11 ([**68, 71**]). *Let G be a polynomial semigroup belonging to $\mathcal{G}$ generated by a (possibly noncompact) family of quadratic polynomials. Then $J(G)$ is connected.*

If G is generated by a single rational map of degree two or more, or if G is a non-elementary Kleinian group, then either $J(G)$ is connected or $J(G)$ has uncountably many connected components. However, regarding general rational semigroups, even in finitely generated polynomial semigroups belonging to $\mathcal{G}$, we have the following examples.

THEOREM 7.12 ([**68, 71**]). *For any $n \in \mathbb{N}$, there exists a polynomial semigroup $G = \langle h_1, \dots, h_{2n} \rangle \in \mathcal{G}$ such that $\sharp(\text{Con}(J(G))) = n$. Moreover, there exists a 3-generator polynomial semigroup $G = \langle h_1, h_2, h_3 \rangle \in \mathcal{G}$ such that $\sharp(\text{Con}(J(G))) = \aleph_0$ (Figure 4).*

REMARK 7.13. For a finitely generated $G \in \mathcal{G}$, the cardinality of $\text{Con}(J(G))$ is related to the "interaction cohomology" in section 8, which was introduced by the author of this article. Moreover, in [**52**], by developing the ideas in [**68, 71**], it was shown that for any $n \in \mathbb{N}$, there exists a 4-generator polynomial semigroup $G \in \mathcal{G}$ such that $\sharp(\text{Con}(J(G))) = n$. In the same [**52**], it was also shown that for any $G \in \mathcal{G}_{dis}$ and any two distinct elements A, B in $\text{Con}((F(G)))$, the closures $\overline{A}$ and $\overline{B}$ are separated by a family of mutually disjoint quasicircles in $J(G)$ which are parameterized by the Cantor set.

8. Interaction cohomology of (backward) self-similar systems

For a finitely generated polynomial semigroup G, we consider the interaction of generators focusing on the backward self-similarity of $J(G)$. In order to do that, we consider the following general subject, and then apply it to the dynamics of polynomial semigroups. In particular, we introduce a new kind of cohomology theory, which was defined by the author of this article.

DEFINITION 8.1 ([**63**]). Let (X, d) be a metric space. Let $h_j : X \to X$ ($j = 1, \dots, m$) be a continuous map. Let L be a nonempty compact subset of X. We say that $\mathfrak{L} = (L, (h_1, \dots, h_m))$ is a **backward self-similar system** if the following two conditions are satisfied.
(1) $L = h_1^{-1}(L) \cup \cdots \cup h_m^{-1}(L)$.
(2) For each $j = 1, \dots, m$ and each $z \in L$, $h_j^{-1}(z) \neq \emptyset$.

DEFINITION 8.2 ([**63**]). For a backward self-similar system $\mathfrak{L} = (L, (h_1, \dots, h_m))$, we use the following notation.

(1) For each $x = (x_1, x_2, \dots) \in \Sigma_m$, we set $L_x := \bigcap_{j=1}^{\infty} h_{x_1}^{-1} \cdots h_{x_j}^{-1}(L)$ $(\neq \emptyset)$.
(2) For each $k \in \mathbb{N}$, we set $\mathcal{U}_k := \{h_{x_1}^{-1} \cdots h_{x_k}^{-1}(L)\}_{(x_1, \dots, x_k) \in \{1, \dots, m\}^k}$ (a finite covering of L).
(3) For each covering $\mathcal{U}$ of L, we denote by $N(\mathcal{U})$ the nerve of $\mathcal{U}$. The vertex set of $N(\mathcal{U}_k)$ is equal to $\{1, \dots, m\}^k$. Let $\varphi_k : N(\mathcal{U}_{k+1}) \to N(\mathcal{U}_k)$ be the simplicial map induced by $(x_1, \dots, x_k, x_{k+1}) \mapsto (x_1, \dots, x_k)$, where $(x_1, \dots, x_k, x_{k+1}) \in \{1, \dots, m\}^{k+1}$. Note that this φ_k is equal to the map induced by the refinement map $h_{x_1}^{-1} \cdots h_{x_{k+1}}^{-1}(L) \in \mathcal{U}_{k+1} \mapsto h_{x_1}^{-1} \cdots h_{x_k}^{-1}(L) \in \mathcal{U}_k$. $\{\varphi_k : N(\mathcal{U}_{k+1}) \mapsto N(\mathcal{U}_k)\}_{k \in \mathbb{N}}$ makes an inverse system of simplicial maps.
(4) For each $p \in \mathbb{N} \cup \{0\}$ and for each $\mathbb{Z}$-module R, we set $\check{H}^p(\mathfrak{L}; R) := \varinjlim_k H^p(N(\mathcal{U}_k); R)$. This is called the **$p$-th interaction cohomology group of $\mathfrak{L}$ with coefficients in R**. Moreover, $H^p(N(\mathcal{U}_k); R)$ is called

the p-**th interaction cohomology group of $\mathfrak{L}$ at the** k-**th stage with coefficients in** R.

(5) There exists a natural homomorphism from $\check{H}^p(\mathfrak{L};R)$ to $\check{H}^p(L;R)$ (p-th Čech cohomology group of L with coefficients in R). This homomorphism is denoted by $\Psi : \check{H}^p(\mathfrak{L};R) \to \check{H}^p(L;R)$. (Note: If for each j, h_j^{-1} is well-defined and is contractive, then Ψ is an isomorphism. However, Ψ is not an isomorphism in general.)

(6) For a general simplicial complex Y, we denote by $|Y|$ the "realization" of Y ([**45**]).

Under the above notation, we have the following.

THEOREM 8.3 ([**63**]). *Let $\mathfrak{L} = (L,(h_1,\ldots,h_m))$ be a backward self-similar system. Suppose that for each $x \in \Sigma_m$, L_x is connected. Then, we have all of the following (1)–(5).*

(1) $\varprojlim_k \mathrm{Con}(|N(\mathcal{U}_k)|) \cong \mathrm{Con}(L)$ *(an isomorphism between sets).*

(2) *"L is connected" $\iff$ "$|N(\mathcal{U}_1)|$ is connected" $\iff$ "for each $i,j \in \{1,\ldots,m\}$, there exists a sequence $\{i_t\}_{t=1}^s \subset \{1,\ldots,m\}$ such that $i_1 = i$, $i_s = j$, $h_{i_t}^{-1}(L) \cap h_{i_{t+1}}^{-1}(L) \neq \emptyset$ $(t = 1,\ldots,s-1)$".*

(3) *Suppose $m = 2$ and L is disconnected. Then $h_1^{-1}(L) \cap h_2^{-1}(L) = \emptyset$ and $\mathrm{Con}(L) \cong \{1,2\}^{\mathbb{N}}$. In particular $\sharp(\mathrm{Con}(L)) > \aleph_0$.*

(4) *Suppose $m = 3$ and L is disconnected. Then $\sharp(\mathrm{Con}(L)) \geq \aleph_0$.*

(5) *Let R be a field. Then $\dim_R \check{H}^0(\mathfrak{L};R) < \infty \iff \sharp(\mathrm{Con}(L)) < \infty$. Moreover, if these conditions hold, then $\dim_R \check{H}^0(\mathfrak{L};R) = \sharp(\mathrm{Con}(L))$.*

REMARK 8.4. The above Theorem 8.3(3) plays a very important role to show the results in Theorem 10.4 in section 10.

We next investigate $\check{H}^1(\mathfrak{L};R)$. Regarding finitely generated polynomial semigroups $\langle h_1,\ldots,h_m\rangle$, we can easily find examples such that $\check{H}^1(\mathfrak{L};R)$ is 0. We consider finding examples such that $\check{H}^1(\mathfrak{L};R) \neq 0$.

THEOREM 8.5 ([**63**]). *Let $m \geq 3$ and let $\mathfrak{L} = (L,(h_1,\ldots,h_m))$ be a backward self-similar system. We assume all of the following (1)–(4).*

(1) *$|N(\mathcal{U}_1)|$ is connected.*

(2) *$(h_1^2)^{-1}(L) \cap (\bigcup_{i:i\neq 1} h_i^{-1}(L)) = \emptyset$.*

(3) *There exist mutually distinct elements $j_1,j_2,j_3 \in \{1,\ldots,m\}$ such that $j_1 = 1$ and for each $k = 1,2,3$, $h_{j_k}^{-1}(L) \cap h_{j_{k+1}}^{-1}(L) \neq \emptyset$ (where $j_4 := j_1$).*

(4) *For each $s,t \in \{1,\ldots,m\}$, if $s,t,1$ are mutually distinct, then $h_1^{-1}(L) \cap h_s^{-1}(L) \cap h_t^{-1}(L) = \emptyset$.*

Then, for each field R, $\dim_R \check{H}^1(\mathfrak{L};R) = \infty$.

REMARK 8.6. In [**63**], it was shown that under the assumption that the system is "postunbranched", then $H^p(N(\mathcal{U}_k);R)$ can be calculated inductively on k. The growth rate of the rank of $H^p(N(\mathcal{U}_k);R)$ as $k \to \infty$ is considered as a quantity which represents the complexity of the system. In order to consider an example, let $h_j(z) = 2(z-p_j)+p_j$, $j = 1,2,3$, where $p_1,p_2,p_3 \in \mathbb{C}$ are three points which make a regular triangle in $\mathbb{C}$. Let $G = \langle h_1,h_2,h_3\rangle$. Then $J(G)$ is the Sierpiński gasket (Figure 2). Applying the above result to $(J(G),(h_1,h_2,h_3))$, we obtain that for each field R, setting $a_k := \dim_R H^1(N(\mathcal{U}_k);R)$, we have $a_1 = 1$ and $a_{k+1} = 3a_k + 1$.

Moreover, applying the above result to a backward self-similar system which is made by subtracting some n-words of $\{h_j\}$, we can show that the invariant set L of the system has infinitely many connected components. The information of the first (or higher) cohomology groups of a system can be related to the information of the number of connected components of the invariant set L of the system.

REMARK 8.7. The "interaction cohomology" can also be defined for a system $(L, (h_1, \dots, h_m))$ such that $L = \bigcup_{j=1}^{m} h_j(L)$ (such a system is called a **forward self-similar system**), and we obtain results similar to the above ([**63**]).

For a finitely generated polynomial semigroup $G = \langle h_1, \dots, h_m \rangle$, $(J(G), (h_1, \dots, h_m))$ is a backward self-similar system. We apply the above results to this system. We first note that by using Proposition 3.17(4) (an application of potential theory), the Riemann-Hurwitz formula, and Lemma 3.15(2), we obtain the following.

PROPOSITION 8.8. *Let $m \in \mathbb{N}$. For each $j = 1, \dots, m$, let h_j be a polynomial of degree two or more. Let $G = \langle h_1, \dots, h_m \rangle$. Then $G \in \mathcal{G}$ if and only if the backward self-similar system $(J(G), (h_1, \dots, h_m))$ satisfies that for each $x \in \Sigma_m$, $(J(G))_x$ is connected.*

COROLLARY 8.9 ([**63**]). *(Corollary of Theorem 8.3 and Proposition 8.8). If $G = \langle h_1, \dots, h_m \rangle \in \mathcal{G}$, then for the backward self-similar system $\mathfrak{L} = (J(G), (h_1, \dots, h_m))$, statements (1)–(5) in Theorem 8.3 hold.*

Combining Theorem 8.5, Leray's theorem ([**25**]), and the Alexander duality theorem ([**45**]), etc., we obtain the following.

THEOREM 8.10 ([**63**]). *Let $G = \langle h_1, \dots, h_m \rangle \in \mathcal{G}$. Suppose that the backward self-similar system $\mathfrak{L} = (J(G), (h_1, \dots, h_m))$ satisfies the assumptions of Theorem 8.5. Then, for each field R, all of the following hold.*

(1) $\dim_R \check{H}^1(\mathfrak{L}; R) = \dim_R(\Psi(\check{H}^1(\mathfrak{L}; R))) = \infty$, *and the natural homomorphism $\Psi : \check{H}^1(\mathfrak{L}; R) \to \check{H}^1(J(G); R)$ is a monomorphism.*

(2) *$F(G)$ $(= \hat{\mathbb{C}} \setminus J(G))$ has infinitely many connected components.*

PROPOSITION 8.11 ([**63**]). *There exists an element $h = (h_1, h_2, h_3, h_4) \in \mathcal{Y}^4$ which satisfies the assumptions of Theorem 8.10. In particular, for this h, setting $G = \langle h_1, \dots, h_4 \rangle$, regarding the system $\mathfrak{L} = (J(G), (h_1, \dots, h_4))$, we have $\dim_R \check{H}^1(\mathfrak{L}; R) = \infty$ (R is a field).*

QUESTION 8.12 (Open). Let $m \in \mathbb{N}$ and let R be a field. Are there any elements $(h_1, \dots, h_m) \in \mathcal{Y}^m$ such that $\langle h_1, \dots, h_m \rangle \in \mathcal{G}$ and such that regarding the system $\mathfrak{L} = (J(\langle h_1, \dots, h_m \rangle), (h_1, \dots, h_m))$, $0 < \dim_R \check{H}^1(\mathfrak{L}; R) < \infty$.

9. Random complex dynamics

In this section, we discuss random complex dynamics. We first introduce the results of Fornaess and Sibony, who were the first researchers to investigate random complex dynamics, and then we introduce the results of R. Brück, M. Büger, S. Reitz et al., who developed the idea of Fornaess and Sibony. After that, we introduce the recent idea of the author of this article, which combines the theory of the dynamics of rational semigroups and that of random complex dynamics. Applying that idea, we present that complex analogues of the devil's staircase appear naturally.

We now introduce the results of Fornaess-Sibony ([**22**]). We use the following notation and setting.

NOTATION AND SETTING. Let W be a domain in $\mathbb{C}$ and let $R : W \times \hat{\mathbb{C}} \to \hat{\mathbb{C}}$ be a holomorphic map. For each $c \in W$, we set $R_c(z) := R(c, z)$. We assume that each R_c is a rational map of degree two or more. For each nonempty compact subset K of W, we set $X(K) = K^{\mathbb{N}}$. Let $\sigma : X(K) \to X(K)$ be the shift map; that is, σ is the map defined by $\sigma(x_1, x_2, \ldots) = (x_2, x_3, \ldots)$. Let $f : X(K) \times \hat{\mathbb{C}} \to X(K) \times \hat{\mathbb{C}}$ be the rational skew product over $\sigma : X(K) \to X(K)$ defined by $f(x, y) = (\sigma(x), R_{x_1}(y))$, where $x = (x_1, x_2, \ldots)$. Moreover, for each $c \in \mathbb{C}$ and each $r > 0$, we set $D(c, r) := \{z \in \mathbb{C} \mid |z - c| < r\}$.

Under the above notation and setting, the following holds.

THEOREM 9.1 ([**22**]). *Suppose that for each $z \in \hat{\mathbb{C}}$, the map $c \mapsto R_c(z)$ is not constant. Let $c_0 \in W$ be a point. Suppose that R_{c_0} has exactly k (≥ 1) attracting cycles $\gamma_1, \gamma_2, \ldots, \gamma_k$. For each $j \in \{1, \ldots, k\}$, let V_j be an open neighborhood of γ_j which is contained in the attracting region of γ_j. Then, there exists a $\delta_0 > 0$ such that for each $0 < \delta < \delta_0$, setting $K = \overline{D(c_0, \delta)}$ and considering the rational skew product $f : X(K) \times \hat{\mathbb{C}} \to X(K) \times \hat{\mathbb{C}}$ in the above "notation and setting", there exist continuous functions $h_1, \ldots, h_k$ on $\hat{\mathbb{C}}$ satisfying all of the following conditions.*

(1) *For each $j \in \{1, \ldots, k\}$ and each $y \in \hat{\mathbb{C}}$, we have $0 \leq h_j(y) \leq 1$ and $\sum_{j=1}^k h_j(y) = 1$.*
(2) *Let λ be the normalized Lebesgue measure on $\overline{D(c_0, \delta)}$. For each $y \in \hat{\mathbb{C}}$, there exist mutually disjoint nonempty open subsets $U_{1,y}, U_{2,y}, \ldots, U_{k,y}$ of $X(K)$ such that $(\bigotimes_{i=1}^{\infty} \lambda)(U_{j,y}) = h_j(y)$ $(\forall j)$, and such that for each $j \in \{1, \ldots, k\}$, for each $x \in U_{j,y}$ and for each sufficiently large $n \in \mathbb{N}$, $f_{x,n}(y) \in V_j$.*
(3) *For almost every point $x \in X(K)$ with respect to $(\bigotimes_{i=1}^{\infty} \lambda)$, the 2-dimensional Lebesgue measure of $J_x(f)$ is equal to 0.*

The above theorem was shown by applying the "no wandering domain theorem" ([**54**]) and by using some method in random dynamics. The proof has many interesting ideas. However, in the above theorem, only small values $\delta > 0$ are dealt with. By developing the idea of Fornaess and Sibony, Brück, Gong et al. investigated the following for the case $W = \mathbb{C}$ and $R(c, z) = z^2 + c$ (Theorems 9.2, 9.3, 9.4).

THEOREM 9.2 ([**10**]). *Let $K = \overline{D(0, 1/4)}$. Let λ be a Borel probability measure on K whose support is equal to K. Then there exists a Borel subset $\mathcal{U}$ of $X(K)$ with $(\bigotimes_{i=1}^{\infty} \lambda)(\mathcal{U}) = 1$ such that for each $x \in \mathcal{U}$ and for each $y \in \overline{D(0, 1/2)}$, $\{f_{x,n}(y)\}_{n \in \mathbb{N}}$ is dense in $\overline{D(0, 1/2)}$.*

NOTATION. Letting $R_c(z) = z^2 + c$, we set $\mathcal{M} := \{c \in \mathbb{C} \mid \{R_c^n(0)\}_{n \in \mathbb{N}}$ is bounded in $\mathbb{C}\}$, and this is called the **Mandelbrot set** ([**37**]). Note that if $\delta > 1/4$, then $\overline{D(0, \delta)} \cap (\mathbb{C} \setminus \mathcal{M}) \neq \emptyset$.

THEOREM 9.3 ([**10, 8**]). *Let $\delta > 1/4, K := \overline{D(0, \delta)}$ and let λ be the normalized 2-dimensional Lebesgue measure on K. Then, we have all of the following.*

(1) *For each $y \in \hat{\mathbb{C}}$, there exists a Borel set $U_y \subset X(K)$ with $(\bigotimes_{i=1}^{\infty} \lambda)(U_y) = 1$ such that for each $x \in U_y$, $f_{x,n}(y) \to \infty$ $(n \to \infty)$. In particular, for*

almost every point $x \in X(K)$ *with respect to* $\bigotimes_{i=1}^{\infty} \lambda$, *the* 2*-dimensional Lebesgue measure of* $J_x(f)$ *is equal to* 0.

(2) *For almost every point* $x \in X(K)$ *with respect to* $\bigotimes_{i=1}^{\infty} \lambda$, $J_x(f)$ *has infinitely many connected components.*

Moreover, regarding connected components of $J_x(f)$, we have the following.

THEOREM 9.4 ([**23**]). *Let* $\delta > 1/4$ *and* $K = \overline{D(0,\delta)}$. *Then there exists a countable intersection* $\mathcal{U}$ *of open dense subsets of* $X(K)$ *such that for each* $x \in \mathcal{U}$, *each connected component of* $J_x(f)$ *consists of one point.*

Note that the proofs of Theorem 9.3, etc. are based on the papers of Büger ([**11, 12**]). In this article, we introduce an important result of them.

THEOREM 9.5 ([**11**]). *Let* $(\pi, Y = X \times \hat{\mathbb{C}}, X)$ *be a trivial* $\hat{\mathbb{C}}$*-bundle and let* $f : Y \to Y$ *be a polynomial skew product over* $g : X \to X$ *such that for each* $x \in X$, $d(x) \geq 2$. *Let* $x \in X$ *be a point and let* U *be a nonempty open subset of* $J_x(f)$. *Then, there exists an* $n \in \mathbb{N}$ *such that* $f_{x,n}(U) = J_{g^n(x)}(f)$.

For other research of random complex dynamics, see [**9, 14, 15, 16, 17, 2, 3, 4, 5, 40, 41**], etc. (We remark that many statements of the main results in [**17**] easily follow from [**43, 58**].)

As mentioned above, regarding random complex dynamics, various results have been obtained. However, only very small perturbations of a single rational map were discussed in [**22**], and only the random dynamics of "quadratic" polynomials was discussed in [**10, 8, 9, 23**]. Such limited situations have been observed by using heavily the assumptions so far. The author of this article considered investigating random complex dynamics more generally and more systematically, combining the study of random complex dynamics with that of the dynamics of rational semigroups as follows.

The setting and notation of random complex dynamics by the author of this article ([72, 66, 67]):

(1) Let τ be a Borel probability measure on Rat. In the following, we consider the (i.i.d.) random dynamical system on $\hat{\mathbb{C}}$ such that at every step we choose a rational map according to the probability distribution τ. (This determines a Markov process on $\hat{\mathbb{C}}$ such that for each $x \in \hat{\mathbb{C}}$ and each Borel subset A of $\hat{\mathbb{C}}$, the transition probability $p(x, A)$ from x to A is equal to $\tau(\{g \in \text{Rat} \mid g(x) \in A\})$.)
(2) We set $X_\tau := (\text{supp}\,\tau)^{\mathbb{N}}$, where supp τ denotes the topological support of τ in Rat.
(3) We set $\tilde{\tau} := \bigotimes_{j=1}^{\infty} \tau$. This is a Borel probability measure on X_τ.
(4) Let G_τ be the rational semigroup generated by the family supp τ of rational maps.
(5) We set $C(\hat{\mathbb{C}}) := \{\varphi : \hat{\mathbb{C}} \to \mathbb{C} \mid \varphi \text{ is continuous}\}$ endowed with the supremum norm. We define an operator $M_\tau : C(\hat{\mathbb{C}}) \to C(\hat{\mathbb{C}})$ by $M_\tau(\varphi)(z) := \int_{\text{supp}\,\tau} \varphi(g(z))\, d\tau(g)$.
(6) Let $\mathcal{M}_1(\hat{\mathbb{C}})$ be the space of all Borel probability measures on $\hat{\mathbb{C}}$ endowed with the weak* topology. Note that $\mathcal{M}_1(\hat{\mathbb{C}})$ is a compact metric space. Moreover, we denote by $M_\tau^* : \mathcal{M}_1(\hat{\mathbb{C}}) \to \mathcal{M}_1(\hat{\mathbb{C}})$ the dual of M_τ (thus $M_\tau^*(\mu)(\varphi) = \int M_\tau(\varphi) d\mu$ for each $\mu \in \mathfrak{M}_1(\hat{\mathbb{C}}), \varphi \in C(\hat{\mathbb{C}})$).

(7) We denote by $F_{\text{meas}}(\tau)$ the set of all $\mu \in \mathcal{M}_1(\hat{\mathbb{C}})$ satisfying the following: "There exists a neighborhood B of μ in $\mathcal{M}_1(\hat{\mathbb{C}})$ such that $\{(M_\tau^*)^n : B \to \mathcal{M}_1(\hat{\mathbb{C}})\}_{n\in\mathbb{N}}$ is equicontinuous on B". Moreover, we set $J_{\text{meas}}(\tau) := \mathcal{M}_1(\hat{\mathbb{C}}) \setminus F_{\text{meas}}(\tau)$.

(8) If $\text{supp}\,\tau \subset \mathcal{Y}$, then for any $z \in \hat{\mathbb{C}}$, we set

$$T_{\infty,\tau}(z) := \tilde{\tau}(\{h = (h_1, h_2, \ldots) \in X_\tau \mid h_n \cdots h_1(z) \to \infty, \text{as } n \to \infty\}).$$

This is nothing else but the probability of tending to $\infty \in \hat{\mathbb{C}}$ starting with the initial value $z \in \hat{\mathbb{C}}$ regarding the random dynamics on $\hat{\mathbb{C}}$ such that at every step we choose a polynomial map according to τ. (Note: If $\infty \in F(G_\tau)$, then for each connected component U of $F(G_\tau)$, $T_{\infty,\tau}|_U$ is a constant function (the constant depends on U).)

REMARK 9.6. If $h \in \text{Rat}$ is of degree two or more and if τ is the Dirac measure at h, then $J_{\text{meas}}(\tau) \neq \emptyset$. In fact, by embedding $\hat{\mathbb{C}}$ into $\mathcal{M}_1(\hat{\mathbb{C}})$ under the map $g \mapsto \delta_g$, we have $J(h) \subset J_{\text{meas}}(\tau)$.

The following is the key to investigating the dynamics of rational semigroups and random complex dynamics.

DEFINITION 9.7 ([**72, 66, 67**]). Let G be a rational semigroup. We set $J_{\text{ker}}(G) := \bigcap_{g\in G} g^{-1}(J(G))$, and this is called the **kernel Julia set** of G.

The following result is fundamental.

THEOREM 9.8 ([**72, 66, 67**], Cooperation principle). *Suppose that* $\text{supp}\,\tau$ *is compact in* Rat *and* $J_{\text{ker}}(G_\tau) = \emptyset$. *Let* $f : X_\tau \times \hat{\mathbb{C}} \to X_\tau \times \hat{\mathbb{C}}$ *be the rational skew product associated with* $\text{supp}\,\tau$. *Then,* $F_{\text{meas}}(\tau) = \mathcal{M}_1(\hat{\mathbb{C}})$, *and for almost every* $\gamma \in X_\tau$ *with respect to* $\tilde{\tau}$, *the 2-dimensional Lebesgue measure of* $J_\gamma(f)$ *is equal to* 0.

From Theorem 9.8, the following follows.

PROPOSITION 9.9 ([**72, 66, 67**]). *Suppose that* $\text{supp}\,\tau$ *is a compact subset of* $\mathcal{Y}$ *and* $J_{\text{ker}}(G_\tau) = \emptyset$. *Then* $T_{\infty,\tau} : \hat{\mathbb{C}} \to [0,1]$ *is continuous on* $\hat{\mathbb{C}}$.

Combining Theorem 9.8, Proposition 9.9, and Theorem 7.10, etc., we obtain the following.

THEOREM 9.10 ([**66, 67**]). *Suppose that* $\text{supp}\,\tau$ *is a compact subset of* $\mathcal{Y}$ *and* $G_\tau \in \mathcal{G}_{\text{dis}}$. *Then, all of the following (1)–(7) hold.*

(1) *For each connected component* U *of* $F(G_\tau)$, *there exists a constant* $C_U \in [0,1]$ *such that* $T_{\infty,\tau}|_U \equiv C_U$.

(2) *(***Continuity***)* $T_{\infty,\tau} : \hat{\mathbb{C}} \to [0,1]$ *is continuous and* $M_\tau(T_{\infty,\tau}) = T_{\infty,\tau}$.

(3) *(***Monotonicity***) Let* $\mathcal{A} := \{U \in \text{Con}(F(G_\tau)) \mid U \text{ is doubly connected}\}$.

   (a) *If* $A_1, A_2 \in \mathcal{A}$, $A_1 <_s A_2$, *then* $C_{A_1} < C_{A_2}$. *In particular, all elements of* $\{C_A \mid A \in \mathcal{A}\}$ *are mutually distinct.*

   (b) *If* $J_1, J_2 \in \text{Con}(J(G_\tau))$ *and* $J_1 <_s J_2$, *then* $\max_{z\in J_1} T_{\infty,\tau}(z) \leq \min_{z\in J_2} T_{\infty,\tau}(z)$.

(4) *For each* $A \in \mathcal{A}$, $T_{\infty,\tau}|_{\hat{K}(G_\tau)} \equiv 0 < C_A < 1 \equiv C_{F_\infty(G_\tau)}$. *Here we denote by* $F_\infty(G_\tau)$ *the connected component of* $F(G_\tau)$ *containing* ∞.

(5) *Let Q be an open subset of $\hat{\mathbb{C}}$. If $Q \cap \Big(\bigcup_{A\in\mathcal{A}} \partial A \cup \partial(F_\infty(G_\tau)) \cup \partial(\hat{K}(G_\tau))\Big) \neq \emptyset$, then $T_{\infty,\tau}|_Q$ is not constant. (From (1),...,(5), $T_{\infty,\tau}$ looks like the "devil's staircase". Such a function is called a* **devil's coliseum**. *See Figures 5, 6.)*

(6) *We have $J_{\ker}(G_\tau) = \emptyset$. In particular, $F_{\text{meas}}(\tau) = \mathcal{M}_1(\hat{\mathbb{C}})$.*

(7) *There exists a unique M_τ^*-invariant Borel probability measure μ on $\hat{K}(G_\tau)$ that satisfies the following. "For each $\varphi \in C(\hat{\mathbb{C}})$,*

$$M_\tau^n(\varphi)(z) \to T_{\infty,\tau}(z) \cdot \varphi(\infty) + (1 - T_{\infty,\tau}(z)) \cdot (\int_{\hat{\mathbb{C}}} \varphi \, d\mu) \quad (n \to \infty)$$

uniformly on $\hat{\mathbb{C}}$". Thus

$$(M_\tau^*)^n(\nu) \to (\int_{\hat{\mathbb{C}}} T_{\infty,\tau} \, d\nu) \cdot \delta_\infty + (\int_{\hat{\mathbb{C}}} (1 - T_{\infty,\tau}) \, d\nu) \cdot \mu \quad (n \to \infty)$$

uniformly on $\mathcal{M}_1(\hat{\mathbb{C}})$. Moreover, the M_τ-invariant subspace of $C(\hat{\mathbb{C}})$ is two-dimensional and it is spanned by the constant function and $T_{\infty,\tau}$. Furthermore, the set of ergodic components of M_τ^*-invariant Borel probability measures on $\hat{\mathbb{C}}$ is equal to $\{\delta_\infty, \mu\}$.

REMARK 9.11. If we obtain some results of the dynamics of rational semigroups, then they are applied to random complex dynamics. Conversely, by using Theorems 9.8, 9.10, and Proposition 9.9, etc., and studying the level sets of $T_{\infty,\tau}$, we can get much information about $J(G)$. In other words, in order to investigate the dynamics of rational semigroups, it is very effective that we once work on the associated random complex dynamics and then we apply the results to the dynamics of original rational semigroups.

10. The space of 2-generator polynomial semigroups

In this section, applying the results in the previous sections, we investigate the details of the space of 2-generator polynomial semigroups. In particular, we consider the space of 2-generator (hyperbolic) polynomial semigroups with bounded postcritical set, and we present that at each parameter in a neighborhood of the closure of the disconnectedness locus (the disconnectedness locus denotes the set of parameters of which the corresponding Julia sets are disconnected), the function T_∞ (which is defined on $\hat{\mathbb{C}}$) of probability of tending to ∞ has many properties on $\hat{\mathbb{C}}$ which are similar to those of the devil's staircase (the Cantor function). Moreover, we present that at each parameter of the closure of the disconnectedness locus, T_∞ can be differentiable by the probability parameter p, and the partial derivative of T_∞ with respect to the probability parameter can be regarded, as a function of the initial values $z \in \hat{\mathbb{C}}$, as a complex analogue of the Takagi function.

DEFINITION 10.1 ([**66**]). We use the following notation.

- We set $\mathcal{B} := \{(h_1, h_2) \in \mathcal{Y}^2 \mid P^*(\langle h_1, h_2\rangle) \text{ is bounded in } \mathbb{C}\}$.
- We set $\mathcal{C} := \{(h_1, h_2) \in \mathcal{Y}^2 \mid J(\langle h_1, h_2\rangle) \text{ is connected}\}$.
- We set $\mathcal{D} := \{(h_1, h_2) \in \mathcal{Y}^2 \mid J(\langle h_1, h_2\rangle) \text{ is disconnected})\}$.
- We set $\mathcal{H} := \{(h_1, h_2) \in \mathcal{Y}^2 \mid \langle h_1, h_2\rangle \text{ is hyperbolic}\}$.
- We set $\mathcal{I} := \{(h_1, h_2) \in \mathcal{Y}^2 \mid J(h_1) \cap J(h_2) \neq \emptyset\}$.
- We set $\mathcal{Q} := \{(h_1, h_2) \in \mathcal{Y}^2 \mid J(h_1) = J(h_2)$, and $J(h_1)$ and $J(h_2)$ are quasicircles$\}$.

LEMMA 10.2. *The sets $\mathcal{H}, \mathcal{B} \cap \mathcal{H}, \mathcal{D} \cap \mathcal{B} \cap \mathcal{H}$ are nonempty and open in $\mathcal{Y}^2$.*

DEFINITION 10.3. For each $h = (h_1, h_2) \in \mathcal{Y}^2$, each $z \in \hat{\mathbb{C}}$, and each $p \in (0, 1)$, we use the following notation.

- We denote by $T(h_1, h_2, p, z)$ the probability of tending to $\infty \in \hat{\mathbb{C}}$ regarding the random dynamics on $\hat{\mathbb{C}}$ such that at every step we choose h_1 with probability p and we choose h_2 with probability $1 - p$. More precisely, setting $\tau_{h_1,h_2,p} := p\delta_{h_1} + (1 - p)\delta_{h_2}$ (δ_{h_i} denotes the Dirac measure concentrated at h_i), we set $T(h_1, h_2, p, z) := T_{\infty,\tau_{h_1,h_2,p}}(z)$. (Note: $z \mapsto T(h_1, h_2, p, z)$ is locally constant on $F(\langle h_1, h_2\rangle)$.)
- Let $\tilde{h} : \Sigma_2 \times \hat{\mathbb{C}} \to \Sigma_2 \times \hat{\mathbb{C}}$ be the rational skew product associated with $h = (h_1, h_2) \in \mathcal{Y}^2$. Let ρ be an $\tilde{h}$-invariant Borel probability measure on $\Sigma_2 \times \hat{\mathbb{C}}$. We define a function $\tilde{p} : \Sigma_2 \times \hat{\mathbb{C}} \to \mathbb{R}$ as $\tilde{p}(z) = p$ if $\pi(z) = (1, \dots)$ and $\tilde{p}(z) = 1 - p$ if $\pi(z) = (2, \dots)$. Moreover, we set $u(h_1, h_2, p, \rho) := -(\int_{\Sigma_2\times\hat{\mathbb{C}}} \log \tilde{p}\, d\rho)/(\int_{\Sigma_2\times\hat{\mathbb{C}}} \log |\tilde{h}'|\, d\rho)$ (this is well-defined when the integral of the denominator converges).
- Let V be an open subset of $\mathbb{C}$. For any function $\varphi : V \to \mathbb{R}$ and any point $y \in V$, if φ is bounded around y, we set

$$\text{Höl}(\varphi, y) := \inf\{\beta \in \mathbb{R} \mid \limsup_{z\to y}(|\varphi(z) - \varphi(y)|/|z - y|^\beta) = \infty\},$$

and this is called the **pointwise Hölder exponent of** φ **at** y. (**Note:** If $\text{Höl}(\varphi, y) < 1$, then φ is not differentiable at y. If $\text{Höl}(\varphi, y) > 1$, then φ is differentiable at y and the derivative is equal to 0.)

Under the above notation, we have the following result obtained by the author of this article.

THEOREM 10.4 ([**66**]). *All of the following (1)–(15) hold.*

(1) *Let $(h_1, h_2) \in \mathcal{D}\cap\mathcal{B}$ and let $G = \langle h_1, h_2\rangle$. Then $h_1^{-1}(J(G)) \cap h_2^{-1}(J(G)) = \emptyset$.*
(2) $\overline{\text{int}(\mathcal{C})} \cap \mathcal{B} \cap \mathcal{H} = \mathcal{C} \cap \mathcal{B} \cap \mathcal{H}$.
(3) *For each $(h_1, h_2) \in (\overline{\mathcal{D}}\cap\mathcal{B}\cap\mathcal{H}) \setminus \mathcal{Q}$, (h_1, h_2) satisfies the open set condition.*
(4) *For each $(h_1, h_2) \in \overline{\mathcal{D}} \cap \mathcal{B} \cap \mathcal{H}$, $\dim_H(J(\langle h_1, h_2\rangle)) < 2$.*
(5) *Let $(h_1, h_2) \in (\overline{\mathcal{D}}\cap\mathcal{B}\cap\mathcal{H}) \setminus \mathcal{Q}$. Then, for each $z \in \hat{\mathbb{C}} \setminus P(\langle h_1, h_2\rangle)$, we have $\dim_H(J(\langle h_1, h_2\rangle)) = Z_{(h_1,h_2)}(z)$. Moreover, there exists an $\epsilon > 0$ and a neighborhood V of (h_1, h_2) in $\mathcal{B} \cap \mathcal{H}$ such that for each $(g_1, g_2) \in V$ and for each $z \in \hat{\mathbb{C}} \setminus P(\langle g_1, g_2\rangle)$, $\dim_H(J(\langle g_1, g_2\rangle)) \leq Z_{(g_1,g_2)}(z) \leq 2 - \epsilon$.*
(6) *$\mathcal{D} \cap \mathcal{Q} = \emptyset$. For each connected component $\mathcal{V}$ of $\mathcal{Y}^2$, $\mathcal{Q} \cap \mathcal{V}$ is included in a proper holomorphic subvariety of $\mathcal{V}$.*
(7) *$((\partial\mathcal{C}) \cap \mathcal{B} \cap \mathcal{H}) \setminus \mathcal{Q}$ is dense in $(\partial\mathcal{C}) \cap \mathcal{B} \cap \mathcal{H}$.*
(8) *Suppose that $h_1 \in \mathcal{Y}$, $\langle h_1\rangle \in \mathcal{G}$, and h_1 is hyperbolic. Moreover, let $d \in \mathbb{N}, d \geq 2$ and suppose that $(\deg(h_1), d) \neq (2, 2)$. Then there exists an element $h_2 \in \mathcal{Y}$ such that $(h_1, h_2) \in ((\partial\mathcal{C}) \cap \mathcal{B} \cap \mathcal{H}) \setminus \mathcal{I}$ and $\deg(h_2) = d$.*
(9) *For each $(h_1, h_2) \in (\mathcal{D} \cap \mathcal{B}) \cup ((\partial\mathcal{C}) \cap \mathcal{B} \cap \mathcal{H})$ and for each $0 < p < 1$, we have $J(\langle h_1, h_2\rangle) = \{z_0 \in \hat{\mathbb{C}} \mid$ for each neighborhood U of z_0, $z \mapsto T(h_1, h_2, p, z)$ is not constant on $U\}$.*
(10) *Let $(h_1, h_2) \in (\partial\mathcal{C}) \cap \mathcal{B} \cap \mathcal{H}$ and let $0 < p < 1$. Then,*

$$\text{“}z \mapsto T(h_1, h_2, p, z) \text{ is continuous on } \hat{\mathbb{C}}\text{”} \iff \text{“}J(h_1) \cap J(h_2) = \emptyset.\text{”}$$

(11) *Let* $(h_1, h_2) \in ((\partial \mathcal{C}) \cap \mathcal{B} \cap \mathcal{H}) \setminus \mathcal{I}$. *Then there exists a neighborhood* V *of* (h_1, h_2) *in* $\mathcal{B} \cap \mathcal{H}$ *such that for each* $(g_1, g_2) \in V$ *and for each* $0 < p < 1$, $z \mapsto T(g_1, g_2, p, z)$ *is continuous on* $\hat{\mathbb{C}}$.

(12) *Let* $(h_1, h_2) \in (\mathcal{D} \cap \mathcal{B}) \cup (((\partial \mathcal{C}) \cap \mathcal{B} \cap \mathcal{H}) \setminus \mathcal{I})$ *and let* $G = \langle h_1, h_2 \rangle$. *Then, for each* $z \in \hat{\mathbb{C}}$, *the function* $p \mapsto T(h_1, h_2, p, z)$ *is real-analytic on* $(0,1)$. *Moreover, for each* $n \in \mathbb{N} \cup \{0\}$, *the function* $(p, z) \mapsto (\partial^n T / \partial p^n)(h_1, h_2, p, z)$ *is continuous on* $(0,1) \times \hat{\mathbb{C}}$. *Moreover, letting* $M_{h_1,h_2,p} : C(\hat{\mathbb{C}}) \to C(\hat{\mathbb{C}})$ *be the operator defined by* $M_{h_1,h_2,p}(\varphi)(z) = p\varphi(h_1(z)) + (1-p)\varphi(h_2(z))$, *the function* $z \mapsto T(h_1, h_2, p, z)$ *is characterized by the unique element* $\varphi \in C(\hat{\mathbb{C}})$ *such that* $M_{h_1,h_2,p}(\varphi) = \varphi, \varphi|_{\hat{K}(G)} \equiv 0, \varphi|_{F_\infty(G)} \equiv 1$. *Here, we denote by* $F_\infty(G)$ *the connected component of* $F(G)$ *containing* ∞. *Furthermore, inductively, for any* $n \in \mathbb{N} \cup \{0\}$, *the function* $z \mapsto (\partial^{n+1} T / \partial p^{n+1})(h_1, h_2, p, z)$ *is characterized by the unique element* $\varphi \in C(\hat{\mathbb{C}})$ *such that*

$$\varphi(z) \equiv M_{h_1,h_2,p}(\varphi)(z) + (n+1)((\partial^n T/\partial p^n)(h_1, h_2, p, h_1(z)) - (\partial^n T/\partial p^n)(h_1, h_2, p, h_2(z))),$$

$\varphi|_{\hat{K}(G) \cup F_\infty(G)} \equiv 0$.

(13) (**Non-differentiability**) *Let* $(h_1, h_2) \in \mathcal{D} \cap \mathcal{B}$, $G = \langle h_1, h_2 \rangle$, *and* $0 < p < 1$. *Regarding the multi-map* $h = (h_1, h_2) \in \mathcal{Y}^2$ *and the weight* $a = (p, 1-p)$, *let* ρ^a *be the Borel probability measure on* $\Sigma_2 \times \hat{\mathbb{C}}$ *in Theorem* 3.22 *and let* $\lambda_p := (\pi_{\hat{\mathbb{C}}})_*(\rho^a)$. *Then,* $\text{supp}\,\lambda_p = J(G)$ *and for almost every point* $z_0 \in J(G)$ *with respect to* λ_p, $(*)$ $\text{Höl}(T(h_1, h_2, p, \cdot), z_0) \le u(h_1, h_2, p, \rho^a) = -(p \log p + (1-p)\log(1-p)) / (p \log(\deg(h_1)) + (1-p)\log(\deg(h_2))) < 1$ *and* $z \mapsto T(h_1, h_2, p, z)$ *is not differentiable at* z_0. *In particular, there exists an uncountable dense subset* A *of* $J(G)$ *such that at every point of* A, *the function* $z \mapsto T(h_1, h_2, p, z)$ *is not differentiable. Moreover, if* $(h_1, h_2) \in \mathcal{D} \cap \mathcal{B} \cap \mathcal{H}$, *then we can replace "*$\le$*" in* $(*)$ *by "*$=$*".*

(14) (**Hölder continuity when the semigroup is hyperbolic**) *Let* $(h_1, h_2) \in \mathcal{D} \cap \mathcal{B} \cap \mathcal{H}$ *and let* $0 < p < 1$. *Then,* $z \mapsto T(h_1, h_2, p, z)$ *is Hölder continuous on* $\hat{\mathbb{C}}$.

(15) *Let* $(h_1, h_2) \in \mathcal{D} \cap \mathcal{B} \cap \mathcal{H}$, $G = \langle h_1, h_2 \rangle$, *and* $0 < p < 1$. *Let* $v = \dim_H(J(G))$ *and let* H^v *be the* v*-dimensional Hausdorff measure. (Note: By* [**60**], $0 < H^v(J(G)) < \infty$.*) Moreover, applying Theorem* 6.10 *to* $h = (h_1, h_2)$, *let* ν *be the Borel probability measure on* $\tilde{J}(\tilde{h})$ *and let* $\alpha \in C(\tilde{J}(\tilde{h}))$ *be the element in Theorem* 6.10. *Let* $\eta := \alpha \cdot \nu$ *(*η *is an* $\tilde{h}$*-invariant Borel probability measure on* $\tilde{J}(\tilde{h})$*). Then, for almost every* $z_0 \in J(G)$ *with respect to* H^v, $\text{Höl}(T(h_1, h_2, p, \cdot), z_0) = u(h_1, h_2, p, \eta)$.

REMARK 10.5. Let $(h_1, h_2) \in \mathcal{D} \cap \mathcal{B} \cap \mathcal{H}$ be an element. By statements (13),(15) in Theorem 10.4, it follows that if p is close enough to 0 or 1, then (1) for almost every $z_0 \in J(\langle h_1, h_2 \rangle)$ with respect to λ_p, $z \mapsto T(h_1, h_2, p, z)$ is not differentiable at z_0, but (2) for almost every $z_0 \in J(\langle h_1, h_2 \rangle)$ with respect to H^v, $z \mapsto T(h_1, h_2, p, z)$ is differentiable at z_0 and the derivative is equal to 0.

REMARK 10.6. The function $z \mapsto T(h_1, h_2, p, z)$ is the complex analogue of the devil's staircase or Lebesgue's singular functions (see Figures 5, 6), and the function $z \mapsto (\partial T / \partial p)(h_1, h_2, p, z)$ is the complex analogue of the Takagi function (see Figure 7).

FIGURE 5. The graph of $z \mapsto T(h_1, h_2, 1/2, z)$, where (h_1, h_2) is the element in Figure 3. A devil's coliseum (a complex analogue of the devil's staircase). $(h_1, h_2) \in \mathcal{D} \cap \mathcal{B} \cap \mathcal{H}$. The set of varying points is equal to Figure 3.

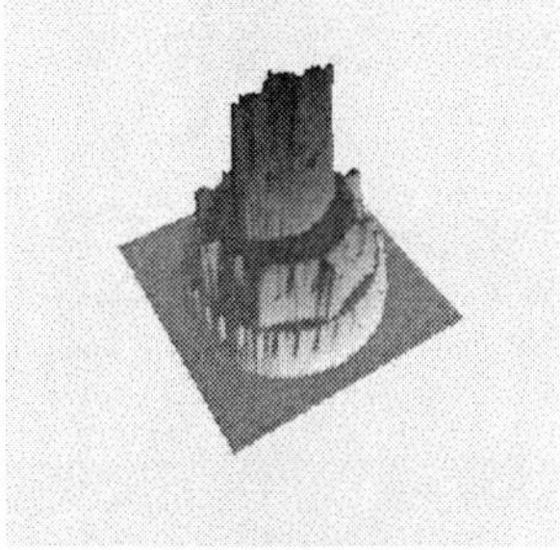

FIGURE 6. Figure 5, upside down.

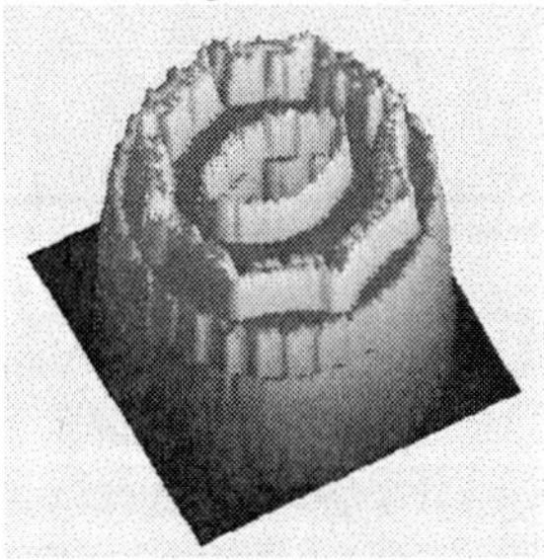

FIGURE 7. The graph of $z \mapsto (\partial T/\partial p)(h_1, h_2, 1/2, z)$, where (h_1, h_2) is the element in Figure 3. A complex analogue of the Takagi function.

REMARK 10.7. In the proof of "$\text{Höl}(T(h_1, h_2, p, \cdot), z_0) = u(h_1, h_2, p, \rho)$ (where ρ is ρ^a or η)" in Theorem 10.4(13),(15), we use the Birkhoff ergodic theorem and the Koebe distortion theorem. In the calculation of $u(h_1, h_2, p, \rho^a)$ by using $p, \deg(h_j)$ in Theorem 10.4(13), we use Remark 3.23(2) and potential theory.

11. Added in translation

In this section, we present some recent results on random complex dynamics. We show that if $J_{\ker}(G_\tau) = \emptyset$, then the chaos of the averaged system disappears (Theorem 11.2). This new phenomenon (which cannot hold in the usual iteration dynamics of a single rational map of degree two or more) holds since many kinds of maps in the system "automatically cooperate" so that they make the chaos disappear although each map has a chaotic part (Julia set). This is called the **cooperation principle**. Moreover, we show that for most Borel probability measures τ on $\mathcal{Y}$ with compact support, we have $J_{\ker}(G_\tau) = \emptyset$ (Lemma 11.5, Proposition 11.6). Thus it turns out that in the random dynamics of complex polynomials on $\hat{\mathbb{C}}$, for most cases, the chaos of the averaged system disappears due to the cooperation of many kinds of maps in the system (Remark 11.7). We now introduce some notation.

DEFINITION 11.1. Let τ be a Borel probability measure on Rat.

(1) Let $\mathcal{U}_\tau$ be the space of all finite linear combinations of unitary eigenvectors of $M_\tau : C(\hat{\mathbb{C}}) \to C(\hat{\mathbb{C}})$, where an eigenvector is said to be unitary if the absolute value of the corresponding eigenvalue is 1.
(2) We set $\mathcal{B}_{0,\tau} := \{\varphi \in C(\hat{\mathbb{C}}) \mid M_\tau^n(\varphi) \to 0 \text{ as } n \to \infty\}$.

We now present a theorem on the disappearance of chaos which occurs by the cooperation of many kinds of maps in the systems.

THEOREM 11.2 ([**72**], Cooperation Principle and Disappearance of Chaos). *Let τ be a Borel probability measure on* Rat *such that* $\operatorname{supp}\tau$ *is compact. Let $f : X_\tau \times \hat{\mathbb{C}} \to X_\tau \times \hat{\mathbb{C}}$ be the rational skew product associated with* $\operatorname{supp}\tau$. *Suppose $J_{\ker}(G_\tau) = \emptyset$ and $J(G_\tau) \neq \emptyset$. Then, we have all of the following* (1)–(8).

(1) $F_{\mathrm{meas}}(\tau) = \mathfrak{M}_1(\hat{\mathbb{C}})$.
(2) $\mathcal{B}_{0,\tau}$ *is a closed subspace of* $C(\hat{\mathbb{C}})$ *and* $C(\hat{\mathbb{C}}) = \mathcal{U}_\tau \oplus \mathcal{B}_{0,\tau}$.
(3) $\dim_{\mathbb{C}} \mathcal{U}_\tau < \infty$.
(4) For each $\varphi \in \mathcal{U}_\tau$ and for each connected component U of $F(G_\tau)$, $\varphi|_U$ is constant.
(5) For any $z \in \hat{\mathbb{C}}$, there exists a Borel set $\mathcal{A}_z \subset (\mathrm{Rat})^{\mathbb{N}}$ with $\tilde{\tau}(\mathcal{A}_z) = 1$ with the following property.
  - *for any $\gamma = (\gamma_1, \gamma_2, \ldots) \in \mathcal{A}_z$, there exists a $\delta = \delta(z, \gamma) > 0$ such that* $\operatorname{diam}\gamma_n \cdots \gamma_1(B(z,\delta)) \to 0$ *as $n \to \infty$, where* diam *denotes the diameter with respect to the spherical distance.*

(6) For $\tilde{\tau}$-a.e. $\gamma = (\gamma_1, \gamma_2, \ldots) \in (\mathrm{Rat})^{\mathbb{N}}$, the 2-dimensional Lebesgue measure of $J_\gamma(f)$ is equal to zero.
(7) There exist at least one and at most finitely many minimal sets of G_τ in $\hat{\mathbb{C}}$, where we say that a nonempty compact subset K of $\hat{\mathbb{C}}$ is a minimal set of G_τ in $\hat{\mathbb{C}}$ if K is minimal in $\{L \subset \hat{\mathbb{C}} \mid \emptyset \neq L \text{ is compact}, \forall g \in G_\tau, g(L) \subset L\}$ with respect to the inclusion.
(8) Let L_τ be the union of minimal sets of G_τ in $\hat{\mathbb{C}}$. Then for each $z \in \hat{\mathbb{C}}$, there exists a Borel set $\mathcal{C}_z \subset (\mathrm{Rat})^{\mathbb{N}}$ with $\tilde{\tau}(\mathcal{C}_z) = 1$ such that for each $\gamma = (\gamma_1, \gamma_2, \ldots) \in \mathcal{C}_z$, $d(\gamma_n \cdots \gamma_1(z), L_\tau) \to 0$ as $n \to \infty$.

REMARK 11.3. Theorem 11.2 describes new phenomena which cannot hold in the usual iteration dynamics of a single $g \in \mathrm{Rat}$ with $\deg(g) \geq 2$.

We now consider the nondifferentiability of nonconstant elements $\varphi \in \mathcal{U}_\tau$ at $J(G_\tau)$. We consider the following setting.

SETTING AND NOTATION. Let $h_1, h_2 \in \mathcal{Y}$ and let $0 < p_1, p_2 < 1$ with $p_1 + p_2 = 1$. Let $\tau := \sum_{i=1}^{2} p_i \delta_{h_i}$ (a Borel probability measure on $\mathcal{Y}$). Regarding the multi-map $h = (h_1, h_2) \in \mathcal{Y}^2$ and the weight $a = (p_1, p_2)$, let ρ^a be the Borel probability measure on $\Sigma_2 \times \hat{\mathbb{C}}$ in Theorem 3.22 and let $\lambda_{p_1} := (\pi_{\hat{\mathbb{C}}})_*(\rho^a)$. For each $(w, y) = ((w_1, w_2, \ldots), y) \in \Sigma_2 \times \mathbb{C}$, we set

$$\mathcal{G}_w(y) := \lim_{n\to\infty} \frac{1}{\deg(h_{w_n} \circ \cdots \circ h_{w_1})} \max\{\log |h_{w_n} \circ \cdots \circ h_{w_1}(y)|, 0\}.$$

Moreover, for each $w = (w_1, w_2, \ldots) \in \Sigma_2$, let $\Omega(w) := \sum_c \mathcal{G}_w(c)$, where c runs over all critical points of h_{w_1} in $\mathbb{C}$.

We now present a theorem.

THEOREM 11.4 ([**72**]). *Under the above setting and notation, we assume that G_τ is hyperbolic, $h_1^{-1}(J(G_\tau)) \cap h_2^{-1}(J(G_\tau)) = \emptyset$, and there exists a point $z \in \mathbb{C}$ such that $\bigcup_{h \in G_\tau} \{h(z)\}$ is bounded in $\mathbb{C}$. Then, we have all of the following statements* (1), (2), (3).

(1) $J_{\ker}(G_\tau) = \emptyset$, $T_{\infty,\tau} \in \mathcal{U}_\tau$ *and* $T_{\infty,\tau}$ *is nonconstant.*

(2) $\dim_H(J(G_\tau)) < 2$.

(3) There exists an uncountable dense Borel subset A of $J(G_\tau)$ with $\lambda_{p_1}(A) = 1$ satisfying all of the following (a) *and* (b).

(a) $\dim_H(A) \geq \dim_H(\lambda_{p_1}) = \frac{\sum_{i=1}^{2} p_i \log \deg(h_i) - \sum_{i=1}^{2} p_i \log p_i}{\sum_{i=1}^{2} p_i \log \deg(h_i) + \int_{\Gamma^{\mathbb{N}}} \Omega(\gamma)\, d\tilde{\tau}(\gamma)} > 0.$

(b) For each $z \in A$ and each nonconstant $\varphi \in \mathcal{U}_\tau$,

$$\text{Höl}(\varphi, z) = u(h_1, h_2, p_1, \rho^a) = \frac{-\sum_{i=1}^{2} p_i \log p_i}{\sum_{i=1}^{2} p_i \log \deg(h_i) + \int_{\Gamma^{\mathbb{N}}} \Omega(\gamma)\, d\tilde{\tau}(\gamma)} < 1$$

and φ is not differentiable at z.

It is very interesting to ask when the kernel Julia set of a rational semigroup is empty. We present some results on this question. The following lemma follows from the forward invariance of $J_{\ker}(G)$ and Montel's theorem.

LEMMA 11.5 ([**72**]). *Let Γ be a compact subset of $\mathcal{Y}$. If there exists an $f_0 \in \mathcal{Y}$ and a nonempty open subset U of $\hat{\mathbb{C}}$ such that $\{f_0 + c \mid c \in U\} \subset \Gamma$, then the polynomial semigroup G generated by Γ satisfies that $J_{\ker}(G) = \emptyset$.*

By using the above lemma, we can show the following.

PROPOSITION 11.6 ([**72**]). *Let τ be a Borel probability measure on $\mathcal{Y}$ with compact support. Let V_1 be any neighborhood of τ in the space of all Borel probability measures on $\mathcal{Y}$ endowed with the topology induced by the weak convergence, and let V_2 be any neighborhood of the element* $\operatorname{supp} \tau$ *in the space of all nonempty compact subsets of $\hat{\mathbb{C}}$ endowed with the Hausdorff metric. Then there exists a Borel probability measure ρ on $\mathcal{Y}$ such that $\rho \in V_1$,* $\operatorname{supp} \rho \in V_2$, $\sharp \operatorname{supp} \rho < \infty$, *and* $J_{\ker}(G_\rho) = \emptyset$.

REMARK 11.7. Lemma 11.5 and Proposition 11.6 imply that from this point of view, for most Borel probability measures τ on $\mathcal{Y}$ with compact support, we have $J_{\ker}(G_\tau) = \emptyset$. Combining this with Theorem 11.2, it follows that in the random dynamics of complex polynomials on $\hat{\mathbb{C}}$, for most cases, the chaos of the averaged system disappears, due to the cooperation of many kinds of maps in the system. It should also be mentioned that, as illustrated in Theorems 10.4, 11.4, and Figure 5, a kind of chaos may still remain in the "C^1" sense, even though the chaos disappears in the "C^0" sense.

It is also very interesting to ask what happens if $J_{\ker}(G_\tau) \neq \emptyset$. We present a result on this problem.

THEOREM 11.8 ([**72**]). *Let τ be a Borel probability measure on* Rat *with compact support such that each element of* supp τ *is of degree two or more. Suppose that G_τ is semihyperbolic and $F(G_\tau) \neq \emptyset$. Then we have the following.* (1) $J_{\ker}(G_\tau) = \emptyset$ *if and only if* $F_{\text{meas}}(\tau) = \mathfrak{M}_1(\hat{\mathbb{C}})$. (2) *If* $J_{\ker}(G_\tau) \neq \emptyset$, *then* $J_{\text{meas}}(\tau) = \mathfrak{M}_1(\hat{\mathbb{C}})$.

(Note: This article is dedicated to the memory of the author's father Seiji Sumi.)

Acknowledgement

The author thanks the referees for many valuable comments. The author also thanks R. Stankewitz for reading several sections of this article.

References

[1] P. Allaart and K. Kawamura, *Extreme values of some continuous nowhere differentiable functions*, Math. Proc. Cambridge Philos. Soc. **140** (2006), no. 2, 269–295. MR2212280 (2007g:26005)

[2] A. Ambroladze, *Ergodic properties of random iterations of analytic functions*, Ergodic Theory Dynam. Systems, **19** (1999), no. 6, 1379–1388. MR1738942 (2000m:37076)

[3] A. Ambroladze and H. Wallin, *Random iteration of Möbius transformations and Furstenberg's theorem*, Ergodic Theory Dynam. Systems, **20** (2000), no. 4, 953–962. MR1779388 (2001h:37112)

[4] I. N. Baker and P. J. Rippon, *On compositions of analytic self-mappings of a convex domain*, Arch. Math., Vol. 55, 380-386 (1990). MR1076065 (91k:30062)

[5] A. F. Beardon, T. K. Carne, D. Minda, and T. W. Ng, *Random iteration of analytic maps*, Ergodic Theory Dynam. Systems, **24** (2004), 659-675. MR2060992 (2005b:37086)

[6] D. Boyd, An invariant measure for finitely generated rational semigroups, Complex Variables Theory Appl. **39** (1999), no. 3, 229–254. MR1717572 (2000e:37050)

[7] D. Boyd, *The immediate basin of attraction of infinity for polynomial semigroups of finite type*, J. London Math. Soc. (2) **69** (2004), no. 1, 201–213. MR2025336 (2004j:37081)

[8] R. Brück, *Connectedness and stability of Julia sets of the composition of polynomials of the form* $z^2 + c_n$, J. London Math. Soc. **61** (2000), 462-470. MR1760694 (2001e:30040)

[9] R. Brück, *Geometric properties of Julia sets of the composition of polynomials of the form* $z^2 + c_n$, Pacific J. Math., **198** (2001), no. 2, 347–372. MR1835513 (2002d:37078)

[10] R. Brück, M. Büger and S. Reitz, *Random iterations of polynomials of the form* $z^2 + c_n$*: Connectedness of Julia sets*, Ergodic Theory Dynam. Systems, **19**, (1999), No.5, 1221–1231. MR1721617 (2000h:37064)

[11] M. Büger, *Self-similarity of Julia sets of the composition of polynomials*, Ergodic Theory Dynam. Systems, **17** (1997), 1289–1297. MR1488318 (98k:58190)

[12] M. Büger, *On the composition of polynomials of the form* $z^2 + c_n$, Math. Ann. **310** (1998), no. 4, 661–683. MR1619744 (99c:30050)

[13] L. Carleson, P. W. Jones and J.-C. Yoccoz, *Julia and John,* Bol. Soc. Bras. Mat. **25, N.1** (1994), 1–30. MR1274760 (95d:30040)

[14] M. Comerford, *Infinitely many grand orbits*, Michigan Math. J. **51** (2003), no. 1, 47–57. MR1960920 (2003k:37061)

[15] M. Comerford, *Conjugacy and counterexample in random iteration*, Pacific J. Math. **211** (2003), no. 1, 69–80. MR2016591 (2004j:37082)

[16] M. Comerford, *A survey of results in random iteration*, Fractal geometry and applications: A jubilee of Benoit Mandelbrot. Part 1, 435–476, Proc. Sympos. Pure Math., 72, Part 1, Amer. Math. Soc., Providence, RI, 2004. MR2112116 (2006b:37082)

[17] M. Comerford, *Hyperbolic non-autonomous Julia sets*, Ergodic Theory Dynam. Systems, **26** (2006), no. 2, 353–377. MR2218765 (2007c:37049)

[18] L. DeMarco and S. L. Hruska, *Axiom A polynomial skew products of* $\mathbb{C}^2$ *and their postcritical sets*, Ergodic Theory Dynam. Sys. **28** (2008), no. 6, 1749–1779. MR2465599 (2009m:37132)

[19] L. DeMarco and S. L. Hruska, Correction to Axiom A polynomial skew products of $\mathbb{C}^2$ and their postcritical sets, preprint, http://www.math.uic.edu/~demarco/correction.pdf.

[20] G. De Rham, *Sur quelques courbes définies par des équations fonctionnelles*, Rend. Sem. Mat. Torino, **16** (1957), 101-113. MR0095227 (20:1733)

[21] K. Falconer, *Techniques in Fractal Geometry*, John Wiley & Sons, 1997. MR1449135 (99f:28013)

[22] J. E. Fornaess and N. Sibony, *Random iterations of rational functions*, Ergodic Theory Dynam. Systems, **11**(1991), 687–708. MR1145616 (93c:58173)

[23] Z. Gong, W. Qiu and Y. Li, *Connectedness of Julia sets for a quadratic random dynamical system*, Ergodic Theory Dynam. Systems, **23** (2003), 1807-1815. MR2032489 (2004j:37087)

[24] Z. Gong and F. Ren, *A random dynamical system formed by infinitely many functions,* Journal of Fudan University, **35** (1996), 387–392. MR1435167 (98k:30032)

[25] R. Gunning, *Introduction to Holomorphic Functions of Several Variables, Vol. III: Homological Theory*, Wadsworth & Brooks/Cole, 1990. MR1059457 (92b:32001c)

[26] T. Harada, *The dynamics of nearly abelian polynomial semigroups at infinity*, Proc. Japan Academy, Vol. 73, Ser. A, No. 3, (1997). MR1453525 (98f:58162)

[27] T. Harada, *The Teichmüller space of rational semigroup*, preprint, http://www.math.okayama-u.ac.jp/~tatsu/prep/index.html

[28] M. Hata and M. Yamaguti, *Takagi function and its generalization*, Japan J. Appl. Math., **1**, (1984), 183-199. MR839313 (87g:26005)

[29] A. Hinkkanen and G. J. Martin, *The Dynamics of Semigroups of Rational Functions I*, Proc. London Math. Soc. (3) **73** (1996), 358–384. MR1397693 (97e:58198)

[30] A. Hinkkanen and G. J. Martin, *Julia Sets of Rational Semigroups* , Math. Z. **222** (1996), no.2, 161–169. MR1429333 (98d:30038)

[31] A. Hinkkanen and G. J. Martin, *Some properties of semigroups of rational functions*, XVIth Rolf Nevanlinna Colloquium (Joensuu, 1995), 53-58, de Gruyter, Berlin, 1996. MR1427070 (97i:30036)

[32] M. Jonsson, *Dynamics of polynomial skew products on* $\mathbb{C}^2$, Math. Ann. **314** (1999), 403-447. MR1704543 (2000f:32025)

[33] M. Jonsson, *Ergodic properties of fibered rational maps*, Ark. Mat. **38** (2000), 281–317. MR1785403 (2002k:37073)

[34] K. Katagata, *On a certain kind of polynomials of degree* 4 *with disconnected Julia set*, Discrete Contin. Dyn. Syst. **20** (2008), no. 4, 975–987. MR2379483 (2009b:37082)

[35] O. Lehto and K. I. Virtanen, *Quasiconformal Mappings in the Plane,* Springer-Verlag, 1973. MR0344463 (49:9202)

[36] R. May, *Stability and Complexity in Model Ecosystems*, 2nd edition, Princeton Univ. Press, Princeton, NJ, 1975.

[37] C. T. McMullen, *Complex Dynamics and Renormalization*, Annals of Mathematics Studies Number 135, Princeton University Press, Princeton, NJ, 1994. MR1312365 (96b:58097)

[38] J. Milnor, *Dynamics in One Complex Variable* (Third Edition), Annals of Mathematical Studies, Number 160, Princeton University Press, Princeton, NJ, 2006. MR2193309 (2006g:37070)

[39] R. Näkki and J. Väisälä, *John Discs,* Expo. Math. **9** (1991), 3–43.

[40] Y. Okuyama, *Nevanlinna theoretical exceptional sets of rational towers and semigroups*, Conform. Geom. Dyn. **10** (2006), 100–116 (electronic). MR2218642 (2007f:37065)

[41] S. Rohde, *Composition of random rational functions*, Complex Variable Theory Appl., 1996, Vol. 29, No. 1, 1-7. MR1381998 (97i:30037)

[42] T. Sekiguchi and Y. Shiota, *A generalization of Hata-Yamaguti's results on the Takagi function*, Japan J. Appl. Math. **8** (1991), 203-219. MR1111613 (92i:26010)

[43] O. Sester, *Hyperbolicité des polynômes fibrés*, Bull. Soc. Math. France **127** (1999), no. 3, 393–428. MR1724402 (2000j:37057)

[44] O. Sester, *Combinatorial configurations of fibered polynomials*, Ergodic Theory Dynam. Systems, **21** (2001), 915-955. MR1836438 (2002d:37074)

[45] E. H. Spanier, *Algebraic Topology*, Springer-Verlag, New York-Berlin, 1981. MR666554 (83i:55001)

[46] R. Stankewitz, *Completely invariant Julia sets of polynomial semigroups,* Proc. Amer. Math. Soc., **127** (1999), No. 10, 2889–2898. MR1600149 (2000a:37032)

[47] R. Stankewitz, *Completely invariant sets of normality for rational semigroups*, Complex Variables Theory Appl. Vol. 40 (2000), 199–210. MR1753707 (2000m:37069)

[48] R. Stankewitz, *Uniformly perfect sets, rational semigroups, Kleinian groups and IFS's,* Proc. Amer. Math. Soc. **128**, (2000), No. 9, 2569–2575. MR1662218 (2000m:37075)

[49] R. Stankewitz, *Density of repelling fixed points in the Julia set of a rational or entire semigroup,* to appear in Journal of Difference Equations and Applications, http://www.bsu.edu/web/rstankewitz/# Publications.

[50] R. Stankewitz, *Density of repelling fixed points in the Julia set of a rational or entire semigroup, II,* to appear in Discrete Contin. Dyn. Syst. Series A, Proceedings of the conference Dynamical Systems II, Denton 2009, http://www.bsu.edu/web/rstankewitz/# Publications.

[51] R. Stankewitz, T. Sugawa and H. Sumi, *Some counterexamples in dynamics of rational semigroups*, Ann. Acad. Sci. Fenn. Math. **29** (2004), 357–366. MR2097238 (2005m:37107)

[52] R. Stankewitz and H. Sumi, *Dynamical properties and structure of Julia sets of postcritically bounded polynomial semigroups*, to appear in Trans. Amer. Math. Soc., http://arxiv.org/abs/0708.3187.

[53] T. Sugawa, *Various domain constants related to uniform perfectness*, Complex Variables Theory Appl. **36** (1998), no. 4, 311–345. MR1670075 (99h:30043)

[54] D. Sullivan, *Quasiconformal homeomorphisms and dynamics. I. Solution of the Fatou-Julia problem on wandering domains*, Ann. of Math. (2) **122** (1985), no. 3, 401–418. MR819553 (87i:58103)

[55] H. Sumi, *On Dynamics of Hyperbolic Rational Semigroups,* J. Math. Kyoto Univ., Vol. 37, No. 4, 1997, 717–733. MR1625944 (99h:30028)

[56] H. Sumi, *On Hausdorff dimension of Julia sets of hyperbolic rational semigroups*, Kodai Math. J., Vol. 21, No. 1, pp. 10–28, 1998. MR1625124 (99h:30029)

[57] H. Sumi, *Skew product maps related to finitely generated rational semigroups,* Nonlinearity, **13**, (2000), 995–1019. MR1767945 (2001g:37060)

[58] H. Sumi, *Dynamics of sub-hyperbolic and semi-hyperbolic rational semigroups and skew products*, Ergodic Theory Dynam. Systems **21** (2001), 563–603. MR1827119 (2002d:37073a)

[59] H. Sumi, *A correction to the proof of a lemma in 'Dynamics of sub-hyperbolic and semi-hyperbolic rational semigroups and skew products'*, Ergodic Theory Dynam. Systems, **21** (2001), 1275–1276. MR1849611 (2002d:37073b)

[60] H. Sumi, *Dimensions of Julia sets of expanding rational semigroups*, Kodai Math. J., Vol. 28, No. 2, 2005, 390–422. MR2153926 (2006c:37050)

[61] H. Sumi, *Semi-hyperbolic fibered rat ional maps and rational semigroups*, Ergodic Theory Dynam. Systems, **26** (2006), 893–922. MR2237476 (2007d:37068)

[62] H. Sumi, *Erratum to 'Semi-hyperbolic fibered rational maps and rational semigroups' (Ergodic Theory and Dynamical Systems, 26 (2006) 893-922)*, Ergodic Theory Dynam. Systems, **28** (2008), 1043-1045. MR2237476 (2007d:37068)

[63] H. Sumi, Interaction cohomology of forward or backward self-similar systems, Adv. Math., **222** No.3 (2009), 729–781. MR2553369

[64] H. Sumi, *Rational semigroups, random complex dynamics, and singular functions on the complex plane* (written in Japanese), Sugaku, Vol. 61, No. 2 (2009), 133-161. MR2560252

[65] H. Sumi, *Topics in dynamics of rational semigroups and fibered rational maps*, RIMS Kokyuroku 1220, pp. 78-122, 2001. (Proceedings paper of a conference.) MR1864533

[66] H. Sumi, *The space of postcritically bounded 2-generator polynomial semigroups with hyperbolicity*, RIMS Kokyuroku 1494, 62–86, 2006. (Proceedings paper of a conference.)

[67] H. Sumi, *Random dynamics of polynomials and devil's-staircase-like functions in the complex plane*, Appl. Math. Comput. **187** (2007), 489-500. (Proceedings paper of a conference.) MR2323605 (2008d:37075)

[68] H. Sumi, *Dynamics of postcritically bounded polynomial semigroups I: Connected components of the Julia sets*, to appear in Discrete Contin. Dyn. Syst. Series A, http://arxiv.org/abs/0811.3664.

[69] H. Sumi, *Dynamics of postcritically bounded polynomial semigroups II: Fiberwise dynamics and the Julia sets*, preprint, http://arxiv.org/abs/1007.0613.

[70] H. Sumi, *Dynamics of postcritically bounded polynomial semigroups III: Classification of semi-hyperbolic semigroups and random Julia sets which are Jordan curves but not quasicircles*, to appear in Ergodic Theory Dynam. Systems, http://arxiv.org/abs/0811.4536.

[71] H. Sumi, *Dynamics of postcritically bounded polynomial semigroups*, preprint 2007, http://arxiv.org/abs/math/0703591.

[72] H. Sumi, Random complex dynamics and semigroups of holomorphic maps, to appear in Proc. London Math. Soc., http://arxiv.org/abs/0812.4483.

[73] H. Sumi and M. Urbański, *The equilibrium states for semigroups of rational maps*, Monatsh. Math., **156** (2009), no. 4, 371–390. MR2486604

[74] H. Sumi and M. Urbański, *Real analyticity of Hausdorff dimension for expanding rational semigroups*, Ergodic Theory Dynam. Systems, **30** (2010), 601–633. http://arxiv.org/abs/0707.2447. MR2599895

[75] H. Sumi and M. Urbański, Measures and dimensions of Julia sets of semi-hyperbolic rational semigroups, preprint 2008, http://arxiv.org/abs/0811.1809.

[76] Y. Sun and C-C. Yang, *On the connectivity of the Julia set of a finitely generated rational semigroup*, Proc. Amer. Math. Soc., **130**, No. 1, (2001), 49–52. MR1855618 (2002e:37065)

[77] T. Takagi, *A simple example of the continuous function without derivative*, The Collected Papers of Teiji Takagi, Iwanami Shoten Pub., 1973, pp. 5-6.

[78] M. Yamaguti, M. Hata, and J. Kigami, Mathematics of fractals. Translated from the 1993 Japanese original by Kiki Hudson. Translations of Mathematical Monographs, 167. American Mathematical Society, Providence, RI, 1997. MR1471705 (98j:28006)

[79] W. Zhou and F. Ren, The Julia sets of the random iteration of rational functions, Chinese Sci. Bulletin, **37**(12) (1992), 969–971.

Department of Mathematics, Graduate School of Science, Osaka University, 1-1, Machikaneyama, Toyonaka, Osaka, 560-0043, Japan

E-mail address: `sumi@math.sci.osaka-u.ac.jp`

Translated by HIROKI SUMI

Amer. Math. Soc. Transl.
(2) Vol. **230**, 2010

Salem Polynomials and the Bimeromorphic Automorphism Group of a Hyper-Kähler Manifold

Keiji Oguiso

1. Introduction

This survey is a continuation of an excellent survey [**Kn2**] by S. Kondo published in 1990. [**Kn2**] concludes with the following suggestion (being original) : "It is shown by [**Fu3**] that the second cohomology group of a Kähler symplectic manifold, which is considered to be a higher dimensional version of a $K3$ surface, has a structure of lattice. It would be interesting to study such a manifold by means of the lattice. Please refer to [**Mu2**] about Kähler symplectic manifolds."

On the other hand, in 2002, C. T. McMullen[1] in complex dynamics brought new insight in studying automorphisms of $K3$ surfaces. Namely, he proposed to study them from the viewpoint of Salem numbers[2] or their minimal polynomials called Salem polynomials. From this view, he found $K3$ surfaces (McMullen's $K3$ surfaces) of Picard number 0 which admit very curious automorphisms of infinite order ([**Mc1**], Section 5). I have also been attracted by his beautiful construction, which shows us a wonderful combination of lattice theory, analytic number theory (the Baker-Feldman theorem), complex geometry (the Siegel-Sternberg theorem) and so on with the theory of $K3$ surfaces through Salem polynomials.

Meanwhile, C. Voisin [**Vo1**] gave an impressive counterexample for a long standing problem (Kodaira's problem): "Is any compact Kähler manifold deformed to a projective manifold under some small deformation?". McMullen's $K3$ surface is also applicable to construct another counterexample (Section 5).

In this survey, after recalling basic facts about Salem polynomials, $K3$ surfaces and symplectic manifolds (which we call hyper-Kähler manifolds in this survey), I would like to explain McMullen's $K3$ surfaces and some new results along the lines of Kondo's suggestion from the point of view of McMullen's new insight.

This article originally appeared in Japanese in Sūgaku **59** (1) (2007), 1–23.

2000 *Mathematics Subject Classification.* Primary 32J25; Secondary 14J28, 32J27, 53C26.

[1]Needless to say, he is one of the superstars in mathematics. He was awarded the Fields medal at ICM Berlin 1998 and he is now a professor at Harvard University.

[2]An algebraic integer named after Raphael Salem (1898-1963) who is also famous for the Salem prize. Salem numbers are important in Diophantine approximation and harmonic analysis and also in knot theory. We shall give a precise definition in Section 2.

2. Salem polynomials and Hodge isometries

2.1. Salem numbers and Salem polynomials.

DEFINITION 2.1. Let $a \in (1, \infty)$ be an algebraic integer.[3] We call a a **Salem number** if the Galois conjugates of a consist of a, $1/a$ and some complex numbers of absolute value 1. We call the minimal polynomial $S(x) \in \mathbf{Z}[x]$ of a Salem number a **Salem polynomial**. A Salem polynomial is nothing but a polynomial $S(x)$ which satisfies the following (S1) and (S2):

(S1) $S(x)$ is an irreducible monic polynomial in $\mathbf{Z}[x]$;

(S2) The set of zeros of $S(x)$ consists of numbers of the following form:

$$a \in (1, \infty)\ ,\ \frac{1}{a} \in (0, 1)\ ,\ \alpha_i\ ,\ \overline{\alpha_i} = \frac{1}{\alpha_i} \in S^1 \setminus \{\pm 1\}\ (1 \le i \le d-1)\ ,$$

where $S^1 := \{z \in \mathbf{C}^\times \mid |z| = 1\}$.

According to Kronecker's theorem, a root of unity is nothing but an algebraic integer whose Galois conjugates are all of absolute value 1. Thus, among all irreducible monic polynomials in $\mathbf{Z}[x]$, cyclotomic polynomials are the simplest ones and Salem polynomials are the next simplest ones, from the viewpoint of the distribution of the absolute values of solutions. On the other hand, if we regard cyclotomic polynomials as something "of finite order", then Salem polynomials are essentially something "of infinite order" (Theorem 2.6).

By (S2), a Salem polynomial is a reciprocal polynomial of even degree. In general, for a reciprocal polynomial $f(x)$ of degree $2d$ in $\mathbf{Z}[x]$, there is a unique polynomial $g(y)$ in $\mathbf{Z}[y]$ such that $f(x) = x^d g(x + (1/x))$. We call this $g(y)$ the trace polynomial of $f(x)$. The roots of $f(x) = 0$ are obtained by finding solutions b of $g(y) = 0$ first and then by solving $x + (1/x) = b$. Thus, a Salem polynomial is nothing but a polynomial whose trace polynomial $R(y)$ satisfies the following (R1) and (R2):

(R1) $R(y)$ is an irreducible monic polynomial in $\mathbf{Z}[y]$;

(R2) $R(y)$ has exactly one zero in $(2, \infty)$ and its other zeros are in $(-2, 2)$.

EXAMPLE 2.2. (1) A Salem trace polynomial of degree 1 is of the form $x - c$ (c is an integer ≥ 3). Thus, a Salem polynomial of degree 2 is of the form $x^2 - cx + 1$.

(2) (cf. [**Mc1**], [**GM**]) $R_3(y) = (y^2 - 4)(y - a) - 1$ ($a \ge 3$) is a Salem trace polynomial. Thus,

$$S_6(x) = x^3 R_3(x + \frac{1}{x}) = x^6 - ax^5 - x^4 + (2a-1)x^3 - x^2 - ax + 1$$

is a Salem polynomial of degree 6. $S_6(x)$ is related to an automorphism of a non-projective Kummer surface (cf. [**Mc1**]).

(3) $R_5(y) = (y+1)^2(y^2-4)(y-1) - 1$ is a Salem trace polynomial of degree 5. Thus, $S_{10}(x) = x^5 R_5(x + \frac{1}{x}) = x^{10} + x^9 - x^7 - x^6 - x^5 - x^4 - x^3 + x + 1$ is a Salem polynomial of degree 10. This polynomial $S_{10}(x)$ is called the Lehmer polynomial. $S_{10}(x)$ is famous for the fact that its Salem number (about 1.17628) is the smallest known Salem number.[4] $S_{10}(x)$ is also closely related to an automorphism of a surface which is a 10 points blowup of $\mathbf{P}^2$, and it plays a distinguished role in the entropy of automorphisms of complex surfaces ([**Mc2**]).

[3] A complex number which is a zero of a monic polynomial in $\mathbf{Z}[x]$.

[4] It is unknown if this is really the smallest among all Salem numbers.

(4) ([**Mc1**])

$$S_{22}(x) = 1 + x - x^3 - 2x^4 - 3x^5 - 3x^6 - 2x^7 + 2x^9 + 4x^{10} + 5x^{11}$$
$$+4x^{12} + 2x^{13} - 2x^{15} - 3x^{16} - 3x^{17} - 2x^{18} - x^{19} + x^{21} + x^{22}$$

is a Salem polynomial[5] of degree 22. Its Salem number is about 1.37289.
$R_{11}(y) = -1 - 8y + 24y^2 + 42y^3 - 54y^4 - 66y^5 + 40y^6 + 42y^7 - 11y^8 - 11y^9 + y^{10} + y^{11}$ is the Salem trace polynomial of $S_{22}(x)$. $S_{22}(x)$ is important in the construction of McMullen's $K3$ surfaces (Theorem 2.5, Section 5). We note that $|S_{22}(\pm 1)| = 1$. This fact is also important there.

There are at most finitely many cyclotomic polynomials of bounded degree. For Salem polynomials, we have:

PROPOSITION 2.3. *Let ϵ be a positive real number and let d be a positive integer. Then, there are only finitely many Salem polynomials of degree $\le d$ and of Salem number $\in (1, 1+\epsilon]$. In particular, Salem numbers of bounded degree and of bounded absolute value are finite. Moreover, given positive real numbers d, B, there are only finitely many Salem polynomials[6] of degree $\le d$ and of trace $\le B$.*

2.2. Salem polynomials and Hodge isometries. Throughout this section, $L := (L, (*, **))$ stands for a nondegenerate lattice of signature $(3, n)$. That is, L is a free $\mathbf{Z}$-module of rank $n + 3$ and $(*, **) : L \times L \longrightarrow \mathbf{Z}$ is an integral-valued symmetric bilinear form of signature[7] $(3, n)$. As we shall see in the next section, the second cohomology group $H^2(X, \mathbf{Z})$ of a $K3$ surface or a hyper-Kähler manifold is a lattice w.r.t. the cup product or the Beauville-Fujiki form.

Let $L_{\mathbf{C}} = L^{1,1} \oplus L^{2,0} \oplus L^{0,2}$ be an (abstract) **Hodge structure** of weight 2 on L, i.e., a decomposition $L_{\mathbf{C}}$ into a direct sum[8] such that

$$L^{2,0} = \mathbf{C}\sigma\ ,\ L^{0,2} := \mathbf{C}\overline{\sigma}\ ;\ (L^{1,1}, \sigma) = (L^{1,1}, \overline{\sigma}) = 0\ ,\ (\sigma, \sigma) = 0\ ,\ (\sigma, \overline{\sigma}) > 0\ .$$

The conditions $(\sigma, \sigma) = 0$, $(\sigma, \overline{\sigma}) > 0$ are called Riemann's relations. A Hodge structure of weight 2 on L is determined by choosing a σ that satisfies Riemann's relations. Thus, the set of Hodge structures of weight 2 on L can be identified with a connected complex manifold, $\mathcal{P}(L) := \{[\sigma] \in \mathbf{P}(L_{\mathbf{C}}) \,|\, (\sigma, \sigma) = 0\ ,\ (\sigma, \overline{\sigma}) > 0\}$ $(\subset \mathbf{P}(L_{\mathbf{C}}))$.

Put $L_{\mathbf{R}}^{1,1} = L^{1,1} \cap L_{\mathbf{R}}$. Then $L^{1,1} = (L_{\mathbf{R}}^{1,1})_{\mathbf{C}}$ and $L_{\mathbf{R}}^{1,1}$ is of signature $(1, n)$. By Schwarz' inequality, the number of connected components of $P(L_{\mathbf{R}}^{1,1}) := \{v \in V \,|\, (v, v) > 0\}$ is 2. We choose $P^+(V)$, one of the two connected components of $P(L_{\mathbf{R}}^{1,1})$, and call it the **positive cone**[9] of L.

[5]We can check (S1) or (R1) by applying the intermediate value theorem for $R_{11}(y)$ and irreducibility by using Kronecker's theorem.

[6]As the number of imaginary roots of a Salem polynomial is $\le d - 2$ and their absolute values are 1, the sum of the imaginary roots is bounded. Thus, for Salem polynomials, bounding traces is equivalent to bounding Salem numbers. Under this condition, the coefficients of Salem polynomials, which are elementary symmetric functions of the roots, are all bounded.

[7]It is the signature of the real symmetric matrix associated to the form, i.e., the numbers of positive, zero negative eigenvalues with multiplicities. When it is nondegenerate, we drop the number of zero eigenvalues.

[8]For a ring k, we put $L_k := L \otimes_{\mathbf{Z}} k$. The k-valued symmetric bilinear form on L_k induced by $(*, **)$ is also denoted by the same symbol $(*, **)$.

[9]When L is the second cohomology group $H^2(M, \mathbf{Z})$ of a Kähler surface or a hyper-Kähler manifold, we usually choose the component that contains the Kähler cone as its positive cone.

An automorphism of a lattice L equipped with a Hodge strcture of weight 2, i.e., an automorphism of L that preserves the bilinear form $(*, **)$ and each $L^{i,j}$, is called a **Hodge isometry** of L. The set of Hodge isometries of L is denoted by $\mathrm{O}_{\mathrm{Hodge}}(L)$. $\mathrm{O}_{\mathrm{Hodge}}(L)$ is a subgroup of the group $\mathrm{O}(L)$ of isometries of the lattice L. The following subgroup of $\mathrm{O}_{\mathrm{Hodge}}(L)$ is a normal subgroup of index 2:

$$\mathrm{O}^{+}_{\mathrm{Hodge}}(L) := \{g \in \mathrm{O}_{\mathrm{Hodge}}(L) \,|\, g(P^{+}(L^{1,1}_{\mathbf{R}})) = P^{+}(L^{1,1}_{\mathbf{R}})\} \,.$$

We call an element of $\mathrm{O}^{+}_{\mathrm{Hodge}}(L)$ a **positive Hodge isometry**.

An automorphism of a $K3$ surface or a bimeromorphic automorphism of a hyper-Kähler manifold acts on its second cohomology group as a positive Hodge isometry. The next important theorem shows that a Salem polynomial as well as a cyclotomic polynomial naturally appear in its characteristic polynomial:

THEOREM 2.4 ([**Mc1**], [**Og4**], [**Og7**]). *Let $g \in \mathrm{O}^{+}_{\mathrm{Hodge}}(L)$. Then the characteristic polynomial $\Phi_g(x) = \det(xI - g)$ of g is either one of the following forms:*

(i) *a product of finitely many cyclotomic polynomials;*

(ii) *a product of finitely many (possibly 0) cyclotomic polynomials and one Salem polynomial (counted with multiplicities).*

[**Mc1**] shows this result when g is the action of an automorphism of a $K3$ surface on the second cohomology group. The formulation here ([**Og4**], [**Og7**]) is a formal generalization, by which we can apply it also for hyper-Kähler manifolds. An essential point is the fact that $L^{1,1}_{\mathbf{R}}$ is of signature $(1, n)$. The proof is then linear-algebraic.

The next theorem, which is quite deep, gives a partial converse of the theorem above:

THEOREM 2.5 ([**Mc1**], [**GM**]). *Let $S(x)$ be (any) Salem polynomial of degree 22 such that $|S(\pm 1)| = 1$ and let $\delta \in S^1$ be a root of $S(x)$ of absolute value 1. Then, there is a pair (M, f) of a $K3$ surface M and an automorphism f of M such that*

(i) *The characteristic polynomial $\det(xI - f^*|H^2(M, \mathbf{Z}))$ of $f^*|H^2(M, \mathbf{Z})$ is $S(x)$, and*

(ii) *$f^*\sigma_M = \delta\sigma_M$. Here $\sigma_M \neq 0$ is a global holomorphic 2-form on M.*

The $\mathbf{Z}$-module $L := \mathbf{Z}[x]/(S(x)) \simeq \mathbf{Z}^{22}$ has an automorphism F defined by $[f(x)] \mapsto [xf(x)]$. The characteristic polynomial of F is $S(x)$. The proof of Theorem 2.5 consists of two parts: (i) an arithmetic part which constructs a nondegenerate bilinear form $(*, **)$ on L such that L is isomorphic to the $K3$ lattice, $U^{\oplus 3} \oplus E_8^{\oplus 2}$, on which F acts as an isometry, and (ii) a geometric part which shows that $F = \iota g^* \iota^{-1}$ for some marked $K3$ surface (M, ι) and an automorphism g of M. Here two of the most important theorems on $K3$ surfaces, surjectivity of the period map and the global Torelli theorem (cf. [**BHPV**]), are used. This theorem is very important in the construction of McMullen's $K3$ surfaces (cf. Section 5).

The next theorem is useful when we study group theoretical aspects of the automorphism group of a $K3$ surface and the bimeromorphic automorphism group of a hyper-Kähler manifold. This will be used in Section 6:

THEOREM 2.6 ([**Og4**], [**Og7**]). *Let L be a free $\mathbf{Z}$-module of finite rank r and let G be an abelian subgroup*[10] *of $\mathrm{GL}(L)$. Suppose that G satisfies the following (i) and (ii):*

[10]We do not assume that it is finitely generated.

(i) *For each* $g \in G$, *its characteristic polynomial* $\Phi_g(x) = \det(xI - g)$ *is a product of at most one Salem polynomial (counted with multiplicities) and some, possibly* 0, *cyclotomic polynomials.*

(ii) *If* $g \in G$ *has eigenvalue* 1, *then* $g = id$.

Then, for each $g \in G$, *its characteristic polynomial* $\Phi_g(x)$ *is either one Salem polynomial or a power of one cyclotomic polynomial. Moreover,* G *is a cyclic group. More precisely, one of the following (iii) or (iv) holds:*

(iii) *If there is* $g \in G$ *such that* $g \neq id$ *and* $\Phi_g(x)$ *is a power of a cyclotomic polynomial, then* G *is isomorphic to* $\mathbf{Z}/n$ *for some positive integer*[11] n *such that* $\varphi(n)|r$.

(iv) *If there is* $g \in G$ *such that* $\Phi_g(x)$ *is a Salem polynomial, then* G *is isomorphic to the infinite cyclic group* $\mathbf{Z}$.

Finiteness of bounded Salem numbers (Proposition (2.3)) is important for the proof. One can also use this theorem to study the representation of the bimeromorphic automorphism group of a hyper-Kähler manifold on the transcendental lattice (cf. [**Og6**], Section 6).

3. $K3$ surfaces and hyper-Kähler manifolds

3.1. Hyper-Kähler manifolds and their deformations. In this section, comparing with the case of $K3$ surfaces, we briefly recall some of the basic facts about hyper-Kähler manifolds. A good reference is [**GHJ**, Part III]. We note that many results here are generalized to symplectic varieties with singularities ([**Na1**], [**Na2**], [**Na3**]).

DEFINITION 3.1. A compact connected complex manifold M that satisfies the following three conditions (i), (ii), (iii) is called an **irreducible holomorphic symplectic manifold** or a **hyper-Kähler manifold**[12]: (i) M is Kähler; (ii) M is simply connected; (iii) there is an everywhere nondegenerate holomorphic 2-form σ_M such that $H^0(M, \Omega_M^2) = \mathbf{C}\sigma_M$.

From (iii), it follows that M is even dimensional, the tangent bundle TM is isomorphic to the cotangent bundle Ω_M^1 and the canonical bundle $K_M := \wedge^{\dim M}\Omega_M^1$ is trivial; i.e., M is a Calabi-Yau manifold in a wide sense, etc. From (i) and (ii), we have $H^0(M, TM) = H^0(M, \Omega_M^1) = 0$ and $H^1(M, TM) = H^1(M, \Omega_M^1) \neq \{0\}$. Kähler classes of M form an open convex cone of $H^{1,1}(M, \mathbf{R})$. We call this cone the **Kähler cone** of M.

EXAMPLE 3.2. (1) A 2-dimensional hyper-Kähler manifold is nothing but a $K3$ **surface**, as Siu [**Si**] has shown that in dimension 2, conditions (ii) and (iii) imply the condition (i).[13] We should note that this is no longer true in higher dimensions: Yoshioka [**Yo1**] constructed a Moishezon manifold that satisfies (ii) and (iii) but not (i), and Guan [**Gu**] constructed a compact complex manifold that satisfies (ii) and (iii) but it is not even in Fujiki's class $\mathcal{C}$.

---

[11] $\varphi(n)$ is the Euler function; i.e., $\varphi(n)$ is the number of integers coprime to n in $[1, n]$.

[12] Our definition here may appear a bit artificial. See [**Be1**] or [**GHJ**] for a more natural definition by means of the holonomy group.

[13] Siu's result together with Miyaoka's results [**Mi1**], [**Mi2**] gave an affirmative answer for Kodaira's conjecture: "A compact complex analytic surface with even first Betti number is always Kähler." Recently, [**La**] and [**Bu**] gave a classfication-free proof for Kodaira's conjecture. See also [**BHPV**].

(2) ([**Fu2**]) Let S be a $K3$ surface. Let $B(S \times S)$ be the manifold obtained by the blowup of $S \times S$ along the diagonal Δ. The involution $\iota : (x, y) \mapsto (y, x)$ on $S \times S$ induces an involution $\tilde{\iota}$ on $B(S \times S)$. The quotient manifold $S^{[2]}$ of $B(S \times S)$ by $\tilde{\iota}$ is a 4-dimensional hyper-Kähler manifold. By construction, $S^{[2]}$ is the Douady space (Hilbert scheme) of 0-dimensional analytic subsets $(Z, \mathcal{O}_Z)$ of length 2 on S.

(3) ([**Be1**]) More generally, the Douady space (Hilbert scheme) $S^{[r]}$ of 0-dimensional analytic subsets $(Z, \mathcal{O}_Z)$ of length $r \geq 2$ on a $K3$ surface S is a $2r$-dimensional hyper-Kähler manifold. From [**Fo**], it is nonsingular and from [**Va**] the condition (i) follows. Put $S^{(r)} := S^r/\Sigma_r$ (Σ_r is the symmetric group of degree r). Then $S^{(r)}$ is the Chow variety consisting of 0-dimensional cycles of length r on S. An element $[(Z, \mathcal{O}_Z)] \in S^{[r]}$ defines an element $\sum_{P \in \mathrm{Supp}\, Z} (\dim_{\mathbf{C}} \mathcal{O}_{Z,P}) \cdot P$ of $S^{(r)}$. This correspondence defines a morphism $S^{[r]} \longrightarrow S^{(r)}$ called the Hilbert-Chow morphism. Conditions (ii) and (iii) for $S^{[r]}$ can be checked by using this morphism. For instance, an everywhere nondegenerate holomorphic 2-form on $S^{[r]}$ is obtained as the extension of the pull-back of an everywhere nondegenerate holomorphic 2-form $\sigma_{S^{(r)}_{\mathrm{reg}}}$ on $S^{(r)}_{\mathrm{reg}}$ which is naturally induced from a nowhere-vanishing holomorphic 2-form σ_S on S. Moreover, $b_2(S^{[r]}) = b_2(S^{(r)}) + 1 = b_2(S) + 1 = 23$.

(4) ([**Be1**]) Let r be a positive integer and T be a 2-dimensional complex torus. Similarly to (3), the Douady space $T^{[r+1]}$ is a nonsingular manifold of dimension $2(r+1)$ with an everywhere nondegenerate holomorphic 2-form. However, $T^{[r+1]}$ is not simply connected. In fact, the composition $a : T^{[r+1]} \longrightarrow T$ of the Hilbert-Chow morphism $T^{[r+1]} \longrightarrow T^{(r+1)}$ and a morphism $T^{(r+1)} \longrightarrow T$ which maps each element to its sum w.r.t. the addition of T is a surjective morphism. This morphism a is also a smooth0 morphism whose fibers are all isomorphic. We denote the equivalence class of fibers by $K_r(T)$. Beauville found that $K_r(T)$ is a $2r$-dimensional hyper-Kähler manifold.[14] We have $b_2(K_r(T)) = 7$ for $r \geq 2$. In particular, $K_r(T)$ is never equivalent to $S^{[r]}$ under a smooth deformation if $r \geq 2$. Up to now, except for O'Grady's examples [**Od2**], [**Od3**] there is no known example of a hyper-Kähler manifold which is not equivalent to $S^{[r]}$ or $K_r(T)$ under smooth deformations (cf. [**Mu1**], [**Yo1**], [**Yo2**]).

A small deformation of a hyper-Kähler manifold M is again a hyper-Kähler manifold, as condition (i) is closed under small deformations by [**KS**].[15] Thus, each fiber of a Kuranishi family $u : (M \subset \mathcal{U}) \longrightarrow (0 \in \mathcal{K})$ of M is a hyper-Kähler manifold. A Kuranishi family is also (locally) universal, as $H^0(M, TM) = 0$. When $M = S$ is a $K3$ surface, the Kuranishi space is a 20-dimensional smooth manifold, which is realized as an open neighbourhood of the origin of (the 20-dimensional vector space) $H^1(S, TS)$, as $H^2(S, TS) \simeq H^{1,2}(S) = 0$.

In general, $H^2(M, TM) \neq 0$. However, we have the following:[16]

THEOREM 3.3 (Bogomolov-Tian-Todorov-Ran-Kawamata). *The Kuranishi space of a hyper-Kähler manifold M is a* $\dim H^1(M, TM) = h^{1,1}(M) (> 0)$-

---

[14] $K_1(T)$ is a Kummer $K3$ surface $\mathrm{Km}\, T$, i.e., the minimal resolution of $T/-1$. For this reason, $K_r(T)$ is sometimes called a generalized Kummer manifold.

[15] It is easy to check that conditions (ii) and (iii) are closed under small deformations.

[16] This important result (unobstructedness) was first shown by Bogomolov [**Bo**], based on a general method of Kodaira-Spencer which solves the Cartan-Maurer equation step by step. Later, the result was generalized for Calabi-Yau manifolds in a wide sense by [**Ti**] and [**To**] by using Yau's solution of Calabi's conjecture. After that, from a new view called T^1-lifting properties, an alternative proof for Calabi-Yau manifolds in a wide sense was given by [**Ra**] and [**Ka2**].

dimensional smooth manifold, which is realized as an open neighbourhood of the origin of $H^1(M, TM)$. In particular, a hyper-Kähler manifold is never rigid.

3.2. The Beauville-Fujiki form. A **Calabi-Yau manifold** (in a narrow sense) is a compact simply connected Kähler manifold M such that $H^0(X, \Omega_X^i) = 0$ $(0 < i < n = \dim X)$ but there is a nowhere-vanishing global holomorphic n-form ω_X. By definition, a $K3$ surface is not only a 2-dimensional hyper-Kähler manifold but also a 2-dimensional Calabi-Yau manifold. However, it seems that $K3$ surfaces share more properties with higher-dimensional hyper-Kähler manifolds than with higher-dimensional Calabi-Yau manifolds. I do not know the essential reason why it is so. At least in technical points of view, this is due to the following.

THEOREM 3.4 (The Beauville-Fujiki form [**Be1**], [**Fu3**], [**GHJ**]). *Let M be a $2d$-dimensional hyper-Kähler manifold and let σ_M be a global holomorphic 2-form on M. Here we normalize*[17] *σ_M so that $\int_M \sigma_M^d \overline{\sigma_M}^d = 1$. We define the real quadratic form $\tilde{q}_M$ on $H^2(M, \mathbf{R})$ by*

$$\tilde{q}_M(\alpha) := \frac{d}{2}\int_M \alpha^2 \sigma_M^{d-1}\overline{\sigma_M}^{d-1} + (1-d)\int_M \alpha\sigma_M^{d-1}\overline{\sigma_M}^{d} \cdot \int_M \alpha\sigma_M^{d}\overline{\sigma_M}^{d-1},$$

$$\alpha \in H^2(M, \mathbf{R}).$$

We denote the real symmetric bilinear from[18] *associated with $\tilde{q}_M$ by $\tilde{b}_M(*, **) : H^2(M, \mathbf{R}) \times H^2(M, \mathbf{R}) \longrightarrow \mathbf{R}$. Then the following (1)-(4) hold.*

(1) *$\tilde{b}_M(*, **)$ is compatible with the Hodge structure on $H^2(M, \mathbf{Z})$ in the following sense:*

$$\tilde{b}_M(H^{1,1}(M), H^{2,0}(M)) = \tilde{b}_M(H^{1,1}(M), H^{0,2}(M)) = 0\ ;$$

$$\tilde{b}_M(\sigma_M, \sigma_M) = \tilde{b}_M(\overline{\sigma_M}, \overline{\sigma_M}) = 0\ ,\ \tilde{b}_M(\sigma_M, \overline{\sigma_M}) > 0\ .$$

(2) *$\tilde{b}_M(*, **)$ is nondegenerate, of signature $(3, b_2(M) - 3)$. More precisely, if η is a Kähler class of M, then $\tilde{b}_M(*, **)$ is positive definite on the 3-dimensional subspace $H_\eta := \langle \eta, \mathrm{Re}\,\sigma_M, \mathrm{Im}\,\overline{\sigma_M}\rangle$ and it is negative definite on the perpendicular subspace $H_\eta^\perp$.*

(3) *There is a real positive number $c > 0$ such that $b_M(*, **) := c\tilde{b}_M(*, **)$ is an integral-valued symmetric bilinear form on $H^2(M, \mathbf{Z})$ and it is primitive in the sense that the greatest common divisor of the values is 1. This new form $b_M(*, **)$ is uniquely determined by M and it is invariant under deformation (whenever fibers are hyper-Kähler manifolds). We call $b_M(*, **) : H^2(M, \mathbf{Z}) \times H^2(M, \mathbf{Z}) \longrightarrow \mathbf{Z}$ the* **Beauville-Fujiki form.**[19]

(4) *There is a universal constant $C > 0$ such that $\alpha^{2d} = C \cdot q_M(\alpha)^d$ for all $\alpha \in H^2(M, \mathbf{C})$.*

Since M is simply connected, $H^2(M, \mathbf{Z})$ is a free $\mathbf{Z}$-module of rank $b_2(M)$. By $b_M(*, **)$, $H^2(M, \mathbf{Z})$ is a nondegenerate lattice (2-dimensional cohomology lattice) of signature $(3, b_2(M) - 3)$. Contrary to the case of a $K3$ surface, $H^2(M, \mathbf{Z})$ is

[17]The method is not unique, but $\tilde{q}_M(*)$ below turns out to be unique.

[18]By definition it is the cup product when M is a $K3$ surface.

[19]Some authors call it the Beauville-Bogomolov form. Bogomolov certainly contributed an important theorem (3.3), but he did nothing for this form. On the other hand, Fujiki found a very important property (4) of this form. For this reason, I call the form the Beauville-Fujiki form here.

neither unimodular nor even, in general. For $S^{[r]}$ and $K_r(T)$, their lattice structures have been computed as follows (here the direct sum decomposition is compatible with the Hodge structure).

EXAMPLE 3.5. (1) ([**Be1**], [**Od1**]) For $S^{[r]}$ $(r \geq 2)$, we have $H^2(S^{[r]}, \mathbf{Z}) \simeq H^2(S, \mathbf{Z}) \oplus \langle -2(r-1) \rangle$. Here, $\langle -2(r-1) \rangle$ is the lattice $(\mathbf{Z}, (*, **))$ of rank 1 defined by $(1,1) = -2(r-1)$.

(2) ([**Yo1**]) For $K_r(T)$ $(r \geq 2)$, we have $H^2(K_r(T), \mathbf{Z}) \simeq H^2(T, \mathbf{Z}) \oplus \langle -2r \rangle$.

Theorem (3.4)(1) is obvious and (2) is essentially the Hodge-Riemann bilinear relation ([**En**], [**GHJ**, Part III]). (3) was first shown by [**Be1**]. There, based on Theorem (3.3), the following important identity w.r.t. $\alpha, \lambda \in H^2(M, \mathbf{C})$,

$$v(\lambda)^2 q_M(\alpha) = q_M(\lambda)\{(2d-1)v(\lambda) \int_M \lambda^{2r-2}\alpha^2 - (2d-2)(\int_M \lambda^{2d-1}\alpha)^2\},$$

$$v(\lambda) := \int_M \lambda^{2d}$$

was shown in the course of his proof of the local Torelli theorem (Theorem 3.10). Choosing an element λ_0 in $H^2(M, \mathbf{Q})$ which is sufficiently close to $\sigma_M + \overline{\sigma_M}$, and then substituting it into the identity above, we find that $q_M(\lambda_0)^{-1} q_M(\alpha)$ $(\alpha \in H^2(M, \mathbf{Q}))$ is a $\mathbf{Q}$-valued quadratic from. From this, (3) follows. The statement (4) was first shown by [**Fu3**], again based on Theorem (3.3). The statement (3) also follows from (4). In [**GHJ**, Part III], there is another proof of (3) which does not use Theorem (3.3) at all.

A surjective morphism $f : M \longrightarrow B$ (from a hyper-Kähler manifold M) is called a **fibration** if its general fiber is connected and the base space B is normal. A d-dimensional smooth submanifold of a $2d$-dimensional hyper-Kähler manifold M is called a **Lagrangian submanifold** if $\sigma_M | N = 0$. For example, any d-dimensional projective space $\mathbf{P}^d$ in M (if it exists) is necessarily a Lagrangian submanifold. A fibration $f : M \longrightarrow B$ whose general fibers are Lagrangian submanifolds is called a **Lagrangian fibration**. The general fibers F of a Lagrangian fibration are necessarily complex tori. This follows from the Liouville-Arnold theorem ([**Be3**], [**Ad**]) for a proper Lagrangian fibration. Or more directly, since F is a Lagrangian submanifold, the tangent bundle TF is dual to the normal bundle$N_{M/F} (\simeq \mathcal{O}_F^{\oplus d})$. Thus TF, and hence Ω_F^1, is trivial. The result follows from this (e.g., consider the Albanese map) (cf. [**CMS**]).

One of the impressive applications of Theorem 3.4 is the following result due to Matsushita.

THEOREM 3.6 (Matsushita [**Ma1**], [**Ma2**]). *Let M be a $2d$-dimensional hyper-Kähler manifold and let $f : M \longrightarrow B$ be a fibration such that $0 < \dim B < \dim M$ and B is projective. Then f is an equidimensional Lagrangian fibration and B is a d-dimensional $\mathbf{Q}$-Fano variety*[20] *of Picard number* 1. *In particular, the general fibers of f are d-dimensional complex tori.*[21]

[20] A $\mathbf{Q}$-Fano variety is a normal projective variety which satisfies: (i) it admits at most $\mathbf{Q}$-factorial klt singularities ([**KMM**], [**KM**]); (ii) the anti-canonical divisor $-K_B$ is ample.

[21] Contrary to the hyper-Kähler case, in the Calabi-Yau case, there appear much more various fiber space structures even if it is 3-dimensional (cf. [**Og1**], [**OS**]).

His proof is a very elegant one which uses only general properties about base spaces of fibrations [**Nk**] and Theorem 3.4. In [**Ma3**], he also studies general singular fibers of a Lagrangian fibration.

We list the following interesting conjectures related to Theorem 3.6. So far, none of them has been solved:

CONJECTURE 3.7. (1) *A nef line bundle L ($\neq 0$) such that $q_M(L) = 0$ is semi-ample. In particular, some multiple of L induces a fibration in Theorem 3.6.*

(2) *The base space B is isomorphic to $\mathbf{P}^d$.*

(3) *The algebraic dimension $a(M)$ of a $2d$-dimensional hyper-Kähler manifold M is either 0, d, or $2d$.*

There are results related to (1) in [**Sa**], [**Ma4**] and some attempt to prove (2) in [**CMS**], [**Ng**]. For (3), $a(M)$ is $2d$ (resp. 0) if $NS(M)$ is hyperbolic (resp. elliptic). So, the case where $NS(M)$ is parabolic is essential in (3). In this case, it is easy to see that $a(M) \le d$. See the next subsection for some of the terminologies used here.

3.3. Néron-Severi lattice and transcendental lattice. As in the previous subsection, we denote by M a $2d$-dimensional hyper-Kähler manifold. The subgroup $c_1(\mathrm{Pic}\,(M))$ of $H^2(M, \mathbf{Z})$ is called the **Néron-Severi lattice** of M and it is denoted by $NS(M)$. The rank of $NS(M)$ is called the **Picard number** of M and it is denoted by $\rho(M)$. The lattice structure of $NS(M)$ is given by $b_M(*, **)$ restricted to $NS(M)$. In the study of M, $NS(M)$ plays very important roles. Despite its definition, $NS(M)$ behaves, under deformations, in a rather delicate way, which is not necessarily algebraic even if M and its deformation are algebraic (cf. Corollary 3.11). The **transcendental lattice** of M, which we will denote by $T(M)$, is the minimal primitive submodule T of $H^2(M, \mathbf{Z})$ such that $\sigma_M \in T_{\mathbf{C}}$. Again, the lattice structure is given by $b_M(*, **)$ restricted to $T(M)$. From the Lefschetz $(1,1)$ theorem and from the fact that $b_M(*, **)$ is nondegenerate on $H^2(M, \mathbf{Z})$, it follows that:

$$NS(M) = H^2(M, \mathbf{Z}) \cap H^{1,1}(M) = \{x \in H^2(M, \mathbf{Z}) | (x, \sigma_M) = 0\}\ ,$$

$$0 \le \rho(M) \le h^{1,1}(M) = b_2(M) - 2\ ;$$

$$NS(M) = T(M)^{\perp}_{H^2(M,\mathbf{Z})}, T(M) = NS(M)^{\perp}_{H^2(M,\mathbf{Z})}; \rho(M) + \mathrm{rank}\, T(M) = b_2(M)\ .$$

PROPOSITION 3.8. *$NS(M)$ and $T(M)$ satisfy one of the following:*
(i) $\mathrm{sgn}\, NS(M) = (1, 0, \rho(M) - 1)$, $\mathrm{sgn}\, T(M) = (2, 0, b_2 - \rho(M) - 2)$;
(ii) $\mathrm{sgn}\, NS(M) = (0, 1, \rho(M) - 1)$, $\mathrm{sgn}\, T(M) = (2, 1, b_2 - \rho(M) - 3)$;
(iii) $\mathrm{sgn}\, NS(M) = (0, 0, \rho(M))$, $\mathrm{sgn}\, T(M) = (3, 0, b_2 - \rho(M) - 3)$.

In the cases (i) and (iii), $NS(M) + T(M) = NS(M) \oplus T(M)$ and this is a finite-index subgroup of $H^2(M, \mathbf{Z})$. In the case (ii), $NS(M) \cap T(M) = \mathbf{Z}v$, $(v^2) = 0$. According to the signature of $NS(M)$, we call M in cases (i), (ii), (iii) **hyperbolic, parabolic, elliptic,** *respectively.*

For a $K3$ surface, the cases (i), (ii), (iii) correspond to the cases of algebraic dimension 2, 1, 0 (cf. Conjecture (3.7)(3)).

The following is a fundamental result, which one can state in terms of lattices

THEOREM 3.9 (Huybrechts [**Hu1**]). *M is projective iff $NS(M)$ is hyperbolic,*[22] *i.e., iff there is a line bundle L such that $q_M([L]) > 0$.*

Theorem 3.9 and Proposition 3.8 are also important in describing groups of bimeromorphic automorphisms of a hyper-Kähler manifold (cf. Sections 4, 6).

3.4. A period map for hyper-Kähler manifolds. Let M be a hyper-Kähler manifold and let Λ be an abstract lattice which is isomorphic to $(H^2(M, \mathbf{Z}), b_M(*, **))$. Thoughout this section, Λ is fixed. We call an isometry $\iota : H^2(M, \mathbf{Z}) \simeq \Lambda$ a marking of M and we call a pair (M, ι) a **marked hyper-Kähler manifold**. Put

$$\mathcal{P} = \mathcal{P}(\Lambda) := \{[\sigma] \in \mathbf{P}(\Lambda_{\mathbf{C}}) \,|\, (\sigma, \sigma) = 0 \ , \ (\sigma, \overline{\sigma}) > 0\} \ .$$

We call $\mathcal{P}$ the **period domain** of M. $\mathcal{P}$ is a connected complex manifold of dimension $b_2(M) - 2 = h^{1,1}(M)$.

Let $f : \mathcal{M} \longrightarrow \mathcal{B}$ be a small deformation of M. For simplicity, we assume that $\mathcal{B}$ is a polydisk. Then, $R^2 f_* \mathbf{Z}_{\mathcal{M}}$ is a constant sheaf, and it defines an isomorphism $\iota : R^2 f_* \mathbf{Z}_{\mathcal{M}} \simeq \Lambda \times \mathcal{B}$ (Theorem 3.4). This ι induces a marking $\iota_b : H^2(\mathcal{M}_b, \mathbf{Z}) \simeq \Lambda$ of each fiber $\mathcal{M}_b$ ($b \in \mathcal{B}$). The image of the sheaf $f_* \Omega^2_{\mathcal{M}/\mathcal{B}} = \mathcal{O}_{\mathcal{B}} \cdot \tilde{\sigma}$ (of relative holomorphic 2-forms) under ι is a holomorphic subbundle of rank 1 of the trivial vector bundle $\Lambda \otimes_{\mathbf{Z}} \mathcal{O}_{\mathcal{B}}$. We then obtain a holomorphic map

$$p_{\mathcal{B}} : \mathcal{B} \longrightarrow \mathcal{P} \ , \ b \mapsto \iota_b(H^{2,0}(\mathcal{M}_b)) = [\iota_b(\tilde{\sigma}_b)] \ .$$

We call this map $p_{\mathcal{B}}$ a **period map** of a family $f : \mathcal{M} \longrightarrow \mathcal{B}$. This map depends on the choice of a marking ι. A period map is a map which geometrically describes how Hodge structures on $H^2(\mathcal{M}_b, \mathbf{Z})$ vary according to the parameter $b \in \mathcal{B}$. Using Theorem 3.4, one can show the following theorem exactly in the same manner as for $K3$ surfaces:

THEOREM 3.10 (Local Torelli Theorem [**Be1**]). *Let $u : \mathcal{U} \longrightarrow \mathcal{K}$ be the Kuranishi family of M and let $p = p_{\mathcal{K}}$ be a period map of u (under some/any fixed marking). Then, the differential $Dp_0 : T_0\mathcal{K} \longrightarrow T_{p(0)}\mathcal{P}$ is an isomorphism. In particular, the period map $p : \mathcal{K} \longrightarrow \mathcal{P}$ gives an isomorphism from a neighbourhood of $0 \in \mathcal{K}$ onto a neighbourhood of $p(0) \in \mathcal{P}$.*

The local Torelli theorem provides us with an effective tool in studying the deformation of hyper-Kähler manifolds, as the period domain is much more explicit (and much easier to treat) than the Kuranishi space. For instance, the following results are applications of Theorem 3.10.

COROLLARY 3.11. *Let $f : \mathcal{M} \longrightarrow \mathcal{B}$ be a nontrivial small deformation of M. Then, the following holds:*[23]

(1) ([**Fu2**]) *The set of $b \in \mathcal{B}$ such that $\mathcal{M}_b$ is projective is dense in $\mathcal{B}$.*

(2) ([**Og3**]) *Let m be the minimum of $\rho(\mathcal{M}_b)$ ($b \in \mathcal{B}$). Then, both the set of $b \in \mathcal{B}$ such that $\rho(\mathcal{M}_b) > m$ and the set of $b \in \mathcal{B}$ such that $\rho(\mathcal{M}_b) = m$ are dense*

---

[22] As in the surface case, this line bundle L is not necessarily ample. The original proof of this theorem has a gap, but a new proof is given by himself. It is also cleanly explained in [**GHJ**, Part III]. In a new proof, the density of projective hyper-Kähler manifolds is not used. Instead, the Nakai-Moishizon type criterion for a Kähler class, due to Demailly and Paun [**DP**], plays a crucial role there.

[23] Calabi-Yau manifolds of dimension ≥ 3 are projective and satisfy $\operatorname{Pic} M \simeq H^2(M, \mathbf{Z})$. Thus, (1) is trivial and (2) and (3) are false for them. The statement (3) is false even for complex tori.

in $\mathcal{B}$. Moreover, if $\mathcal{B}$ is 1-dimensional, then the set of $b \in \mathcal{B}$ such that $\rho(\mathcal{M}_b) > m$ is a countable, dense subset of $\mathcal{B}$.

(3) (*cf.* [**Og3**]) *Let ρ be an integer*[24] *such that $0 \le \rho \le b_2(M) - 2 = h^{1,1}(M)$. Then, in the Kuranishi space $\mathcal{K}$ of M, the set of $b \in \mathcal{K}$ such that $\rho(\mathcal{M}_b) = \rho$ is dense in $\mathcal{K}$. Moreover, the set of $b \in \mathcal{K}$ such that $\rho(\mathcal{M}_b) \ge \rho$ is a countable union of closed analytic subsets of $\mathcal{K}$ of codimension ρ.*

(4) ([**HLOY**]) *Let Δ be a unit disk. Then, there are two projective families of $K3$ surfaces $f_i : \mathcal{M}_i \longrightarrow \Delta$ $(i = 1, 2)$ such that there is a sequence $\{t_n\} \subset \Delta$ which converges to 0 and satisfies $(\mathcal{M}_1)_{t_n} \simeq (\mathcal{M}_2)_{t_n}$, but $(\mathcal{M}_1)_0 \not\simeq (\mathcal{M}_2)_0$ at the limit; i.e., limits of isomorphic projective $K3$ surfaces happen to be nonisomorphic.*

Let us show Corollary 3.10(1) for the Kuranishi family $u : \mathcal{U} \longrightarrow \mathcal{K}$ of M. Choose $\mathcal{K}$ sufficiently small. We then identify $0 \in \mathcal{K}$ with a neighbourhood of $p(0) \in \mathcal{P}$ by the period map p. Let U be a given neighbourhood of $p(0)$ and let η be a Kähler class of M. As $b_M(\sigma_M, \eta) = 0$, the point $p(0)$ is in the hyperplane defined by $(x, \iota_0(\eta)) = 0$ (in the period domain). As $\iota_0(\eta) \in \Lambda_{\mathbf{R}}$, there is a point $a \in \Lambda_{\mathbf{Q}}$ which is very close to $\iota_0(\eta)$. We have $(a^2) > 0$. Moreover, the hyperplane defined by $(x, a) = 0$ (in the period domain) passes through a point in U, say $p(t)$. That is, $(p(t), a) = 0$. Thus, a suitable multiple of $\iota_t^{-1}(a)$, say A, is an element of $NS(\mathcal{U}_t)$ such that $q_{\mathcal{U}_t}(A) > 0$. Thus $\mathcal{U}_t$ is projective by Theorem 3.9. One can also show (2) and (3) in a similar manner. For (4), besides Theorem 3.10, we also use a counting formula ([**HLOY**]) of the Fourier-Mukai partners of a projective $K3$ surface (cf. [**Ka4**]).

The following result due to Huybrechts is fundamental:

THEOREM 3.12 (Surjectivity of period map [**Hu1**]). *Let $\mathcal{P}$ be the period domain of M. Then, for each $[\sigma] \in \mathcal{P}$, there is a marked hyper-Kähler manifold (M', ι') such that $\iota([\sigma_{M'}]) = [\sigma]$ and M' is equivalent to M under smooth deformations.*

An important problem called the global Torelli problem is a problem to study relations between M and M' having the same period, i.e., such that there is a Hodge isometry between $H^2(M, \mathbf{Z})$ and $H^2(M', \mathbf{Z})$. As is well known, we have the following very satisfactory answer for $K3$ surfaces.

THEOREM 3.13 (Global Torelli theorem for $K3$ surfaces [**PSS**],[**BR**],[**BHPV**]). *Let S and S' be $K3$ surfaces. Then, the following holds.*[25]

(1) *If there is a Hodge isometry between $H^2(S, \mathbf{Z})$ and $H^2(S', \mathbf{Z})$, then $S \simeq S'$.*

(2) *A Hodge isometry φ from $H^2(S', \mathbf{Z})$ to $H^2(S, \mathbf{Z})$ that preserves the Kähler cones is uniquely induced by an isomorphism from S to S'. That is, there is a unique isomorphism $f : S \longrightarrow S'$ such that $\varphi = f^*$.*

Let M and M' be hyper-Kähler manifolds of the same dimension ≥ 4. It was expected (cf. [**Mu2**]) that if $H^2(M, \mathbf{Z})$ and $H^2(M', \mathbf{Z})$ are Hodge isometric, then M and M' would be bimeromorphic (cf. Theorem 6.4 for the reason why not isomorphic but just bimeromorphic). However, a counterexample was given by Namikawa [**Na4**]. There is an abelian surface such that $NS(A) = \mathbf{Z}h$ $((h^2) = 6)$ and $A \not\simeq A^*$, where $A^* := \mathrm{Pic}^0 A$. In [**Na4**], he showed that the second cohomology

[24] Note that $0 \le \rho(\mathcal{M}_b) \le h^{1,1}(M)$.

[25] The known proof shows a seemingly stronger statement (2) first and then reduces (1) to (2) by using the fact that the Kähler cone is the fundamental domain of the action of the Weyl group $W(S)$, generated by the reflections arizing from (-2)-curves, on $H^{1,1}(S, \mathbf{R})$ (cf. [**BHPV**]).

lattices of $K_2(A)$ and $K_2(A^*)$ are Hodge isometric, but $K_2(A)$ and $K_2(A^*)$ are not bimeromorphic. Also, in [**Mk**], counterexamples for deformations of $S^{[r]}$ of a $K3$ surface S are given.

4. The automorphism group of a $K3$ surface

The automorphism group of a $K3$ surface is a traditional subject which has been studied since the 19th century as symmetries of Kummer surfaces or semi-invariants of quartics in four variables (cf. [**Ko**]). It remains an attractive subject which is studied from various viewpoints. The current method is more or less based on the following theorem, which is a restatement of Theorem 3.13(2).

THEOREM 4.1. *Let S be $K3$ surface. Then:*

(1) *The correspondence $f^{-1} \mapsto f^*$ gives an isomorphism between* $\mathrm{Aut}\, S$ *and the group of isometries of $H^2(S, \mathbf{Z})$ that preserves the Kähler cone.*

(2) *Assume further that S is projective. Then, the correspondence $f^{-1} \mapsto (f^*|NS(S), f^*|T(S))$ gives an isomorphism between* $\mathrm{Aut}\, S$ *and the group $\{(g_N, g_T) \in \mathrm{O}(NS(S), A(S)) \times \mathrm{O}_{\mathrm{Hodge}}(T(S)) \mid g_N|NS(S)^*/NS(S) = g_T|T(S)^*/T(S)\}$. Here, $\mathrm{O}(NS(S), A(S))$ is the group of isometries of $NS(S)$ that preserve the ample cone, and $\mathrm{O}_{\mathrm{Hodge}}(T(S))$ is the group of Hodge isometries of $T(S)$ with Hodge structure induced naturally from the one on $H^2(S, \mathbf{Z})$. The last equality is the one under the natural identification of the discriminant groups $NS(S)^*/NS(S)$ and $T(S)^*/T(S)$ (derived from the fact that $H^2(S, \mathbf{Z})$ is unimodular and the fact that $NS(S)$ and $T(S)$ are both primitive in $H^2(S, \mathbf{Z})$). Moreover, $\mathrm{O}_{\mathrm{Hodge}}(T(S))$ is a cyclic group of order $2n$, where n is a positive integer such that $\varphi(2n) | \mathrm{rank}\, T(S)$.*

COROLLARY 4.2. *Let S be a projective $K3$ surface. Then:*

(1) ([**PSS**]) $\mathrm{Aut}\, S$ *is isomorphic to $O(NS(S))/W(S))$ up to a finite kernel and a finite cokernel. In particular,* $\mathrm{Aut}\, S$ *is finite iff so is $\mathrm{O}(NS(S))/W(S)$. Here $W(S)$ is the Weyl group of $NS(S)$, i.e., the normal subgroup of $\mathrm{O}(NS(S))$ generated by the reflections $r_c(x) = x + (r, c)c$ corresponding to (-2)-elements c of $NS(S)$.*

(2) ([**St**]) $\mathrm{Aut}\, S$ *is finitely generated.*

Corollary (4.2)(1) follows from Theorem (4.1)(2). By Corollary (4.2)(1), $\mathrm{Aut}\, S$ is a group which extends a finite group by some quotient of an arithmetic subgroup of an algebraic group $O(NS(S)_{\mathbf{C}})$ defined over $\mathbf{Q}$. Therefore, Corollary (4.2)(2) follows from the Borel-Harish-Chandra theorem [**BHC**].[26] Corollary (4.2)(2) is a positive result which supports a possible description of $\mathrm{Aut}\, S$ in terms of finitely many generators and their relations. There are highly nontrivial concrete calculations in [**Vi**], [**Ko**], [**KK**].

For a while, we assume that S is projective. Then, $NS(S)$ is hyperbolic (Proposition 3.8) and $1 \le \rho(S) \le 20$. Let us consider the relation between $\mathrm{Aut}\, S$ and the Picard number $\rho(S)$. The next result follows from Theorem 4.1 and Corollary 4.2:

EXAMPLE 4.3. (1) Assume that $\rho(S) = 1$. Then $\mathrm{Aut}\, S = \{id\}$ if S is of degree ≥ 4 and $\mathrm{Aut}\, S \simeq \mathbf{Z}/2$ if S is of degree 2 ([**Ni1**]). Here degree means the self-intersection number of the ample generator of $NS(S)$.

(2) Assume that $\rho(S) = 2$. Then $\mathrm{Aut}\, S$ is isomorphic to $\{id\}$ or $\mathbf{Z}$ modulo extension by a finite cyclic group ([**PSS**]).

---

[26] [**St**] gives a somewhat different proof.

When $\rho(S) \geq 3$, the following is known:

THEOREM 4.4. (1) ([**SI**]) *If* $\rho(S) = 20$, *then* $\operatorname{Aut} S$ *is an infinite group.*

(2) ([**Ni2**]) *The isomorphism classes of* $NS(S)$ *of* $K3$ *surfaces* S *such that* $\operatorname{Aut} S$ *is finite and* $\rho(S) \geq 3$ *are finite. Moreover* $\rho(S) \leq 19$ *in this case.*[27]

In [**SI**], they showed the assertion (1) geometrically. They actually showed that S with $\rho(S) = 20$ always admits a Jacobian fibration of positive Mordell Weil rank (see Section 6 for terminologies). In a series of his deep studies (see [**Ni2**] and the references therein), Nikulin classified the isomorphism classes of even hyperbolic lattices L such that $\operatorname{rank} L \geq 3$ and $O(L)/W(L)$ is finite. Combining this with Corollary 4.1(1), he obtained the assertion (2).

According to Corollary 3.11(3), $K3$ surfaces become more special if the Picard numbers become larger. Both Example 4.3 and Theorem 4.4 then support an intuitively acceptable phrase, "Automorphism groups become larger if manifolds become more special". The following theorem shows that this phrase is mostly true for *projective* $K3$ surfaces (cf. Section 5).

THEOREM 4.5 ([**Og3**]). *Let* $f : \mathcal{M} \longrightarrow \Delta$ *be a nontrivial projective family of* $K3$ *surfaces over the unit disk* Δ. *Let* m *be the minimum of* $\rho(\mathcal{M}_t)$ $(t \in \Delta)$ *and let* $\mathcal{S} := \{\, t \in \Delta \mid \rho(\mathcal{M}_t) > m \,\}$.[28] *Then, there are a (possibly empty) finite subset* $\mathcal{F} \subset \mathcal{S}$, *(possibly infinite) group* G *and a positive integer* N *such that*

(1) *for each* $t \in \Delta \setminus \mathcal{F}$, *there is an injective homomorphism* $\iota_t : G \subset \operatorname{Aut} \mathcal{M}_t$;

(2) *for all* $t \in \Delta \setminus \mathcal{S}$, $[\operatorname{Aut} \mathcal{M}_t : \iota_t(G)] < N$;

(3) *there is a dense subset* $\mathcal{D}$ *of* $\mathcal{S}$ *such that* $\operatorname{Aut} \mathcal{M}_t$ *is infinite for each* $t \in \mathcal{D}$.

The case where $\mathcal{F} \neq \emptyset$ actually happens. Let $f : \mathcal{M} \longrightarrow \Delta$ be a sufficiently general family of quartic $K3$ surfaces. A general fiber $\mathcal{M}_t$ $(t \in \Delta \setminus \mathcal{S})$ is of Picard number 1 and therefore $\operatorname{Aut} \mathcal{M}_t = \{id\}$. Thus, $G = \{1\}$ in this family. The assertion (3) claims that nevertheless there densely exist fibers $\mathcal{M}_t$ whose automorphism groups are infinite.[29]

Next, we shall describe automorphism groups of nonprojective $K3$ surfaces. As a result, they are rather moderate groups which are very close to finitely generated abelian groups:

PROPOSITION 4.6 ([**Og6**]). *Let* M *be a nonprojective* $K3$ *surface. If* $NS(M)$ *is parabolic, then* $\operatorname{Aut} M$ *is an almost abelian group of rank* $\leq \rho(M) - 1$. *If* $NS(M)$ *is elliptic, then* $\operatorname{Aut} M$ *is an almost abelian group of rank* ≤ 1.

Here an almost abelian group of rank r is a group which is isomorphic to $\mathbf{Z}^r$ modulo a finite kernel and a finite cokernel. From Corollary 4.2(2) and Proposition 4.6, we obtain:

COROLLARY 4.7 ([**Og6**]). *The automorphism group of a complex* $K3$ *surface is finitely generated.*

---

[27] There is a projective $K3$ surface S such that $\rho(S) = 19$ and $|\operatorname{Aut} S| < \infty$ ([**Ni1**], [**Kn1**]). When $\rho(S) = 2$, the isomorphism classes of $NS(S)$ such that $\operatorname{Aut} S$ is infinite and the isomorphism classes of $NS(S)$ such that $\operatorname{Aut} S$ is finite are both infinite.

[28] $\mathcal{S}$ is a countable dense subset of Δ.

[29] In other words, the automorphism groups of quartic $K3$ surfaces can be drastically changed if one changes the coefficients slightly. For this reason, the assertion (3) is a negative assertion which suggests some difficulties in describing automorphism groups explicitly and systematically.

It is unknown if the automorphism group is finitely presented or not. It is also unknown if a similar result holds for a $K3$ surface in positive characteristic or not.

Finally, we shall consider an automorphism of a $K3$ surface from the point of view of **entropy**. This observation will also provide a good preparation to Section 5. This view is brought by Cantat [**Ca1**], [**Ca2**], [**Ca3**] and McMullen [**Mc1**], [**Mc2**], and it is a relatively new view in the theory of complex surfaces.

Let M be a compact Kähler manifold and let g be an automorphism of M. The entropy of g is a nonnegative real number which measures a sort of complexity or randomness of g or the orbit of a point under $\langle g \rangle$. From a "general meaning" of entropy, one can imagine that the behaviour is moderate if the entropy is 0 and it becomes more and more complicated if the entropy becomes more and more positive (cf. Proposition 4.8). There are three basic ways to define entropy precisely: topological, metrical, and cohomological. It is known that these three approaches yield the same value of entropy for an automorphism of a compact Kähler manifold ([**Gr**], [**Ym**], [**Fr**]). The cohomological definition (here we are only concerned with whether it is zero or positive) is as follows: Consider the action $g^*|H^*(M, \mathbf{Z})$ of an automorphism g of M on the cohomology ring $g^*|H^*(M, \mathbf{Z})$. g is of **positive entropy**.[30]

From the point of view of entropy, we obtain the following:

PROPOSITION 4.8 ([**Og5**]). *Let S be a (not necessarily projective) $K3$ surface and let g be an automorphism of S. Then,*

(1) *g is of positive entropy iff there is a point P such that the orbit $\langle g \rangle \cdot P$ is Zariski dense in S.*

(2) *g is of null entropy iff either g is of finite order or g preserves some elliptic fibration on S.*

PROPOSITION 4.9 ([**Og4**], [**Og5**]). *Let S be a (necessarily projective) $K3$ surface with $\rho(S) = 20$. Then* $\mathrm{Aut}(S)$ *has a noncommutative free subgroup*[31] $G \simeq \mathbf{Z} * \mathbf{Z}$ *satisfying the following (i), (ii), (iii):*

(i) *G has an element of positive entropy.*

(ii) *Let $P \in S$ be a sufficiently general point of S. Then the orbit $G \cdot P$ is dense in S w.r.t. classical topology* [**Ca3**].

(iii) *Let P be as above. Then the orbit $G \cdot P$ is not measurable w.r.t. the invariant measure $\sigma_S \overline{\sigma_S}$ of S.*

So, when $\rho(S) = 20$, i.e., the most special cases, $\mathrm{Aut}\, S$ is (not only infinite but also) highly nonabelian and its geometrical behaviour is rather complicated.

The following unsolved problem posed by McMullen [**Mc1**] is very interesting also from the view of Proposition 4.8:

---

[30] By the classification of Kähler surfaces, if a surface S has an automorphism of positive entropy, then S is isomorphic to either a $K3$ surface, a complex torus, an Enriques surface, their blowing up, or a rational surface which is a blowing up of $\mathbf{P}^2$ in at least 10 points (possibly infinitesimally near points) [**Ca1**], [**Ca2**] if g^* has an eigenvalue with absolute value > 1. Otherwise, g is of **null entropy**. g is of null entropy iff the eigenvalues of $g^*|H^*(M)$ are all of absolute value 1.

[31] The fact that S admits 2 different Jacobian fibrations of positive Mordell-Weil rank reduces the existence of G to Proposition 6.9. It follows from Proposition 4.8 that this G satisfies (i). The assertion (ii) is due to Cantat. The assertion (iii) is a $K3$ version of the so-called Banach-Tarski paradox. Similar to the Banach-Tarski paradox in other examples, our proof is based on the shape of the graph of G, which grows self-similarly (cf. [**DS**]).

PROBLEM 4.10. Let S be a projective $K3$ surface and let g be an automorphism of S of positive entropy. Then, is there a point $P \in S$ such that the orbit $\langle g \rangle \cdot P$ is dense in S w.r.t. the classical topology?

McMullen pairs (Section 5) show that the projectivity in Problem (4.10) is inevitable. So, for an affirmative answer, besides the fact that S is a $K3$ surface, one will need to find/study something based on the projectivity, such as the orbit of a point which is highly transcendental from the field of definition of S and g, in an essential way.

5. McMullen's $K3$ surfaces

We call n complex numbers $\alpha_i \neq 0$ $(1 \leq i \leq n)$ **multiplicatively independent** if they have no nontrivial relation, i.e., only $(m_1, m_2, \ldots, m_n) = (0, 0, \ldots, 0)$ is an integer solution of $\alpha_1^{m_1}\alpha_2^{m_2}\cdots\alpha_n^{m_n} = 1$. For instance, $\alpha_1 \neq 0$ is multiplicatively independent iff α_1 is not a root of unity.

DEFINITION 5.1. Let $(\alpha_i)_{i=1}^d \in (S^1)^d$. Then $(\alpha_i)_{i=1}^d$ defines an automorphism F of a polydisk Δ^d by $F^*(z_i)_{i=1}^d = (\alpha_i z_i)_{i=1}^d$. We call a pair (Δ^d, F) a **Siegel polydisk** if in addition $(\alpha_i)_{i=1}^d$ are multiplicatively independent.[32]

Let M be a d-dimensional complex manifold and let f be an automorphism of M. Suppose that D is a domain of M and $f(D) = D$. Then $f_D : D \longrightarrow D$ is an automorphism of D. We also say that f has a **Siegel polydisk** D if the pair (D, f_D) is biholomorphic to (some) Siegel polydisk (Δ^d, F) defined above. If f has a Siegel polydisk, then f is of infinite order. As the domain D is preserved by f, the group $\langle f \rangle$ cannot have any dense orbit in M w.r.t. the classical topology, when M is compact (cf. Problem 4.10).

Let M be a $K3$ surface and let f be an automorphism of M. We call the pair (M, f) a **McMullen pair** if f has a Siegel polydisk.

According to Corollary 3.11(3), $K3$ surfaces with Picard number 0 are the most general ones. In the view of Theorem 4.5, one might imagine that such $K3$ surfaces would never admit high symmetries. At least I did so. Contrary to this intuition, McMullen [**Mc1**] has found the following surprising[33]

THEOREM 5.2 (McMullen [**Mc1**]). *There is a McMullen pair (M, f) of Picard number* 0. *f is of positive entropy but there is no dense orbit under $\langle f \rangle$. Moreover, if we put $f^*\sigma_M = \delta\sigma_M$, then δ is not a root of* 1.

Using a McMullen pair (M, f), one can easily construct another counterexample for Kodaira's problem in the Introduction:

PROPOSITION 5.3. *Let (M, f) be a McMullen pair. Blow up $X := M \times M$ first along the intersection points of the diagonal Δ and the graph Γ_f of f and blow up further along the proper transforms of Δ and Γ. The resulting manifold Y is a* 4*-dimensional simply connected compact nonprojective Kähler manifold with no nontrivial small deformation. In particular, no small deformation of Y is projective.*

---

[32]By definition, F preserves a d-dimensional real torus $(S^1)^d$ centered at the origin and of radius $(r_i)_{i=1}^d$. For each $P \in (S^1)^d$, the orbit $\langle F \rangle P$ is a dense subset of $(S^1)^d$.

[33]It was also a big surprise for me that one can obtain such a concrete picture for (some) $K3$ surface of Picard number 0, as there is no explicit way to write down $K3$ surfaces of Picard number 0.

PROOF. It is clear that Y is a 4-dimensional simply connected compact non-projective Kähler manifold. As exceptional sets are stable under deformation, any small deformation Y_t of Y (where t is parameter of the base space of a given small deformation) induces a deformation $(X_t, \Delta_t, \Gamma_t)$, where $\Delta_t, \Gamma_t \subset X_t$, of (X, Δ, Γ) ([**FN**]). This induces a deformation (M_t, f_t) ($f_t \in \operatorname{Aut} M_t$) of the pair (M, f). Put $f_t^*\sigma_{M_t} = \chi_t\sigma_t$. Then χ_t is an eigenvalue of $f_t^*|H^2(M_t, \mathbf{Z})$, so that it cannot vary continuously. Thus, χ_t is constant, i.e., $\chi_t = \chi_0 = \delta$. As δ is not a root of 1, it follows that M_t is nonprojective. On the other hand, any *nontrivial* small deformation of a $K3$ surface densely contains projective ones (Corollary 3.11(1)). Thus, $M_t \simeq M$ for all t. As $\operatorname{Aut} M$ is discrete, this implies that $(M_t, f_t) \simeq (M, f)$, whence, $Y_t \simeq Y$ for all t. □

The construction here is nothing but Voisin's construction [**Vo1**] applied (not for complex tori but) for a McMullen pair (M, f). Her original result [**Vo1**] is much stronger. In fact, for each $d \geq 6$, she constructed a d-dimensional compact Kähler manifold V such that the cohomology ring $H^*(V, \mathbf{Q})$ is never isomorphic to a cohomology ring of a projective manifold.[34]

In the rest of this section, we shall explain the geometric part of the proof of Theorem 5.2.

5.1. The Baker-Feldman theorem and applications.

THEOREM 5.4 (Baker-Feldman [**Fe**], [**Ba**]). (1) *Let n, d, A, $B \geq 2$ be positive integers and let $(m_i)_{i=1}^n \in \mathbf{Z}^n$.*

Let $S := S(n, d, A, (m_i)_{i=1}^n)$ be the set consisting of $(\alpha_i)_{i=1}^n \in \mathbf{C}^n$ such that $\alpha_i \neq 0$ are algebraic numbers[35] *of height $\leq A$ and of degree $\leq d$ such that $\log \alpha_i$ with branch*[36] *m_i are linearly independent over $\mathbf{Q}$.*

Let $T := T(n, d, B)$ be the set consisting of algebraic numbers $(\beta_i)_{i=0}^n \in \mathbf{C}^{n+1}$ of degree $\leq d$ and of height $\leq B$.

Then there is a positive real number C that depends on n, d, A, $(m_i)_{i=0}^n$, but does not depend on B, such that $|\beta_0 + \beta_1 \log \alpha_1 + \cdots + \beta_n \log \alpha_n| > B^{-C}$ for all $(\alpha_i)_{i=1}^n \in S$ and for all $(\beta_i)_{i=0}^n \in T \setminus \{0\}$.

(2) *In particular,*[37] *if α_i $(1 \leq i \leq n)$ are algebraic numbers such that $\log \alpha_i$ are linearly independent over $\mathbf{Q}$, then 1, $\log \alpha_i$ $(1 \leq i \leq n)$ are linearly independent also over $\overline{\mathbf{Q}}$.*

---

[34]In [**Vo2**], she also constructed a counterexample for Kodaira's problem in the bimeromorphic version. See also [**Hu3**].

[35]For an algebraic number α, there is a unique (up to ± 1) irreducible polynomial $f(x) \in \mathbf{Z}[x]$ such that $f(\alpha) = 0$. We call the degree of $f(x)$ the degree of α, and the maximum of the absolute values of coefficients of $f(x)$, the height of α.

[36]For a complex number $\alpha \neq 0$, there is unique β_0 such that $e^{\beta_0} = \alpha$ and $-\pi < \operatorname{Im} \beta_0 \leq \pi$. We call this β_0 the principal value of natural logarithm of α. Any β such that $e^\beta = \alpha$, i.e., natural logarithm of α is then of the form $\beta = \beta_0 + 2\pi\sqrt{-1}m$, where m is an integer. We call the integer m the branch of the natural logarithm β of α. We write the natural logarithm with prescribed branch by $\log \alpha$.

[37](2) is an obvious conclusion of (1). Historically, Baker showed (2) first without using (1). Based on his argument, Feldman [**Fe**] then obtained a quantitative version like (1). The formulation of (1) here is the one which was furthermore sharpened by Baker [**Ba**]. For a series of works concerning analytic number theory including this theorem, Baker was awarded the Fields medal in 1970. In [**Sh**], a direct proof of (2) is explained in Japanese. In [**Og7**], Baker's proof of (1) is also explained in Japanese.

This is a very deep theorem which includes the transcendence of fundamental numbers as special cases:

COROLLARY 5.5. (1) (**Hermite**) *e is a transcendental number.*
(2) (**Lindemann**) *π is a transcendental number.*
(3) (**Gelfand-Schneider**) *Any number of the form $\alpha^\beta (= e^{\beta \log \alpha})$, where $\alpha \neq 0, 1$ is an algebraic number and β is an irrational algebraic number, is a transcendental number (an affirmative solution of the 7th problem of Hilbert). For instance $2^{\sqrt{2}}$ is a transcendental number.*

PROOF. $\log e = 1$ is linearly independent over $\mathbf{Q}$, but 1 and 1 are linearly dependent over $\overline{\mathbf{Q}}$. Thus e is a transcendental number by the contrapositive of Theorem 5.4(2). Let us show that π is a transcendental number. By taking a branch 1, we have $\log 1 = 2\pi\sqrt{-1}$. Clearly $2\pi\sqrt{-1}$ is linearly independent over $\mathbf{Q}$. Let β be an algebraic number. Then, so is $-2\sqrt{-1}\beta$. Then $-2\sqrt{-1}\beta \cdot 1 + 1 \cdot 2\pi\sqrt{-1} \neq 0$ by Theorem 5.4(2). That is, $\pi \neq \beta$; i.e., π is a transcendental number. The proof of (3) is similar. □

McMullen has found a surprising connection between Theorem 5.4 and complex geometry (Corollary 5.9). The following Corollary 5.7 is crucial in this connection.

DEFINITION 5.6. We call complex numbers $(\lambda_i)_{i=1}^n$ **jointly Diophantine** if there are positive real numbers c and m such that

$$|\lambda_1^{k_1} \cdots \lambda_n^{k_n} - 1| > c(\max_{1 \leq i \leq n}|k_i|)^{-m}$$

for all $(0) \neq (k_i)_{i=1}^n \in \mathbf{Z}^n$.

Jointly Diophantine numbers are multiplicatively independent. The converse is also true for algebraic numbers on S^1:

COROLLARY 5.7 ([**Mc1**]). *Let $(\lambda_i)_{i=1}^n$ be algebraic numbers with absolute value 1. Then, $(\lambda_i)_{i=1}^n$ are jointly Diophantine provided that $(\lambda_i)_{i=1}^n$ are multiplicatively independent.*

PROOF. Taking branch 1, we have $\log 1 = 2\pi\sqrt{-1}$. We choose principal values as $\log \lambda_i$ $(1 \leq i \leq n)$. As $(\lambda_i)_{i=1}^n$ are multiplicatively independent, $2\pi\sqrt{-1}$, $\log \lambda_1$ ... $\log \lambda_n$ are linearly independent over $\mathbf{Q}$. Thus, by Theorem 5.4, there is a positive number C such that $|k_0 2\pi\sqrt{-1} + k_1 \log \lambda_1 + \cdots + k_n \log \lambda_n| > (\max_{0 \leq i \leq n}|k_i|)^{-C}$ for all $0 \neq (k_i)_{i=0}^n \in \mathbf{Z}^{n+1}$. As $|\lambda_i| = 1$, we have $\log \lambda_i = 2\sqrt{-1}\theta_i$ $(-\pi < \theta_i \leq \pi)$. Let $\theta (-\pi < \theta \leq \pi)$ be the polar angle of $z := \prod_{i=1}^n \lambda_i^{k_i}$. From the inequality above, there is a positive number M, depending only on $(\theta_i)_{i=1}^n$, such that $|\theta| \geq (M \cdot \max_{1 \leq i \leq n}|k_i|)^{-C}$. As $|z| = 1$, we have $|z - 1| = 2|\sin(\theta/2)|$. Moreover, $|\sin(\theta/2)| \geq (2/\pi)|\theta/2|$. Thus we obtain the following inequality, from which the assertion follows:

$$|\prod_{i=1}^n \lambda_i^{k_i} - 1| = 2|\sin \frac{\theta}{2}| \geq \frac{2}{\pi}|\theta| > \frac{2}{\pi}(M \cdot \max_{1 \leq i \leq n}|k_i|)^{-C} .$$

□

5.2. Local linearization of an automorphism: the Siegel-Sternberg theorem. Let M be a d-dimensional complex manifold and let f be an automorphism of M. Assume that f fixes a point $P \in M$, i.e., $f(P) = P$. In this subsection, we consider behaviours of f around P. f is called **locally linearizable** at P if there is a local coordinate $(z_i)_{i=1}^d$ at P such that $f^*(z_i)_{i=1}^d = A(z_i)_{i=1}^d$ for some constant matrix A. If, in addition, one can choose a coordinate so that A is a diagonal matrix, we say that f is **locally diagonalizable** at P. In this case, the diagonal of A consists of the eigenvalues (with multiplicities) of the differential $Df_P : T_P M \longrightarrow T_P M$ of f at P.

As is well known, f is locally diagonalizable if f is of finite order. On the other hand, an automorphism of $\mathbf{C}^2$ given by $g^*x = \omega x + y^2$, $g^*y = \omega^2 y$ (where ω is a primitive 3rd root of 1) is not locally linearizable at $(0, 0)$. This is because, even though g is of infinite order, the eigenvalues ω, ω^2 of the differential of g at $(0, 0)$ are "algebraically too close":

THEOREM 5.8 (Siegel-Sternberg[38] [**Se**], [**Sb**]). *Let $\lambda_i (1 \le i \le d)$ be the eigenvalues of Df_P (counted with multiplicities). Assume that $(\lambda_i)_{i=1}^d$ are jointly Diophantine. Then f is locally diagonalizable at P.*

COROLLARY 5.9 ([**Mc1**]). *Let $\lambda_i (1 \le i \le d)$ be the eigenvalues of Df_P (counted with multiplicities). Assume that $(\lambda_i)_{i=1}^d$ are multiplicatively independent algebraic numbers such that $(\lambda_i)_{i=1}^d \in (S^1)^d$. Then f has a Siegel polydisk centered at P.*

PROOF. From Corollary (5.7), $(\lambda_i)_{i=1}^d$ are jointly Diophantine. Thus, by Theorem (5.8), f is locally diagonalizable at P under some local coordinate $(z_i)_{i=1}^d$. Let $\epsilon > 0$ be a sufficiently small real number. As the eigenvalues are of absolute value 1, the domain defined by $|z_i| < \epsilon$ $(1 \le i \le d)$ is a Siegel polydisk of f centered at P. □

5.3. Proof of McMullen's theorem. The geometric part of the proof can be summarized as in the following:

PROPOSITION 5.10. *Let M be a $K3$ surface and let f be an automorphism of M. Put $f^*\omega_M = \delta\omega_M$ and $\tau := \delta + \delta^{-1}$. Assume that:*

(1) The characteristic polynomial of $f^*|H^2(M, \mathbf{Z})$ is a Salem polynomial of trace -1;

(2) τ is a real number such that $\tau > 1 - 2\sqrt{2}$. Moreover, τ has a Galois conjugate τ' such that $\tau < 1 - 2\sqrt{2}$.

Then (M, f) is a McMullen pair of Picard number 0.

In fact, using Theorem 2.5, we can construct a pair (M, f) of a $K3$ surface and its automorphism from the Salem polynomial $S_{22}(x)$ in Example 2.2(4). This pair satisfies conditions (1) and (2) in Proposition 5.10. In what follows, we shall show Proposition 5.10 by dividing it into steps.

Step 1. $\rho(M) = 0$. In particular, M contains no curve.[39]

PROOF. As $S(x)$ is irreducible, one has $H^2(M, \mathbf{Z}) = T(M)$. Thus, $\rho(M) = 0$. As M is Kähler, this implies the last assertion. □

Step 2. The set of the fixed points $M^f := \{P \in M \,|\, f(P) = P\}$ is a one point set, say $\{P\}$, counted with multiplicities. Let α and β be the eigenvalues of the

---

[38]Siegel proved it in dimension 1 and later Sternberg [**Sb**] generalized to any dimension. The property of jointly Diophantine is essential at the final step of the proof concerning the convergence of the formal power series solution. The name of Siegel polydisk comes from this theorem.

[39]A 1-dimensional closed subvariety possibly admitting singularities is called a curve.

differential Df_P counted with multiplicities. Then α and β satisfy the following equations (in particular, α and β are algebraic numbers):

$$\alpha + \beta = \frac{1 + \delta + \delta^2}{1 + \delta}\ , \quad \alpha\beta = \delta\ .$$

PROOF. By Step 1, M^f is a finite set. Then, by the topological Lefschetz trace formula and by the condition (1), the number of fixed points $L(f)$ is $1 - 1 + 1 = 1$ counted with multiplicities. By $f^*\omega_M = \delta\omega_M$, we have $\alpha\beta = \delta$. As $L(f) = 1$, neither α nor β is 1. Thus, by the holomorphic Lefschetz trace formula, one has

$$\frac{1}{(1-\alpha)(1-\beta)} = \sum_{P \in M^f} \frac{1}{\det(I - Df_P)} = \sum_{i \geq 0} (-1)^i \mathrm{tr}\, f^*|H^i(\mathcal{O}_M) = 1 + \frac{1}{\delta}\ .$$

This implies the result. □

By virtue of Corollary 5.9, Step 2 and the next Step 3 complete the proof:

Step 3. $|\alpha| = |\beta| = 1$. Moreover, α and β are multiplicatively independent.

PROOF. From $f^*\omega = \delta\omega$, we have $0 \neq (\omega, \overline{\omega}) = (f^*\omega, f^*\overline{\omega}) = |\delta|^2(\omega, \overline{\omega})$. Thus, $|\delta| = 1$. Moreover, $\delta \neq \pm 1$ by condition (1). Combining this with condition (2), one has $\tau = \delta + \delta^{-1} = \delta + \overline{\delta} \in \mathbf{R}$ and $1 - 2\sqrt{2} < \tau < 2\ (< 1 + 2\sqrt{2})$. We notice the following: □

LEMMA 5.11. *For a real number x, the following (i) and (ii) are equivalent:* (i) $-2 \leq -2 + \frac{(x+1)^2}{x+2} \leq 2$; (ii) $1 - 2\sqrt{2} \leq x \leq 1 + 2\sqrt{2}$.

Let us return back to the proof of Step 3. Put $A := \alpha^2/\delta$, $B := \beta^2/\delta$. From the equalities in Step 2, we have $A + B + 2 = \frac{(\alpha+\beta)^2}{\delta} = \frac{(\tau+1)^2}{\tau+2}$ and $AB = \frac{(\alpha\beta)^2}{\delta^2} = 1$. Thus, A and B are the solutions of $x^2 - kx + 1 = 0$ ($k = -2 + \frac{(\tau+1)^2}{\tau+2}$). By Lemma 5.11 and condition (2), one has $-2 < k < 2$. Thus $|A| = |B| = 1$. As $|\delta| = 1$, it follows that $|\alpha| = |\beta| = 1$.

Let us show the last statement in Step 3. By condition (2), there is an isomorphism $\sigma : \mathbf{Q}(\tau) \longrightarrow \mathbf{Q}(\tau')$ of fields such that $\sigma(\tau) = \tau'$. This isomorphism can be extended to an automorphism $\overline{\mathbf{Q}}$, which we denote by the same symbol σ. Put $\alpha' = \sigma(\alpha)$, $\beta' = \sigma(\beta)$ and $\delta' = \sigma(\delta)$. Then:

LEMMA 5.12. $|\delta'| = 1$ *and* $|\alpha'| = |\beta'|^{-1} \neq 1$.

PROOF. Note that $\alpha'\beta' = \delta'$, $\alpha' + \beta' = \frac{1+\delta'+(\delta')^2}{1+\delta'}$. As $S(\delta) = 0$, we have $S(\delta') = 0$ as well. That is, δ' is also a zero of the Salem polynomial $S(x)$. So, if $|\delta'| \neq 1$, then δ' would be a positive real number. However, then $\tau' = \delta' + \frac{1}{\delta'} > 0 > 1 - 2\sqrt{2}$, a contradiction. Thus, $|\delta'| = 1$. Let A, B and k be as before. Put $A' = \sigma(A)$, $B' = \sigma(B)$ and $k' = \sigma(k)$. Since $A' = \frac{(\alpha')^2}{\delta'}$ and $B' = \frac{(\beta')^2}{\delta'}$, the numbers A' and B' are the solutions of $x^2 - k'x + 1 = 0$. As $\tau' < 1 - 2\sqrt{2}$, it follows from Lemma 5.11 that $k' < -2$ or $k' > 2$. Thus, $|A'| \neq 1$ or $|B'| \neq 1$. Hence $|\alpha'| \neq 1$ or $|\beta'| \neq 1$. As $\alpha'\beta' = \delta'$, we then have $|\alpha'| = |\beta'|^{-1} \neq 1$. This completes the proof of Lemma 5.12.

Now we can complete the proof of Step 3. Suppose $\alpha^m\beta^n = 1$ ($(m, n) \in \mathbf{Z}^2$). By mapping this equality by σ and using $|\beta'|^{-1} = |\alpha'|$, we obtain $1 = |(\alpha')^m(\beta')^n| = |\alpha'|^{m-n}$. As $|\alpha'| \neq 1$, we have $m = n$. Thus $1 = \alpha^m\beta^n = \delta^n$. As δ is a zero of a Salem polynomial, it is not a root of 1. Thus $n = 0$. Hence $m = n = 0$. □

6. Groups of bimeromorphic automorphisms of a hyper-Kähler manifold

Any bimeromorphic map between $K3$ surfaces is biholomorphic. However, in higher dimensions, this is no longer true even for hyper-Kähler manifolds.

EXAMPLE 6.1 (Mukai flop). Let M be a hyper-Kähler manifold of dimension $2d \geq 4$ such that $P := \mathbf{P}^d \subset M$. Let $\tilde{M}$ be the blowup of M along P and let $E \subset \tilde{M}$ be the exceptional divisor. As P is a Lagrangian submanifold of M, the normal bundle $N_{M/P} = (TM|P)/TP$ is isomorphic to the cotangent bundle Ω_P^1. Thus, $E = \mathbf{P}(N_{M/P}) \simeq \mathbf{P}(\Omega_P^1)$. By Euler's exact sequence, the latter space is isomorphic to the hypersurface $\{(x, [h]) | h(x) = 0\}$ in $P \times P^*$, where P^* is the dual projective space of P. In this expression, the natural morphism $E \longrightarrow P$ coincides with the first projection $\mathrm{pr}_1 : E \longrightarrow P$. The second projection $\mathrm{pr}_2 : E \longrightarrow P^*$ gives another $\mathbf{P}^{d-1}$-bundle structure on E. The restriction of the normal bundle $N_{M/P} \simeq \mathcal{O}_E(E)$ to the fibers $F \simeq \mathbf{P}^{d-1}$ of pr_2 is isomorphic to $\mathcal{O}_F(-1)$. Thus, one can blow down $\tilde{M}$ to a smooth manifold, say M^+, along the fibers of pr_2. In this way, we obtain a bimeromorphic map $M \cdots \rightarrow M^+$. We call this bimeromorphic map or M^+ the **Mukai flop** of M along P ([**Mu1**]). As M^+ and M are isomorphic in codimension 1, we see that M^+ is again a hyper-Kähler manifold provided that M^+ is Kähler.[40] It is also known that any birational map between 4-dimensional projective hyper-Kähler manifolds is a composition of Mukai flops and a biholomorphic map ([**WW**]).

EXAMPLE 6.2. Let S be a quartic $K3$ surface containing a unique line l. Then the line through any two general points $P, Q \in S$ meets S at two more points, say P' and Q'. Then, by the correspondence $\{P, Q\} \mapsto \{P', Q'\}$, we obtain a birational map $\iota : S^{[2]} \cdots \rightarrow S^{[2]}$. This ι is an involution which is holomorphic on $S^{[2]} \setminus l^{[2]} \simeq S^{[2]} \setminus \mathbf{P}^2$. In fact, ι coincides with the Mukai flop of $S^{[2]}$ along $l^{[2]}$ ([**Be2**]). In this example, both the source and the target are $S^{[2]}$, but the map ι itself is not an isomorphism.

For this reason, it is more natural to consider not only biholomorphic maps and automorphisms but also bimeromorphic maps and bimeromorphic automorphisms when we study higher-dimensional varieties including hyper-Kähler manifolds.

Let $f : M' \cdots \rightarrow M$ be a bimeromorphic map between hyper-Kähler manifolds.[41] As both K_M and $K_{M'}$ are trivial, f is isomorphic in codimension 1. Thus, we can define the pull-back $f^* : H^2(M, \mathbf{Z}) \longrightarrow H^2(M', \mathbf{Z})$ and f^* is an isomorphism as a $\mathbf{Z}$-module. In general, bimeromorphic manifolds are neither homeomorphic nor of isomorphic cohomology ring,[42] even if they are isomorphic in codimension 1. However, this is not the case for hyper-Kähler manifolds.

THEOREM 6.3 ([**Hu1**]). *Assume that two hyper-Kähler manifolds M and M' are bimeromorphic. Then, they are equivalent under smooth deformation. In particular, M and M' are diffeomorphic and their cohomology rings are isomorphic.*

---

[40] There is an example such that M^+ is not Kähler [**Yo1**].

[41] For simplicity, we always choose f so that the domain where f is holomorphic is maximum.

[42] See [**Og2**] for an impressive example.

For this reason, it is not so easy to find examples of hyper-Kähler manifolds which are bimeromorphic but not isomorphic (cf. [**GHJ**]). Explicit examples of bimeromorphic nonisomorphic hyper-Kähler manifolds are given by [**De**](4-dimensional nonprojective ones), [**Yo1**] (8-dimensional projective ones) and [**HT**], [**Sa**](4-dimensional projective ones).

In the rest of this section, we consider the bimeromorphic automorphism group $\mathrm{Bir}\, M$ of a hyper-Kähler manifold M. The following theorem is fundamental:

THEOREM 6.4 ([**Fu1**], [**En**], [**Hu1**], [**Hu2**]). *Let $f : M' \cdots \to M$ be a bimeromorphic map between hyper-Kähler manifolds M and M'. Then $f^* : H^2(M, \mathbf{Z}) \longrightarrow H^2(M', \mathbf{Z})$ is a positive Hodge isometry. Moreover, if f^* maps at least one Kähler class η of M to a Kähler class of M', then f is biholomorphic.*

The next theorem is an important theorem by which one can well approximate the group $\mathrm{Bir}\, M$ by the induced (linear) action on $H^2(M, \mathbf{Z})$:

THEOREM 6.5 ([**Hu1**]). *The correspondence $f^{-1} \mapsto f^*$ induces a group homomorphism $\rho : \mathrm{Bir}\, M \longrightarrow \mathrm{O}^+_{\mathrm{Hodge}}(H^2(M, \mathbf{Z}))$. Moreover, the kernel is a finite group.*[43]

PROOF. The first assertion follows from Theorem 6.4. Let us show the last assertion. Put $G = \mathrm{Ker}\, \rho$. As $G(\eta) = \eta$ for a Kähler class η of M, from Theorem 6.4, G is a subgroup of $\mathrm{Aut}\, M$. By Yau's theorem [**Ya1, Ya2**], η is induced by a Ricci flat Kähler metric g which is uniquely determined by η. As $G \subset \mathrm{Aut}\, M$ and G preserves η, the group G also preserves the metric g. Thus, G is a subgroup of $\mathrm{O}(M, g)$, which is compact.[44] On the other hand, as $H^0(M, TM) = 0$, the group $\mathrm{Aut}\, M$ is a discrete group. Thus, G is a discrete subgroup of a compact group, and therefore, a finite group. □

Theorems 6.6 and 6.7 below concern the group-theoretical nature of bimeromorphic automorphism groups of hyper-Kähler manifolds. Theorem 6.6 is also a higher-dimensional analogue of Proposition 4.6:

THEOREM 6.6 ([**Og6**]). *Let M be a nonprojective hyper-Kähler manifold. Then, $\mathrm{Bir}\, M$ is an almost abelian group of rank $\le \max(1, \rho(M) - 1)$. In particular, both $\mathrm{Bir}\, M$ and $\mathrm{Aut}\, M$ are finitely generated. More precisely, $\mathrm{Bir}\, M$ is of rank at most $\rho(M) - 1$ when $NS(M)$ is parabolic, and $\mathrm{Bir}\, M$ is either a finite group or an extension of a finite group by $\mathbf{Z}$ when $NS(M)$ is elliptic. Moreover, if $NS(M)$ is elliptic and $b_2(M) - \rho(M)$ is odd, then $\mathrm{Bir}\, M$ is a finite group.*

THEOREM 6.7 (cf. [**Og4**], [**Og7**]). *Let M be a projective hyper-Kähler manifold and let G be a subgroup of $\mathrm{Bir}\, M$. Then G satisfies either (i) or (ii) below.*[45]

(i) *G is an almost abelian group of rank $\le \max(1, \rho(M) - 2)$.*

(ii) *G has a noncommutative free subgroup isomorphic to $\mathbf{Z} * \mathbf{Z}$.*

The following two propositions give sufficient conditions for (i) and (ii) to happen in geometric terms:

---

[43]Let T be a 2-dimensional complex torus. There is a natural action of the group of $(r+1)$-torsion points of T on the generalized Kummer variety $K_r(T)$ $(r \ge 2)$. This group acts trivially on $H^2(K_r(T), \mathbf{Z})$ ([**Be2**]).

[44]The topology is the one induced by the compact-open topology of $\mathrm{Diff}\, M$.

[45](i) and (ii) are exclusive. There is a group which satisfies neither (i) nor (ii). For instance, the group of unipotent upper triangular matrices of size $n \ge 3$ with integer entries is such a group.

PROPOSITION 6.8 ([**Og5**]). *Let M and G be as in Theorem 6.7. If all elements of G are of null entropy, then G is almost abelian.*

We call a fibration $f : M \longrightarrow B$ a **Jacobian fibration** if f admits a globally defined rational section and general fibers of f are abelian varieties. Let us regard one rational section as 0 of general fibers. Then, the set of rational sections of f, which naturally acts on general fibers as parallel translations, gives rise to an abelian subgroup of $\mathrm{Bir}\, M$. We call this group the **Mordell-Weil group** of f and denote it by $\mathrm{MW}(f)$.

PROPOSITION 6.9 (Crescent Theorem[46] [**Og4**]). *Let M be a projective hyper-Kähler manifold. Assume that M admits two different Jacobian fibrations $f_i : M \longrightarrow B_i$ $(i = 1, 2)$ of positive Mordell-Weil rank. Choose $g_i \in \mathrm{MW}(f_i)$ such that g_i is of infinite order. Then, the subgroup $\langle g_1, g_2 \rangle$ of $\mathrm{Bir}\, M$ has a subgroup isomorphic to $\mathbf{Z} * \mathbf{Z}$.*

For instance, $S^{[r]}$ satisfies this condition if S is a $K3$ surface with $\rho(S) = 20$ ([**Og5**]).

To prove Proposition 6.9, we show geometrically that $\langle g_1, g_2 \rangle$ is not almost abelian. We then use Theorem 6.7. For Proposition 6.8, we use Lie's theorem about nilpotent algebraic groups [**Hm**]. In our proof of Theorems 6.6 and 6.7, besides basic properties of Salem polynomials (Section 2), the following theorem by Tits (Tits' alternative) plays a crucial role:

THEOREM 6.10 ([**Ts**]). *Let k be a field of characteristic 0. Let G be a subgroup of $\mathrm{GL}(n, k)$. Then, one of the following holds:*[47]
(i) *G has a solvable subgroup of finite index;*
(ii) *G has a subgroup isomorphic to $\mathbf{Z} * \mathbf{Z}$.*

In what follows, we shall give an outline of the proof of Theorem 6.7. Recall that the transcendental lattice $T(M)$ is the smallest primitive sublattice of $H^2(M, \mathbf{Z})$ such that $\sigma_M \in T(M)_{\mathbf{C}}$. Then, by Theorem 2.4, the natural action of G on $T(M)$ satisfies the conditions (i) and (ii) in Theorem 2.6. Moreover, as M is projective, $T(M)$ is of signature $(2, n)$. By using this, we see that G satisfies (iii) in Theorem 2.6. From this fact and Theorem 6.5, the kernel of the natural homomorphism $G \longrightarrow \mathrm{O}(NS(M))$ is a finite group. In this way, we reduce Theorem 6.7 to the following:

PROPOSITION 6.11. *Let L be a hyperbolic lattice of rank r and let G be a subgroup of $\mathrm{O}(L)$. Then, one of the following holds:*
(i) *G is an almost abelian group of rank $\le \max(r-2, 1)$;*
(ii) *G has a subgroup isomorphic to $\mathbf{Z} * \mathbf{Z}$.*

By Tits' theorem (Theorem 6.10), this proposition follows from the next lemma:

LEMMA 6.12. *Let L and G be as in Proposition 6.11. Then G is an almost abelian group of rank $\le \max(r-2, 1)$, provided that G has a solvable subgroup of finite index.*

---

[46]During my stay in Korea, Professor Kawamata noticed that angles of crescent are different in Japan and in Korea. The first version of this theorem was inspired by this discovery.

[47](i) and (ii) are exclusive. The infinite direct product of the monster group satisfies neither (i) nor (ii). This example is due to Professor Maeno.

In what follows, we shall show this lemma by induction on the rank r of L. If $r = 1$, then the result is clear. Assuming that the result is true for $\mathrm{rank}\, L \le r - 1$, we shall show the case $\mathrm{rank}\, L = r$. We may freely replace G by a subgroup of finite index. Thus, we may assume that G is solvable and that G preserves a positive cone of L. Let S be the Zariski closure of G in the complex algebraic group $\mathrm{O}(L_{\mathbf{C}})$ and let S^0 be the identity component of S. As G is solvable, so are S and S^0. In addition, S^0 is a finite-index subgroup of S. Thus, by replacing G by $G \cap S^0$, we may assume that G is a subgroup of a connected solvable group S^0. By the Lie-Kolchin theorem about connected solvable algebraic groups [**Hm**], S^0 has a common eigenvector, say, $0 \ne v \in L_{\mathbf{C}}$. In particular, for each $g \in G$, one has $g(v) = \alpha(g)v$ for some $\alpha(g) \in \mathbf{C}$. The correspondence $g \mapsto \alpha(g)$ defines a group homomorphism $\alpha : G \longrightarrow \mathbf{C}^{\times}$. Let T be the minimal primitive sublattice of L such that $v \in T_{\mathbf{C}}$. By the minimality of T, we have $GT = T$, $GT^{\perp} = T^{\perp}$. Moreover, as L is hyperbolic, T satisfies one of the following:

(a) T is elliptic; (b) T is parabolic; (c) T is hyperbolic and $\mathrm{rk} T < \mathrm{rk} L$; (d) $T = L$.

In case (a), $T^{\perp}$ is hyperbolic, and in case (c), $T^{\perp}$ is elliptic. Recall that the group of isometries of an elliptic lattice is a finite group. So we obtain the result in cases (a) and (c), by applying the hypothesis of induction for $T^{\perp}$ in case (a) and for T in case (c). In case (b), an isotropic element $u \in T$ is also G-invariant. Using this fact, we can obtain the result in case (b) rather easily. The essential case is case (d). As $L = T$ in case (d), applying Theorem 2.4, we see that the action of G on L satisfies the conditions (i), (ii) in Theorem 2.6. Thus, by Theorem 2.6, G is an almost abelian group of rank at most 1. This completes the proof.

We conclude this survey with the following unsolved interesting problem:

PROBLEM 6.13. Let M be a hyper-Kähler manifold of dimension ≥ 4.

(1) Are $\mathrm{Bir}\, M$ and $\mathrm{Aut}\, M$ finitely generated?

(2) Is $[\mathrm{Bir}\, M : \mathrm{Aut}\, M] < \infty$ (cf. [**Hu2, Ka1, Ka3**])?

(3) Let ρ be the map in Theorem (6.5). What kind of finite groups can be $\mathrm{Ker}\, \rho$ (cf. [**Be2**], [**Mk**])?

Aknowledgement

I would like to express my deep thanks to the referees and Professors Noam D. Elkies and De-Qi Zhang for careful reading and many valuable comments. I would like to express my thanks to Professors A. Fujiki, Y. Kawamata, J.H. Keum, S. Kondo, T. Maeno, K. Nishioka, T. Terasoma, K. Yoshioka for helpful discussions.

References

[Ad] V. I. Arnold, *Mathematical methods of classical mechanics*, Graduate Texts in Mathematics, 60. Springer-Verlag, New York, 1989. MR997295 (90c:58046)

[Ba] A. Baker, *Transcendental number theory*, Cambridge Univ. Press, London, 1975. MR0422171 (54:10163)

[BHPV] W. Barth, K. Hulek, C. Peters, A. Van de Ven, *Compact complex surfaces*, Second edition. Springer-Verlag, Berlin, 2004. MR2030225 (2004m:14070)

[Be1] A. Beauville, *Variétés Kähleriennes dont la première classe de Chern est nulle*, J. Diff. Geom. **18** (1983) 755–782. MR730926 (86c:32030)

[Be2] A. Beauville, *Some remarks on Kähler manifolds with $c_1 = 0$*, In: Classification of algebraic and analytic manifolds (Katata, 1982), Progr. Math. **39** (1983) 1–26, Birkhäuser Boston, Boston, MA. MR728605 (86c:32031)

[Be3] A. Beauville, *Systèmes hamiltoniens complètement integrables associés aux surfaces K3*: In Problems in the theory of surfaces and their classification (Cortona, 1988), 25–31, Sympos. Math., XXXII, Academic Press, London, 1991. MR1273370 (95c:14052)

[Bo] F. Bogomolov, *Hamiltonian Kähler manifolds*, Soviet. Math. Dokl. **19** (1978) 1462–1465. MR514769 (80c:32024)

[BHC] A. Borel, Harish-Chandra, *Arithmetic subgroups of algebraic groups*, Ann. of Math. (2) **75** (1962) 485–535. MR0147566 (26:5081)

[Bu] N. Buchdahl, *On compact Kähler surfaces*, Ann. Inst. Fourier **49** (1999) 287–302. MR1688136 (2000f:32029)

[BR] D. Burns, M. Rapoport, *On the Torelli problem for Kählerian K3 surfaces*, Ann. Sci. Ecole Norm. Sup. **8** (1975) 235–273. MR0447635 (56:5945)

[Ca1] S. Cantat, *Dynamique des automorphismes des surfaces projectives complexes*, C. R. Acad. Sci. Paris Ser. I Math. 328 (1999), no. 10, 901–906. MR1689873 (2000c:32061)

[Ca2] S. Cantat, *Dynamique des automorphismes des surfaces K3*, Acta Math. 187 (2001) 1–57. MR1864630 (2003h:32026)

[Ca3] S. Cantat, *Sur la dynamique du groupe d'automorphismes des surfaces K3*, Transform. Groups 6 (2001), no. 3, 201–214. MR1854708 (2002k:37039)

[CMS] K. Cho, Y. Miyaoka, N.I. Shepherd-Barron, *Characterizations of projective space and application to complex symplectic manifolds*, Adv. Stud. Pure Math. **35** (2002) 1–88. MR1929792 (2003m:14080)

[De] O. Debarre, *Un contre-exemple au théorème de Torelli pour les variétés symplectiques irréductibles*, C.R. Acad. Sc. Paris **299** (1984) 681–684. MR770463 (87c:14039)

[DS] P. Deligne, D. Sullivan, *Division algebras and the Hausdorff-Banach-Tarski paradox*, Enseign. Math. **29** (1983) 145–150. MR702737 (85c:22008)

[DP] J.-P. Demailly, M. Paun, *Numerical characterization of the Kahler cone of a compact Kahler manifold*, Ann. of Math. (2) **159** (2004) 1247–1274. MR2113021 (2005i:32020)

[En] I. Enoki, *Compact Ricci-flat Kähler manifolds*, Adv. Stud. Pure Math. **18** (1990) 229–256. MR1145250 (92k:53078)

[Fe] H.I. Feldman, *An improvement of the estimate of a linear form in the logarithms of algebraic numbers*, Math. USSR Sb. **6** (1968) 393–406.

[Fo] J. Forgarty, *Algebraic families on an algebraic surface*, Amer. J. Math. **90** (1968) 511–521. MR0237496 (38:5778)

[Fr] S. Friedland, *Entropy of algebraic maps*, In the proceedings of the conference in honor of Jean-Pierre Kahane, J. Fourier Anal. Appl. (1995) 215–228. MR1364887 (97k:28040)

[Fu1] A. Fujiki, *A Theorem on Bimeromorphic Maps of Kähler Manifolds and its Applications*, Publ. Res. Inst. Math. Sci. **17** (1981) 735–754. MR642659 (84a:32042)

[Fu2] A. Fujiki, *On primitively symplectic compact Kähler V-manifolds of dimension four* In: Classification of algebraic and analytic manifolds (Katata, 1982), Progr. Math. **39** (1983) 71–250, Birkhäuser Boston . MR728609 (85k:32040)

[Fu3] A. Fujiki, *On the de Rham cohomology group of a compact Kähler symplectic manifold* In: Algebraic geometry, Sendai, 1985, Adv. Stud. Pure Math. **10** (1987) 105–165. MR946237 (90d:53083)

[FN] A. Fujiki and S. Nakano, *Supplement to "On the inverse of monoidal transformation"*, Publ. Res. Inst. Math. Sci. **7** (1971/72) 637–644. MR0294712 (45:3780)

[Gr] M. Gromov, *On the entropy of holomorphic maps*, Enseign. Math. **49** (2003) 217–235. MR2026895 (2005h:37097)

[GM] B. Gross, C.T. McMullen, *Automorphisms of even unimodular lattices and unramified Salem numbers*, J. Algebra **257** (2002) 265–290. MR1947324 (2003j:11071)

[GHJ] M. Gross, D. Huybrechts and D. Joyce, *Calabi-Yau manifolds and related geometries*, Springer-Verlag, 2003. MR1963559 (2004c:14075)

[Gu] D. Guan, *Examples of compact holomorphic symplectic manifolds which are not Kähler, III*, Int. J. Math. **6** (1995) 709–718. MR1351162 (97i:32034)

[HT] B. Hassett, Y. Tschinkel, *Abelian fibrations and rational points on symmetric products*, Int. J. Math. **11** (2000) 709–718. MR1809306 (2002a:14010)

[HLOY] S. Hosono, B.H. Lian, K. Oguiso, S.-T. Yau, *Fourier-Mukai number of a K3 surface*, Algebraic structures and moduli spaces, 177–192, CRM Proc. Lecture Notes, 38, Amer. Math. Soc., Providence, RI, 2004. MR2096145 (2005i:14046)

[Hm] J. E. Humphreys, *Linear algebraic groups*, Graduate Texts in Mathematics, **21** Springer-Verlag, New York-Heidelberg, 1975. MR0396773 (53:633)

[Hu1] D. Huybrechts, *Compact hyper-Kähler manifolds: Basic results*, Invent. Math. **135** (1999) 63–113. Erratum: "Compact hyper-Kähler manifolds: Basic results" Invent. Math. **152** (2003) 209–212. MR1664696 (2000a:32039); MR1965365 (2004c:32041)

[Hu2] D. Huybrechts, *The Kähler cone of a compact hyper-Kähler manifold*, Math. Ann. **326** (2003) 499–513. MR1992275 (2004e:53068)

[Hu3] D. Huybrechts, *Projectivity of Kaehler manifolds - Kodaira's problem (after C. Voisin)*, math.AG/0512039.

[Ka1] Y. Kawamata, *Crepant blowing-up of 3-dimensional canonical singularities and its application to degeneration of surfaces*, Ann. of Math. (2) **127** (1988) 93–163. MR924674 (89d:14023)

[Ka2] Y. Kawamata, *Unobstructed deformations. A remark on a paper of Z. Ran*, J. Algebraic Geom. **1** (1992) 183–190. MR1144434 (93e:14016)

[Ka3] Y. Kawamata, *On the cone of divisors of Calabi-Yau fiber spaces*, Internat. J. Math. **8** (1997) 665–687. MR1468356 (98g:14043)

[Ka4] Y. Kawamata, *Algebraic geometry and derived categories*, Sugaku Expositions **58** (2006), 64–85. (in Japanese) MR2208306 (2007b:14033)

[KMM] Y. Kawamata, K. Matsuda, K. Matsuki, *Introduction to the minimal model problem*, Adv. Stud. Math. **10** (1987) 283–360. MR946243 (89e:14015)

[Ke] J.H. Keum, *Automorphisms of Jacobian Kummer surfaces*, Compositio Math. **107** (1997) 269–288. MR1458752 (98e:14039)

[KK] J.H. Keum, S. Kondo, *The automorphism groups of Kummer surfaces associated with the product of two elliptic curves*, Trans. Amer. Math. Soc. **353** (2001) 1469–1487. MR1806732 (2001k:14075)

[Ko] K. Kodaira, *On stability of compact submanifolds of complex manifolds*, Amer. J. Math. **85** (1963) 79–84. MR0153033 (27:3002)

[KM] J. Kollár, S. Mori, Birational geometry, Iwanami Shoten (1998) (in Japanese).

[Kn1] S. Kondo, *Algebraic $K3$ surfaces with finite automorphism groups*, Nagoya Math. J. 116 (1989), 1–15. MR1029967 (91b:14050)

[Kn2] S. Kondo, *Quadratic forms and $K3$, Enriques surfaces* [translation of Sugaku **42** (1990) 346–360], Sugaku Expositions **6** (1993) 53–72 (1993). MR1083944 (92b:14018)

[Kn3] S. Kondo, *The automorphism group of a generic Jacobian Kummer surface*, J. Algebraic Geom. **7** (1998) 589–609. MR1618132 (99i:14043)

[KS] K. Kodaira and D. Spencer, *On deformations of complex analytic structures III. Stability theorems for complex analytic structures*, Ann. of Math. (2) **71** (1960) 43–76. MR0115189 (22:5991)

[La] A. Lamari, *Currents kähleriens et surfaces compactes*, Ann. Inst. Fourier **49** (1999) 263–285. MR1688140 (2000d:32034)

[Mk] E. Markman, *Integral constraints on the monodromy group of the hyper-Kähler resolution of a symmetric product of a $K3$ surface*, math.AG/0601304.

[Ma1] D. Matsushita, *On fibre space structures of a projective irreducible symplectic manifold*, Topology **38** (1999) 79–83. Addendum: "On fibre space structures of a projective irreducible symplectic manifold", Topology **40** (2001) 431–432. MR1644091 (99f:14054); MR1808227 (2002d:32031)

[Ma2] D. Matsushita, *Equidimensionality of Lagrangian fibrations on holomorphic symplectic manifolds*, Math. Res. Lett. **7** (2000) 389–391. MR1783616 (2001f:32041)

[Ma3] D. Matsushita, *On singular fibres of Lagrangian fibrations over holomorphic symplectic manifolds*, Math. Ann. **321** (2001) 755–773. MR1872528 (2002i:32017)

[Ma4] D. Matsushita, *On nef reductions of projective irreducible symplectic manifolds*, math.AG/0601114.

[Mc1] C. T. McMullen, *Dynamics on $K3$ surfaces: Salem numbers and Siegel disks*, J. Reine Angew. Math. **545** (2002) 201–233. MR1896103 (2003a:37057)

[Mc2] C. T. McMullen, *Dynamics on blowups of the projective plane*, Publ. Math. Inst. Hautes Études Sci. No. 105 (2007), 49–89. MR2354205 (2008m:37076)

[Mi1] Y. Miyaoka, *Extension theorem for Kähler metrics*, Proc. Japan Acad. **50** (1974) 407–410. MR0463483 (57:3432)

[Mi2] Y. Miyaoka, *Kähler metrics on elliptic surfaces*, Proc. Japan Acad. **50** (1974) 533–536. MR0460730 (57:723)

[Mu1] S. Mukai, *Symplectic structure of the moduli space of sheaves on an abelian or $K3$ surface*, Invent. Math. **77** (1984) 101–116. MR751133 (85j:14016)

[Mu2] S. Mukai, *Moduli of vector bundles on $K3$ surfaces and symplectic manifolds* (Japanese) Sugaku Expositions **1** (1988) 139–174. Sugaku **39** (1987) 216–235. MR922020 (89h:32057)

[Ng] Y. Nagai, *Dual fibration of projective Lagrangian fibrations*, preprint.

[Nk] N. Nakayama, *The singularity of the canonical model of compact Kähler manifolds*, Math. Ann. **280** (1988) 509–512. MR936326 (89g:32045)

[Na1] Y. Namikawa, *Deformation theory of singular symplectic n-folds*, Math. Ann. **319** (2001) 597–623. MR1819886 (2002b:32025)

[Na2] Y. Namikawa, *Extension of 2-forms and symplectic varieties*, J. Reine Angew. Math. **539** (2001) 123–147. MR1863856 (2002i:32011)

[Na3] Y. Namikawa, *Projectivity criterion of Moishezon spaces and density of projective symplectic varieties*, Internat. J. Math. **13** (2002) 125–135. MR1891205 (2003b:32025)

[Na4] Y. Namikawa, *Counter-example to global Torelli problem for irreducible symplectic manifolds*, Math. Ann. **324** (2002) 841–845. Erratum: "Counter-example to global Torelli problem for irreducible symplectic manifolds" Math. Ann. **324** (2002), no. 4, 847. MR1942252 (2003k:14008a); MR1942253 (2003k:14008b)

[Ni1] V.V. Nikulin, *On the quotient groups of the automorphism group of hyperbolic forms by the subgroups generated by 2-reflections*, J. Soviet Math. **22** (1983) 1401–1476.

[Ni2] V.V. Nikulin, *Discrete reflection groups in Lobachevsky spaces and algebraic surfaces*, Proceedings of the International Congress of Mathematicians, Vols. 1, 2 (Berkeley, Calif., 1986), 654–671, Amer. Math. Soc., Providence, RI, 1987. MR934268 (89d:11032)

[Od1] K. O'Grady, *Desingularized moduli spaces of sheaves on a $K3$*, J. Reine Angew. Math. **512** (1999) 49–117. MR1703077 (2000f:14066)

[Od2] K. O'Grady, *Desingularized moduli spaces of sheaves on a $K3$*, J. Reine Angew. Math. **512** (1999) 49–117. MR1703077 (2000f:14066)

[Od3] K. O'Grady, *A new six-dimensional irreducible symplectic variety*, J. Algebraic Geom. **12** (2003) 435–505. MR1966024 (2004c:14017)

[Og1] K. Oguiso, *On algebraic fiber space structures on a Calabi-Yau 3-fold. with an appendix by Noboru Nakayama*, Internat. J. Math. **4** (1993) 439–465. MR1228584 (94g:14019)

[Og2] K. Oguiso, *Two remarks on Calabi-Yau Moishezon threefolds*, J. Reine Angew. Math. **452** (1994) 153–161. MR1282199 (95f:32032)

[Og3] K. Oguiso, *Local families of $K3$ surfaces and applications*, J. Algebraic Geom. **12** (2003) 405–433. MR1966023 (2004c:14072)

[Og4] K. Oguiso, *Tits alternative in hyperkähler manifolds*, Math. Research Letters **13** (2006) 307–316. MR2231119 (2007e:14023)

[Og5] K. Oguiso, *Automorphisms of hyper-Kähler manifolds in the view of topological entropy*, Contemp. Math. AMS **422** (2007) 173–185. MR2296437 (2008a:14055)

[Og6] K. Oguiso, *Bimeromorphic automorphism groups of non-projective hyper-Kähler manifolds - a note inspired by C. T. McMullen*, J. Differential Geom. **78** (2008) 163–191. MR2406267 (2009g:32037)

[Og7] K. Oguiso, *Salem polynomial and $K3$ surface*, A chapter in a book, new development in algebraic surfaces by S. Mukai, S. Kondo, K. Matsuki and K. Oguiso (in Japanese, in preparation).

[OS] K. Oguiso, J. Sakurai, *Calabi-Yau threefolds of quotient type*, Asian J. Math. **5** (2001) 43–77. MR1868164 (2002i:14043)

[PSS] I. Piatetskii-Shapiro, I.R. Shafarevich, *A Torelli theorem for algebraic surfaces of type $K3$*, Math. USSR Izv. **5** (1971) 547–588. MR0284440 (44:1666)

[Ra] Z. Ran, *Deformations of manifolds with torsion or negative canonical bundles*, J. Algebraic Geom. **1** (1992) 279–291. MR1144440 (93e:14015)

[Sa] J. Sawon, *Lagrangian fibrations on Hilbert schemes of points on $K3$ surfaces*, J. Algebraic Geom. **16** (2007) 477–497. MR2306277 (2008d:14008)

[Se] C. L. Siegel, *Iterations of analytic functions*, Ann. of Math. (2) **43** (1942) 607–612. MR0007044 (4:76c)

[SI] T. Shioda, H. Inose, *On singular $K3$ surfaces*, In Complex analysis and algebraic geometry, Iwanami-Shoten (1977) 119–136. MR0441982 (56:371)

[Sh] U. Shiokawa, *Irrational number and transcendental number*, Morihoku Shuppan.

[Si] Y.-T. Siu, *Every $K3$ surface is Kähler*, Invent. Math. **73** (1983) 139–150. MR707352 (84j:32036)

[St] H. Sterk, *Finiteness results for algebraic $K3$ surfaces*, Math. Z. **189** (1985) 507–513. MR786280 (86j:14038)

[Sb] S. Sternberg, *Infinite Lie groups and formal aspects of dynamical systems*, J. Math. Mech. **10** (1961) 451–474. MR0133400 (24:A3234)

[Ti] G. Tian, *Smoothness of the universal deformation space of compact Calabi-Yau manifolds and its Peterson-Weil metric*, Mathematical Aspects of String Theory, World Scientific, Singapore (1987) 629–646. MR915841

[Ts] J. Tits, *Free subgroups in linear groups*, J. Algebra **20** (1972) 250–270. MR0286898 (44:4105)

[To] A. N. Todorov, *The Weil-Peterson geometry of moduli space of $SU(n \geq 3)$ (Calabi-Yau) manifolds*, Comm. Math. Phys. **126** (1989) 325–346. MR1027500 (91f:32022)

[Va] J. Varouchas, *Kähler spaces and proper open morphisms*, Math. Ann. **283** (1989) 13–52. MR973802 (89m:32021)

[Vi] E.B. Vinberg, *The two most algebraic $K3$ surfaces*, Math. Ann. **265** (1985) 1–21. MR719348 (85k:14020)

[Vo1] C. Voisin, *On the homotopy types of compact Kähler and complex projective manifolds*, Invent. Math. **157** (2004) 329–343. MR2076925 (2005e:32034)

[Vo2] C. Voisin, *On the homotopy types of Kähler manifolds and the birational Kodaira problem*, J. Differential Geom. **72** (2006) 43–71. MR2215455 (2007e:32020)

[WW] J. Wierzba, J.A. Wiśniewski, *Small contractions of symplectic 4-folds*, Duke Math. J. **120** (2003) 65–95. MR2010734 (2004g:14020)

[Ya1] S. T. Yau, *On the Ricci curvature of a compact Kähler manifold and the complex Monge-Ampère equation. I*, Comm. Pure Appl. Math. **31** (1978) 339–411. MR480350 (81d:53045)

[Ya2] S. T. Yau, *Calabi's conjecture and some new results in algebraic geometry*, Proc. Nat. Acad. Sci. U.S.A. **74** (1977) 1798–1799. MR0451180 (56:9467)

[Ym] Y. Yomdin, *Volume growth and entropy*, Israel J. Math. **57** (1987) 285–300. MR889979 (90g:58008)

[Yo1] K. Yoshioka, *Moduli spaces of stable sheaves on abelian surfaces*, Math. Ann. **321** (2001) 817–884. MR1872531 (2002k:14020)

[Yo2] K. Yoshioka, *Moduli spaces of vector bundles on algebraic surfaces*, Sugaku **56** (2004), 225–247. (in Japanese) MR2086113

Department of Mathematics, Osaka University, Toyonaka 560-0043 Osaka, Japan
E-mail address: `oguiso@math.sci.osaka-u.ac.jp`

Translated by KEIJI OGUISO

Amer. Math. Soc. Transl.
(2) Vol. **230**, 2010

Tensor Categories in Operator Algebras

Shigeru Yamagami

We shall review tensor categories from the viewpoint of operator algebras. For this purpose, two approaches seem to be conceivable: one is to discuss tensor categories which arise in operator algebras or more explicitly in subfactor theory, and the other is to deal with tensor categories themselves with operator algebraic structures assumed therein. Of course both of these are woven together and not easy to distinguish clearly, but we focus here on the second approach because there are several review articles already ([**39, 40**]) from the former viewpoint.

Although our coverage of contents is fairly narrow, we do include an account of general terminology and notation on tensor categories for the (Japanese) reader's accessibility in view of the difficulty in finding writings on tensor categories in Japanese, which prevented the author, however, from including topics such as finite tensor categories that are most actively studied presently.

For a question such as 'what is a tensor category?', the author once used to answer "it is something like Tannaka duals of compact groups". After several occasions at meetings, however, it was found that the Tannaka duality theorem itself is not so much well known. Based on this experience, after explanations on general usage of terminology, some classical results on Tannaka duality are reviewed in some detail. Other materials are selected so as to cover analytical apsects in tensor categories. As a result, lots of topics are left untouched in the text. To compensate this fault, though it is not complete at all, we have added several lines to give a guide to related fields for further references.

1. Tensor categories

Let us begin with our notation on categories which is used in this article.

Objects in a category $\mathcal{C}$ are denoted by capital letters such as X, Y or V, W with the set of morphisms from X to Y denoted by $\mathrm{Hom}_{\mathcal{C}}(X,Y)$ or simply $\mathrm{Hom}(X,Y)$ if there is no ambiguity. We also use the notation $\mathrm{End}(X) = \mathrm{Hom}(X,X)$ on occation. For an object X, 1_X stands for the identity morphism. The mapping notation $f : X \to Y$ is also used to designate a morphism f in $\mathrm{Hom}(X,Y)$ with X, Y referred to as the initial and final objects of f respectively.

This article originally appeared in Japanese in Sūgaku **59** (1) (2007), 56–74.

2000 *Mathematics Subject Classification.* Primary 46L10; Secondary 18D10, 22D25, 46L05, 46L37, 46M05, 46M15.

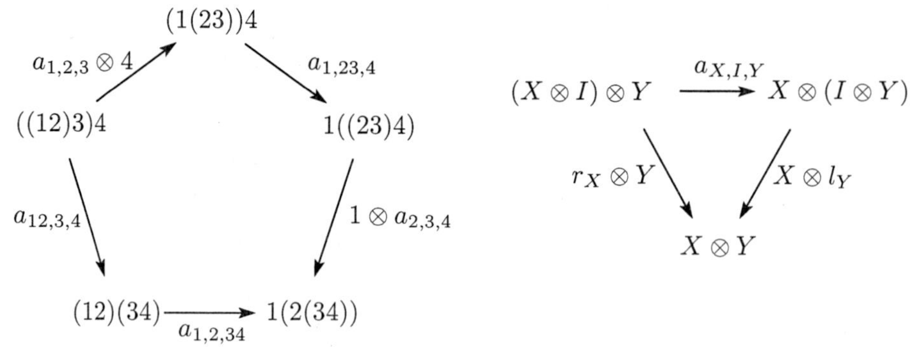

FIGURE 1

In what follows, we shall work with such categories for which $\mathrm{Hom}(X, Y)$ are vector spaces over the complex number field $\mathbb{C}$ and compositions of morphisms are bilinear. A category in this class is called a **linear category**. Functors between linear categories are assumed to preserve linearity.

A linear category $\mathcal{C}$ is called a **tensor category** if it is furnished with

(i) a bifunctor $\otimes : \mathcal{C} \times \mathcal{C} \to \mathcal{C}$,
(ii) a natural equivalence $a_{X,Y,Z} : (X \otimes Y) \otimes Z \to X \otimes (Y \otimes Z)$ between trifunctors and
(iii) a special object I (called the unit object) with natural isomorphisms $l_X : I \otimes X \to X$, $r_X : X \otimes I \to X$

which satisfy the following equations.

Pentagonal equation: Given four objects X_j $(j = 1, 2, 3, 4)$, the left diagram in Figure 1 commutes.

Triangular equation: Given two objects X, Y, the right diagram in Figure 1 commutes.

In the first diagram, a simplified notation is used such as $(X_1 \otimes (X_2 \otimes X_3)) \otimes X_4 = (1(23))4$, $a_{1,23,4} = A_{X_1, X_2 \otimes X_3, X_4}$ and $X \otimes l_Y = 1_X \otimes l_Y$.

We remark here that, although the notation is the same for ordinary tensor products, no identity such as $(X \otimes Y) \otimes Z = X \otimes (Y \otimes Z)$ is assumed here. Instead we relate four tensor products through appropriate isomorphisms, which are required to satisfy a kind of compatibility condition. Note also that we assume no isomorphism between two tensor products $X \otimes Y$, $Y \otimes X$.

If one drops off the condition on linearity in the definition of tensor categories, the notion of **monoidal category** is obtained.

In other literature, our monoidal categories are referred to as tensor categories, whereas the name of tensor category is reserved for standing for symmetric tensor categories (see below for its definition).

THEOREM 1.1 (MacLane's coherence theorem, [**48**]). *In a tensor category $\mathcal{C}$, two morphisms of common initial and final objects are identical if they are obtained from finitely many compositions of structural isomorphisms such as $a_{X,Y,Z}$, l_X, r_X and their tensor product ampliations.*

In other words, if isomorphisms of the form

$$((X_1 \otimes X_2) \otimes X_3) \otimes X_4 \to X_1 \otimes (X_2 \otimes (X_3 \otimes X_4))$$

and

$$(X \otimes I) \otimes Y \to X \otimes (I \otimes Y)$$

give rise to well-defined identifications, then all other identifications of tensor product objects are well-defined. Clearly, this is an analogue of associativity in algebraic systems but renders a proper light on the usual identification of tensor products of vector spaces such as $(U \otimes V) \otimes W = U \otimes (V \otimes W)$.

As a benefit of the above result, we may safely abbreviate parentheses in tensor products of objects; $a_{X,Y,Z}$ and l_X, r_X can be identified with identity morphisms (and it causes no contradiction).

EXAMPLE 1.2. (i) Given a group G, let $\mathcal{R}ep(G)$ be the category of finite-dimensional representations of G. Then $\mathcal{R}ep(G)$ is a tensor category by the natural identification of tensor product representations. Here the unit object is given by the trivial representation of G. The construction is immediately extended to the tensor category of a finite-dimensional corepresentations of a bialgebra (especially of a Hopf algebra).

(ii) Associated to a monoid G (a monoid being a semigroup with unit), we can introduce a tensor category whose objects are labeled by elements in G and hom-sets are given by $\mathrm{Hom}(g,h) = \mathbb{C}\delta_{g,h}$ with the tensor operations specified by the product in G.

(iii) Given a unital $\mathbb{C}$-algebra A, the tensor category $\mathcal{E}nd(A)$ (the tensor category of endomorphisms of A) is defined as follows: Objects in $\mathcal{E}nd(A)$ are unit-preserving endomorphisms of A and hom-sets are set to be

$$\mathrm{Hom}(\rho, \sigma) = \{x \in A; x\rho(y) = \sigma(y)x, \forall y \in A\}$$

(the composition of morphisms being the product in A) with the tensor product operation defined by $\rho \otimes \sigma = \rho \circ \sigma$ (the composition of endomorphisms) and $x \otimes y = x\rho(y) = \rho'(y)x \in \mathrm{Hom}(\rho \otimes \sigma, \rho' \otimes \sigma')$. In this tensor category, the associativity of tensor products holds strictly; $(\rho \otimes \sigma) \otimes \tau = \rho \otimes (\sigma \otimes \tau)$, and so on.

(iv) Consider again a unital $\mathbb{C}$-algebra A. The category $\mathcal{B}imod$ of A-A bimodules is a tensor category with tensor products given by A-module tensor products. Here the associativity is given by the natural isomorphisms $(X \otimes_A Y) \otimes_A Z \cong X \otimes_A (Y \otimes_A Z)$ between triple tensor products.

For functors between tensor categories, it is reasonable to consider those which preserve the structure of tensor products.

A functor $F : \mathcal{C} \to \mathcal{C}'$ from a tensor category $\mathcal{C}$ into another tensor category $\mathcal{C}'$ is called a **tensor functor** if we are given two families of natural isomorphisms $m^F_{X,Y} : F(X \otimes Y) \to F(X) \otimes F(Y)$ and $u^F : F(I) \to I'$ which make the following diagrams commutative:

$$\begin{array}{ccc} F(X\otimes Y\otimes Z) & \xrightarrow{m^F_{X\otimes Y,Z}} & F(X\otimes Y)\otimes F(Z) \\ {\scriptstyle m^F_{X,Y\otimes Z}}\big\downarrow & & \big\downarrow{\scriptstyle m^F_{X,Y}\otimes F(Z)} \\ F(X)\otimes F(Y\otimes Z) & \xrightarrow[F(X)\otimes m^F_{Y,Z}]{} & F(X)\otimes F(Y)\otimes F(Z), \end{array}$$

$$\begin{array}{ccc} F(I\otimes X) & \xrightarrow{m^F_{I,X}} & F(I)\otimes F(X) \\ {\scriptstyle F(l_X)}\big\downarrow & & \big\downarrow{\scriptstyle u^F\otimes 1} \\ F(X) & \xleftarrow[l_{F(X)}]{} & I'\otimes F(X), \end{array} \qquad \begin{array}{ccc} F(X\otimes I) & \xrightarrow{m^F_{X,I}} & F(X)\otimes F(I) \\ {\scriptstyle F(r_X)}\big\downarrow & & \big\downarrow{\scriptstyle 1\otimes u^F} \\ F(X) & \xleftarrow[r_{F(X)}]{} & F(X)\otimes I'. \end{array}$$

Two tensor functors $F, G : \mathcal{C} \to \mathcal{C}'$ are said to be tensor-equivalent if there are natural isomorphisms $\varphi_X : F(X) \to G(X)$ which make the following diagrams commutative:

$$\begin{array}{ccc} F(X\otimes Y) & \xrightarrow{m^F_{X,Y}} & F(X)\otimes F(Y) \\ {\scriptstyle \varphi_{X\otimes Y}}\big\downarrow & & \big\downarrow{\scriptstyle \varphi_X\otimes\varphi_Y} \\ G(X\otimes Y) & \xrightarrow[m^G_{X,Y}]{} & G(X)\otimes G(Y), \end{array} \qquad \begin{array}{ccc} F(I) & \xrightarrow{u^F} & I' \\ {\scriptstyle \varphi_I}\big\downarrow & & \big\| \\ G(I) & \xrightarrow[u^G]{} & I'. \end{array}$$

Apparently tensor functors are analogues of homomorphisms in algebraic systems and the composite functor $F \circ G$ is a tensor functor in a canonical way as can be expected whenever the composition has meaning.

PROPOSITION 1.3 (cf. Saavedra Rivano [**61**]). *For a tensor functor $F : \mathcal{C} \to \mathcal{C}'$, the following conditions are equivalent.*

(i) *We can find a tensor functor $G : \mathcal{C}' \to \mathcal{C}$ such that $F \circ G$ and $G \circ F$ are tensor-equivalent to the identity functor.*
(ii) *For objects $X, Y \in \mathcal{C}$, $F : Hom(X, Y) \to Hom(F(X), F(Y))$ is an isomorphism of vector spaces and, moreover, for any object $Z' \in \mathcal{C}'$, we can choose an object $Z \in \mathcal{C}$ so that $F(Z) \cong Z'$.*

Two tensor categories are said to be '**isomorphic**' if the equivalent conditions in the above proposition are satisfied. (According to the orthodox terminology in category theory, this should be referred to as being 'equivalent'.)

As can be read off from the above-mentioned examples, the notion of a tensor category is too vast without further qualifications, and therefore we usually impose more conditions on them for meaningful investigations.

Let us begin with a property which incorporates the existence of inverse elements in groups.

For objects X, Y in a tensor category $\mathcal{C}$, consider the condition that we can find morphisms of the form $\epsilon : X \otimes Y \to I$ and $\delta : I \to Y \otimes X$ in such a way that both of the compositions

$$X \xrightarrow{1_X\otimes\delta} X\otimes Y\otimes X \xrightarrow{\epsilon\otimes 1_X} X, \qquad Y \xrightarrow{\delta\otimes 1_Y} Y\otimes X\otimes Y \xrightarrow{1_Y\otimes\epsilon} Y$$

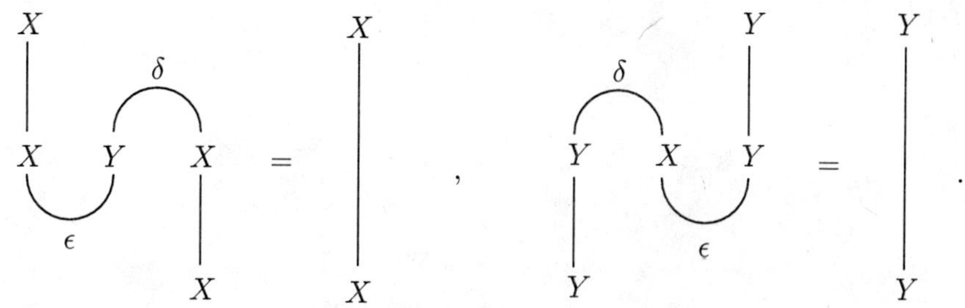

FIGURE 2

are identities. If this is satisfied, Y is called a **right dual** of X (X is a **left dual** of Y) and denoted by $Y = {}^*X$ ($X = Y^*$). By definition, ${}^*(X^*) \cong X \cong ({}^*X)^*$ and the right (or left) dual of an object, if any exists, is unique up to isomorphisms.

The uniqueness is analogous to that of invertible elements in groups but for dual objects is only determined up to isomorphisms, whence it is safe to write $Y \cong X^*$.

Indeterminancy is also applied to morphisms ϵ and δ. It would be, however, worth pointing out that, once one of ϵ and δ is specified, the other is uniquely determined. In view of this fact, we say that a morphism $\epsilon : X \otimes Y \to I$ is **non-degenerate** if it is accompanied by δ ($X \cong Y^*$, in particular).

A tensor category is **rigid** if any object admits both left and right duals.

EXAMPLE 1.4.

(i) In the tensor category of vector spaces, a vector space V has a left (or right) dual if and only if V is finite-dimensional. As a result, the tensor category of finite-dimensional vector spaces is rigid. Furthermore, two finite-dimensional vector spaces V, W are duals of each other if and only if they have the same dimension. In this case, a linear map $\epsilon : V \otimes W \to \mathbb{C}$ is accompanied with $\delta : \mathbb{C} \to W \otimes V$ if and only if ϵ is nondegenerate as a bilinear form.

(ii) The tensor category of a monoid G is rigid if and only if G is a group.

The identities which ensure the relation of dual objects can be visually (and effectively) expressed by one-dimensional objects in the plane as in Figure 2.

The following facts, although immediate from the definition, are quite useful in studying structures of tensor categories.

PROPOSITION 1.5 (Frobenius reciprocity). *If an object Z admits a left dual Z^*, we have the following isomorphisms for arbitrary objects X, Y:*

$$Hom(X \otimes Z, Y) \cong Hom(X, Y \otimes Z^*), \quad Hom(X, Z \otimes Y) \cong Hom(Z^* \otimes X, Y)$$

(cf. Figure 3). *Likewise, if we assume a right dual ${}^*Z$,*

$$Hom(Z \otimes X, Y) \cong Hom(X, {}^*Z \otimes Y), \quad Hom(X, Y \otimes Z) \cong Hom(X \otimes {}^*Z, Y).$$

Suppose that, to each object X, its left dual X^* is chosen together with the duality morphisms $\epsilon_X : X^* \otimes X \to I$, $\delta_X : I \to X \otimes X^*$. In this situation, associated to a morphism $f : X \to Y$, its **transposed morphism** $f^t : Y^* \to X^*$ is defined by the composition

$$Y^* \xrightarrow{1_{Y^*} \otimes \delta_X} Y^* \otimes X \otimes X^* \xrightarrow{1_{Y^*} \otimes f \otimes 1_{X^*}} Y^* \otimes Y \otimes X^* \xrightarrow{\epsilon_Y \otimes 1_{X^*}} X^*$$

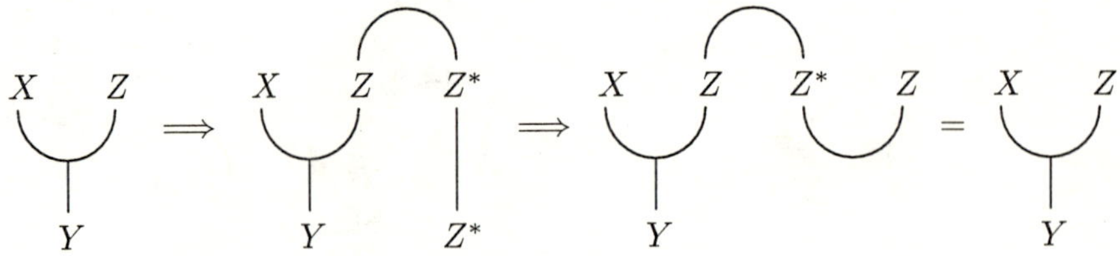

FIGURE 3

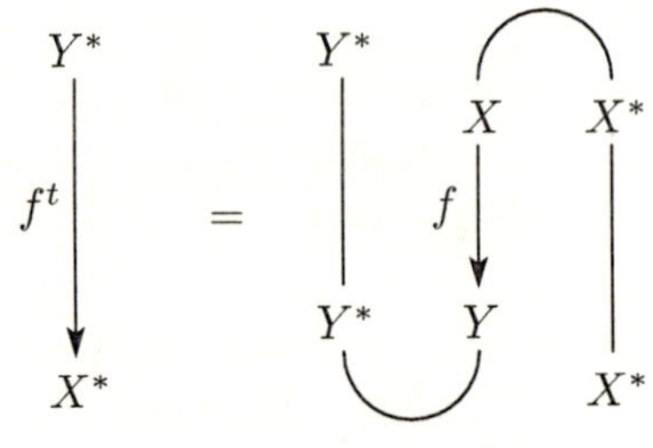

FIGURE 4

(Figure 4), and we have a contravariant functor $X \mapsto X^*$, $f \mapsto f^t$. By the uniqueness of (left) dual objects, we get the isomorphism $(X \otimes Y)^* \to Y^* \otimes X^*$, which turns out to preserve tensor products. Thus a contravariant tensor functor is obtained. If one starts with another choice of left duals and dual morphisms, then the obtained contravariant tensor functor changes to an equivalent one. Clearly a similar construction is applied for right duals.

PROPOSITION 1.6 (Maltsiniotis [**49**]). *In a rigid tensor category, the following conditions are equivalent.*

(i) *The functors of taking left transposed morphisms and right transposed morphisms are equivalent as contravariant tensor functors; i.e., we can find a multiplicative natural isomorphism* $(X^*, f^t) \cong ({}^*X, {}^tf)$.
(ii) *The composition of a left transposed functor with itself is tensor equivalent to the identity functor.*
(iii) *The composition of a right transposed functor with itself is tensor equivalent to the identity functor.*

By an **involution** of a tensor category $\mathcal{C}$, we shall mean a contravariant tensor functor $\{{}^tX, {}^tf\}$ from $\mathcal{C}$ into $\mathcal{C}$ together with a tensor equivalence between $\{{}^t({}^tX), {}^t({}^tf)\}$ and the identity functor.

PROPOSITION 1.7 (Barrett-Westbury [**3**]). *In a tensor category with an involution, two morphisms of common initial and final objects are identical if they are obtained as finitely many repetitions of structural isomorphisms of tensor category and involution.*

We next discuss the commutativity of tensor categories. A **braiding** in a tensor category $\mathcal{C}$ is a natural family of isomorphisms $b_{X,Y} : X \otimes Y \to Y \otimes X$ (X and Y being objects of $\mathcal{C}$) which makes the following diagrams commutative (hexagonal equation).

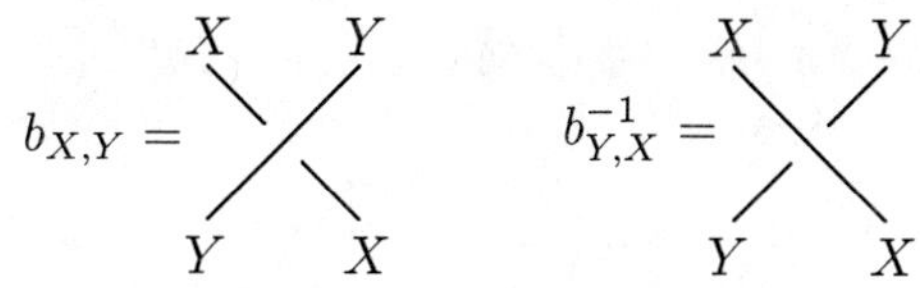

FIGURE 5

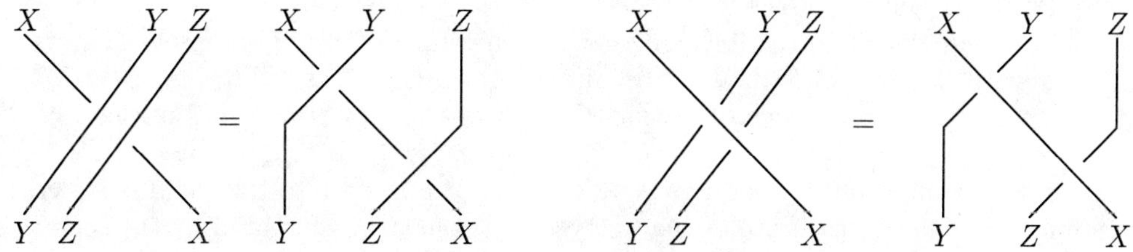

FIGURE 6

Here the associativity isomorphisms $a_{X,Y,Z}$ are revived explicitly to display 'hexagon'.

$$\begin{array}{ccccc} (X\otimes Y)\otimes Z & \xrightarrow{a_{X,Y,Z}} & X\otimes(Y\otimes Z) & \xrightarrow{b_{X,Y\otimes Z}} & (Y\otimes Z)\otimes X \\ {\scriptstyle b_{X,Y}\otimes 1}\downarrow & & & & \downarrow{\scriptstyle a_{Y,Z,X}} \\ (Y\otimes X)\otimes Z & \xrightarrow[a_{Y,X,Z}]{} & Y\otimes(X\otimes Z) & \xrightarrow[1\otimes b_{X,Z}]{} & Y\otimes(Z\otimes X), \end{array}$$

$$\begin{array}{ccccc} (X\otimes Y)\otimes Z & \xrightarrow{a_{X,Y,Z}} & X\otimes(Y\otimes Z) & \xrightarrow{b^{-1}_{Y\otimes Z,X}} & (Y\otimes Z)\otimes X \\ {\scriptstyle b^{-1}_{Y,X}\otimes 1}\downarrow & & & & \downarrow{\scriptstyle a_{Y,Z,X}} \\ (Y\otimes X)\otimes Z & \xrightarrow[a_{Y,X,Z}]{} & Y\otimes(X\otimes Z) & \xrightarrow[1\otimes b^{-1}_{Z,X}]{} & Y\otimes(Z\otimes X). \end{array}$$

From the definition, the inverse morphisms $b^{-1}_{Y,X} : X\otimes Y \to Y\otimes X$ also define a braiding. If a braiding satisfies $b^{-1}_{Y,X} = b_{X,Y}$, it is called a **permutation symmetry**. In this case, we shall often use the notation $s_{X,Y}$ instead of $b_{X,Y}$.

A tensor category is said to be **braided** if it is furnished with a braiding. When the braiding is a permutation symmetry, it is referred to as being **symmetric**. In accordance with the terminology on algebraic systems, one may call it commutative.

To see the reason why 'braids' come into, encode the basic isomorphisms $b_{X,Y}$, $b^{-1}_{Y,X}$ into planar objects as shown in Figure 5 Then the pentagonal equations are reduced to the braid isotopy relations as shown in Figure 6.

THEOREM 1.8 (Joyal-Street [**33**]). *In a braided tensor category, an isomorphism which is obtained from finitely many repetitions of braiding isomorphisms depends only on the isotopy type of the associated braids.*

COROLLARY 1.9 (MacLane [**48**]). *In a symmetric tensor category, isomorphisms obtained from repetitions of permutation symmetry depend only on the* *choice of their initial and final objects. In other words, we can do the identification $X \otimes Y = Y \otimes X$ consistently.*

Next we shall impose restrictions on the underlying linear category. An object X of a linear category $\mathcal{C}$ is said to be **simple** or **irreducible** if $\mathrm{End}(X) = \mathbb{C}1_X$. A linear category $\mathcal{C}$ is defined to be **semisimple** if any object of $\mathcal{C}$ is isomorphic to a direct sum of finitely many simple objects. In tensor categories, we further require the simplicity of the unit object I. With this assumption, the condition of rigidity is equivalent to the following weaker one: In a semisimple tensor category, an object X having a left dual ${}^*X$ admits a right dual X^* as well.

Given a semisimple tensor category $\mathcal{C}$, let S be the set of isomorphism classes of simple objects in $\mathcal{C}$. As in the case of the tensor category of group representations, the way in which tensor products $X \otimes Y$ are decomposed into irreducible objects induces an algebra structure in the free vector space $\mathbb{C}[S]$ generated by the set S:

$$xy = \sum_{z \in S} N_{x,y}^z z, \qquad N_{x,y}^z = \dim \mathrm{Hom}(Z, X \otimes Y) = \dim \mathrm{Hom}(X \otimes Y, Z)$$

(X, Y, Z being representative objects of $x, y, z \in S$, respectively).

The algebra $\mathbb{C}[S]$ is called the Grothendieck ring of the tensor category $\mathcal{C}$. Moreover, if $\mathcal{C}$ is rigid, it is a *-algebra by the *-operation induced from the correspondence $X \mapsto X^*$ and we have the duality identity $N_{x^*,y}^1 = \delta_{x,y}$ ($x, y \in S$ and 1 is the class represented by the unit object). We can derive other identities such as $N_{x,y}^z = N_{x^*,z}^y$ from this.

A *-algebra having a *-invariant distinguished basis with the structure constants being nonnegative integers is called a **fusion rule algebra** if it satisfies the duality identity.

Thus, the Gronthendieck ring of a semisimple tensor category is an example of a fusion rule algebra. Note here that, when the tensor category is braided, the associated fusion rule algebra is commutative. Originally the notion of a fusion rule algebra arises in this context in quantum physics ([**23**]).

EXAMPLE 1.10.

(i) The group algebra $\mathbb{C}[G]$ of a discrete group G.

(ii) The (complexification of) a representation ring of a compact group G. When specialized to SU(2), we have the Clebsch-Gordan algebra, which is defined to be $S = \{x_0 = 1, x_1, x_2, \dots\}$ with

$$x_j x_k = x_{|j-k|} + x_{|j-k|+2} + \cdots + x_{j+k-2} + x_{j+k}.$$

(iii) The Verlinde algebra: $S = \{x_0 = 1, x_1, x_2, \dots, x_n\}$ with

$$x_j x_k = \begin{cases} x_{|j-k|} + x_{|j-k|+2} + \cdots + x_{j+k-2} + x_{j+k} & \text{if } j+k \leq n, \\ x_{|j-k|} + x_{|j-k|+2} + \cdots + x_{2n-j-k-2} + x_{2n-j-k} & \text{if } j+k > n. \end{cases}$$

Although the fusion rule algebra provides us with good information on tensor categories, it is not enough to determine the category itself. For example, it is well known that there are two nonisomorphic groups of order 8, the dihedral group and the quaternion group, which have isomorphic representation rings.

If one splits the isomorphism information of semisimple tensor categories into the part of the fusion rule algebra and the remaining part, then the remaining part turns out to concern nonlinear equations originated from the pentagonal identities, which is highly difficult to solve explicitly. Just for a few classes of fusion rule algebras, tensor categories are determined completely.

EXAMPLE 1.11. (i) For the case of the group algebra $\mathbb{C}[G]$, the possible (nonisomorphic) tensor categories are parametrized by elements in the third cohomology group $H^3(G; \mathbb{C}^\times)$ of G.

(ii) For the fusion rule algebra obtained from the representation ring of the compact group $\mathrm{SU}(N)$, tensor categories are completely classified: they are isomorphic to the quantum deformation of representation categories up to twisting by elements in $H^3(\mathbb{Z}_N; \mathbb{C}^\times)$ ([**43**]).

(iii) Let G be a finite group and consider the fusion rule algebra which is obtained from the group algebra $\mathbb{C}[G]$ by adding an extra element x with the fusion rule extended by

$$gx = xg = x, \qquad x^2 = \sum_{g \in G} g.$$

Then such a fusion rule algebra comes from a semisimple tensor category if and only if G is commutative and, if this is the case, we have a complete description of possible tensor categories ([**69**]).

Finally we discuss some analytical structures in tensor categories. Recall that a *-algebra of bounded linear operators on a Hilbert space is called a (concrete) C*-algebra (respectively W*-algebra or von Neumann algebra) if it is closed under the norm-topology (respectively under the weak operator-topology). A C*-algebra is also abstractly characterized as a Banach *-algebra with the norm satisfying the so-called C*-condition $\|a^*a\| = \|a\|^2$. If one requires such a structure on hom-sets of a linear category, the notion of a C*-category is obtained. A C*-category is known to admit a faithful embedding into the C*-category of Hilbert spaces. A C*-category is called a **C*-tensor category** (or tensor C*-category) if it is furnished with a compatible tensor category structure (the compatibility means that relevant isomorphisms are unitary).

EXAMPLE 1.12.

(i) The C*-tensor category of (continuous) unitary representations of a locally compact group.
(ii) The C*-tensor category of finite-dimensional unitary representations of a compact quantum group.
(iii) The tensor category of a discrete group is a C*-tensor category.
(iv) The C*-tensor category of *-endomorphisms of a C*-algebra.
(v) Given a W*-algebra A, the C*-tensor category of Hilbert spaces on which A acts continuously (relative to the weak operator topology) in a bimodule fashion. More explanations can be found later on W*-bimodules.

2. Fiber functors and Tannaka-Krein duality

Group representations are good sources of tensor categories and constitute a major class of them, which has the special feature that objects are realized as vector spaces with hom-sets given by linear maps. This kind of structure goes back to studies on the duality of nonabelian groups, which was initiated by T. Tannaka in the case of compact groups and further developed in the context of abstract categories by algebraic geometrists. It has since been also a source of quantum groups through the representation theory of Hopf algebras.

Here we shall look into Tannaka type theorems in connection with tensor categories.

In the tensor category of finite-dimensional (co)representations of a Hopf algebra, we have kernels, cokernels and images associated to morphisms (intertwiners), which can be paraphrased in a categorical language; i.e., it is a so-called abelian category. As already remarked, this type of tensor category has the property that objects are realized by vector spaces with morphisms given by linear maps between vector spaces. In other words, tensor categories in this class have realizations in the tensor category $\mathcal{V}ec$ of finite-dimensional vector spaces.

Generally, given a tensor category $\mathcal{C}$ with the underlying linear category abelian, a tensor functor from $\mathcal{C}$ to $\mathcal{V}ec$ is called a **fiber functor** if it is faithful and exact. A fiber functor can be interpreted as a kind of representation of a tensor category.

THEOREM 2.1 (Saavedra Rivano-Ulbrich [**61, 71**]). *The tensor category of finite-dimensional corepresentations of a Hopf algebra is characterized as a rigid tensor category admitting a fiber functor. Hopf algebras are determined up to isomorphisms by pairs of a rigid tensor category and a fiber functor.*

Abstract tensor categories are not enough to determine Hopf algebras. This sort of indeterminancy happens even for representations of finite groups ([**15**]). On the other hand, tensor categories arising from Hopf algebras are known to constitute only a small portion of tensor categories.

THEOREM 2.2 (Woronowicz [**77**]). *The tensor category of finite-dimensional unitary representations of a compact quantum group is characterized as a semisimple rigid C^*-tensor category realized in $\mathcal{H}ilb$. Here $\mathcal{H}ilb$ denotes the C^*-tensor category of finite-dimensional Hilbert spaces.*

THEOREM 2.3 (Tannaka-Krein). *The tensor category of finite-dimensional unitary representations of a compact group is characterized as a semisimple symmetric rigid C^*-tensor category realized in $\mathcal{H}ilb$.*

As an algebraic group version of this classical Tannaka duality theorem, we have

THEOREM 2.4 (Saavedra Rivano-Deligne-Milne [**61, 12**]). *The tensor category of finite-dimensional representations of a proalgebraic group is characterized up to isomorphisms as a symmetric abelian rigid tensor category realized in $\mathcal{V}ec$.*

For Tannaka duals, i.e., for tensor categories of finite-dimensional representations of compact groups, we have the following more intrinsic characterizations. Given an object X in a symmetric tensor category, define a morphism $s_j \in \mathrm{End}(X^{\otimes n})$ $(j = 1, 2, \dots, n-1)$ in its n-fold tensor product $X^{\otimes n} = \overbrace{X \otimes \cdots \otimes X}^{n\text{-times}}$ by

$$s_1 = s_{X,X} \otimes 1_{X^{n-2}}, \quad s_2 = 1_X \otimes s_{X,X} \otimes 1_{X^{n-3}}, \quad \dots, \quad s_{n-1} = 1_{X^{n-2}} \otimes s_{X,X}.$$

Then these satisfy the relations

$$s_j^2 = 1, \quad s_i s_{i+1} s_i = s_{i+1} s_i s_{i+1}, \quad s_j s_j = s_j s_i \ (|i-j| \geq 2),$$

whence they give rise to a homomorphism from $\mathbb{C}[S_n]$ into $\mathrm{End}(X^{\otimes n})$ ($\mathbb{C}[S_n]$ being the group algebra of the symmetric group S_n). Thus we can define an idempotent

$P_n \in \mathrm{End}(X^{\otimes n})$ as the image of the antisymmetrizer

$$\frac{1}{n!}\sum_{\sigma\in S_n} \mathrm{sign}(\sigma)\sigma \in \mathbb{C}[S_n].$$

We use $\wedge^n X$ to stand for the subobject of $X^{\otimes n}$ corresponding to P_n.

THEOREM 2.5 (Deligne [**11**]). *Let $\mathcal{C}$ be a rigid symmetric abelian tensor category satisying the condition that, given any object X, we can find a positive integer n satisfying $\wedge^n X = 0$. Then there exists a fiber functor $\mathcal{C} \to \mathcal{V}ec$ preserving the permutation symmetry.*

Let X be an object in a symmetric C*-tensor category $\mathcal{C}$. An object $Y \in \mathcal{C}$ is said to be **conjugate** to X if $Y = X^*$ (i.e., Y is a left dual of X) and we can choose a pair of rigidity morphisms $\epsilon : X^* \otimes X \to I$, $\delta : I \to X \otimes X^*$ so that $\delta^* = \epsilon \circ s_{X,X^*}$.

In the Tannaka dual of a compact group, it is easy to check that the complex conjugate representation gives a conjugate object in the above sense.

THEOREM 2.6 (Doplicher-Roberts [**13**]). *A symmetric C*-tensor category $\mathcal{C}$ is embedded into the C*-tensor category $\mathcal{H}ilb$ as a C*-tensor category in such a way that it preserves the permutation symmetry if any object in $\mathcal{C}$ admits a conjugate object. Furthermore, the embedding is unique up to natural equivalences.*

In other words, there is a one-to-one correspondence between (isomorphism classes of) compact groups and (isomorphism classes of) symmetric C*-tensor categories admitting conjugate objects.

3. Rigid C*-tensor categories

We shall here review the basic properties which will be obtained when rigidity is combined with C*-structure. By applying such categorical results to the C*-tensor category of W*-bimodules, we can easily derive many combinatorial results in Jones index theory.

PROPOSITION 3.1 (Longo-Roberts [**46**]). *Consider a C*-tensor category $\mathcal{C}$ where the unit object is simple (or irreducible). If an object X of $\mathcal{C}$ admits a dual object X^*, then End(X) is finite-dimensional.*

Since $\mathrm{End}(X)$ is a C*-algebra, finite-dimensionality means that it is isomorphic to a direct sum of finitely many finite-dimensional matrix algebras. Thus rigid C*-tensor categories are semisimple (by adding subobjects associated to orthogonal projections if necessary).

In what follows, we shall exclusively work with C*-tensor categories with simple unit objects. Recall that, for an object X, X^* denotes one of the left duals of X. In a C*-tensor category, it is natural to require an involution to be compatible with C*-structure in the sense that $(f^t)^* = (f^*)^t$ and relevant structural isomorphisms (such as $(X \otimes Y)^* \to Y^* \otimes X^*$) are unitary.

As an analogue to the fact that the rigidity-based involution is reduced to the ordinary duality relation in the case of the tensor category $\mathcal{V}ec$ of finite-dimensional vector spaces, we formulate here the relevant information into the form of definition ([**79, 82**], cf. [**3**]): By a duality system in a C*-tensor category, we shall mean a unitary involution (X^*, f^t) together with a family of morphisms $\{\epsilon_X : X^* \otimes X \to I\}$ fulfilling the following conditions.

(i) Compatibility

$$\begin{array}{ccc} Y^* \otimes X & \xrightarrow{1\otimes f} & Y^* \otimes Y \\ {\scriptstyle f^t\otimes 1}\downarrow & & \downarrow{\scriptstyle \epsilon_Y} \\ X^* \otimes X & \xrightarrow[\epsilon_X]{} & I. \end{array}$$

(ii) Multiplicativity $\epsilon_{X\otimes Y} = \epsilon_Y \circ (1_{Y^*} \otimes \epsilon_X \otimes 1_Y)$.

(iii) Neutrality

$$\epsilon_X(1 \otimes a)\epsilon_X^t = \epsilon_{X^*}(a \otimes 1)\epsilon_{X^*}^t, \quad a \in \mathrm{End}(X).$$

(iv) Positivity $\epsilon_X^t = \epsilon_X^*$ for $X \in \mathcal{C}$.

THEOREM 3.2 (Yamagami [**82**]). *Every rigid C*-tensor category $\mathcal{C}$ satisfying the condition End(I) $= \mathbb{C}$ (I being the unit object) admits one and only one duality system up to unitary isomorphisms.*

As a simple application of the existence and the uniqueness of duality systems, let us introduce the quantum dimension in a rigid C*-tensor category $\mathcal{C}$: For an object $X \neq 0$, its quantum dimension $d(X) > 0$ is defined by

$$d(X)1_I = \epsilon_X \epsilon_X^*.$$

THEOREM 3.3 (Longo-Roberts [**46**]). *The quantum dimension $d(X)$ satisfies the following properties.*

(i) *If* $X \cong Y$, $d(X) = d(Y)$.
(ii) $d(X \oplus Y) = d(X) + d(Y)$.
(iii) $d(X \otimes Y) = d(X)d(Y)$.
(iv) $d(X) = d(X^*)$.
(v) $d(X) \in [2, +\infty) \cup \{2\cos(\pi/n); n = 2, 3, 4, \dots\}$.

Properties (i)–(iv) are equivalently stated in the form that the quantum dimension gives a multiplicative linear functional on the fusion rule algebra $\mathbb{C}[S]$. The restriction (v) on the value of the quantum dimension is then a direct consequence of these.

Originally these kinds of characteristic parameters were introduced under the name of statistical dimension during studies of algebraic quantum field theory, which is later identified by R. Longo with the square root of Jones' index in subfactor theory ([**31**]).

4. Amenability in fusion rule algebras

As an analytical aspect of fusion rule algebras, we shall review the notion of amenability. We remark that the integrality of structure constants is not essential to this end and we can work with the so-called **discrete hypergroup** equally well. Here, by a discrete hypergroup, we shall mean an algebraic structure similar to the fusion rule algebra with the integrality condition on structure constants dropped off (non-negativity as well as uniform separation between positive coefficients and 0 being kept).

Given a fusion rule algebra $\mathbb{C}[S]$ with S its base set, define a linear functional N on $\mathbb{C}[S]$ by

$$N\left(\sum_{s\in S} c(s)\, s\right) = c(1).$$

Then N has the trace property $N(ab) = N(ba)$ and satisfies the positivity $N(c^*c) = \sum_{s\in S} |c(s)|^2$. The associated GNS-representation is denoted by L and is referred to as the left regular representation of the fusion rule algebra $\mathbb{C}[S]$.

By a **dimension function** of $\mathbb{C}[S]$, we shall mean a multiplicative linear functional $d : \mathbb{C}[S] \to \mathbb{C}$ of $\mathbb{C}[S]$ satisfying $d(s) = d(s^*) > 0$ $(s \in S)$.

THEOREM 4.1 (Hiai-Izumi [**27**]). *If a fusion rule algebra $\mathbb{C}[S]$ admits a dimension function d, then the left regular representation is bounded and satisfies the inequality $\|L(s)\| \leq d(s)$ $(s \in S)$.*

With the help of Perron-Frobenius theory, we can prove the following.

THEOREM 4.2 (Arad-Blau-Sunder [**1, 66**]). *In a finite-dimensional fusion rule algebra, one and only one dimension function exists.*

In the infinite-dimensional case, neither existence nor uniqueness holds. For the nonexistence, we just remark that there are plenty of fusion rule algebras with unbounded L (cf. [**80**]). For the nonuniqueness, the following is well known.

EXAMPLE 4.3. In the Clebsch-Gordan algebra (i.e., the representation ring of the compact group SU(2)), dimension functions are parametrized by q-integers (different dimension functions for different $q \geq 1$).

THEOREM 4.4 (Hiai-Izumi [**27**]). *For a dimension function d of a fusion rule algebra $\mathbb{C}[S]$, the following conditions are equivalent.*

(i) *We can find a sequence $\{\xi_n\}_{n\geq 1} \subset \ell^2(S)$ of unit vectors satisfying*

$$\lim_{n\to\infty} \|L(s)\xi_n - d(s)\xi_n\| = 0, \quad \forall s \in S.$$

(ii) *If an element $c = \sum_{s\in S} c(s)s$ in $\mathbb{C}[S]$ satisfies $c(s) \geq 0$ for any $s \in S$, then $\|L(c)\| = d(c)$.*

DEFINITION 4.5. A dimension function is said to be **amenable** if it satisfies the equivalent conditions in the above theorem. A fusion rule algebra is said to be **amenable** if it admits an amenable dimension function. Note that an amenable dimension function is, if any exists, unique from the property (ii) in the theorem.

There are other conditions which paraphrase the amenability, such as the Følner condition, Kesten condition, and so on ([**27**]). Amenability of finite-dimensional fusion rule algebras is an immediate consequence of the Følner condition. If the group algebra of a (discrete) group G is considered to be a fusion rule algebra, it has the unique dimension function $d(g) = 1$ $(g \in G)$, which is known to be amenable if and only if G is amenable (cf. [**27**]).

THEOREM 4.6 (Yamagami [**80**]). *A commutative fusion rule algebra is amenable if and only if the associated regular representation is bounded.*

EXAMPLE 4.7. For the Tannaka dual of the compact quantum group $\mathrm{SU}_q(N)$, its fusion rule algebra is known to be independent of the deformation parameter q ([**59**]). The tensor-categorical dimension function is, however, the one associated with the q-integers, which is amenable in our sense if and only if $q = 1$ (cf. [**25**]).

5. C*-tensor categories of bimodules

In operator algebras, tensor categories arise most naturally from bimodules. Among bimodules, we are here concerned with the case of W*-algebras, which is intimately related with Jones' subfactor theory as well as physical applications.

As for relevant definitions of the basics on W*-algebras, we had several occasions to read in this "ronsetsu" series ([**29, 39, 40, 56**]) and we will not repeat them here. For studies of C*-algebraic bimodules, we just refer to [**35, 36, 37**]. The tensor categories behind them do not seem to be exploited much yet.

In what follows, Hilbert spaces are assumed to be separable; i.e., they are isometrically isomorphic to $\ell^2(\mathbb{N})$. Consider the situation where W*-algebras A, B act (continuously) on a Hilbert space X in a bimodule fashion. We use the notation ${}_AX_B$ to express the bimodule of this type. By the self-duality of Hilbert spaces (the so-called Riesz' lemma) the dual Hilbert space X^* is canonically identified with the complex conjugate $\overline{X}$ of X and we have the bimodule ${}_BX^*_A$ (the dual bimodule of ${}_AX_B$) naturally. Obviously we have the duality on bimodules in the form that the dual bimodule of ${}_BX^*_A$ is naturally isomorphic to the original bimodule ${}_AX_B$.

We next explain the regular representations of W*-algebras to be used to introduce tensor products of bimodules.

For a general W*-algebra A, we can naturally construct the bimodule $L^2(A)$ (also called the standard representation) which corresponds to algebraic regular representations, but we need the elaborate machinery of noncommutative integration to extract the Hilbert space structure from A. When A admits a (continuous) linear functional $\phi : A \to \mathbb{C}$ such that $\phi(a^*b)$ is positive definite and $\phi(ab) = \phi(ba)$ for $a, b \in A$ (called a trace), then the completion of A with respect to the inner product $(a|b) = \phi(a^*|b)$ turns out to be (isomorphic to) $L^2(A)$. The left and right multiplications by elements of A are then continuously extended to actions on $L^2(A)$ resulting in the bimodule ${}_AL^2(A)_A$. Apparently the bimodule obtained this way depends on the choice of ϕ but its reliance is functorial thanks to the Radon-Nikodym type theorem, which enables us to identify them categorically and drop ϕ off (see [**67**], for example).

Returning to the definition of tensor products of W*-bimodules, pick up bimodules ${}_AX_B$, ${}_BY_C$. Then owing to measure-theoretical features of W*-algebras, we can define a Hilbert space $X \otimes_B Y$ as the completion of the algebraic tensor product $X \odot_B Y$ of B-modules, which turns out to be an A-C bimodule naturally. (This explanation is inaccurate because the vector space to be completed is not $X \odot_B Y$ but something like $X \odot_B L^{-2}(B) \odot_B Y$. Here -2 in L^{-2} is not a typo. The middle $L^{-2}(B)$ indicates the fact that $L^2(B)$ appears twice in $X \odot_B Y$ and we are forced to cancel one of them to obtain a natural inner product.)

Since W*-modules are projective, we can reduce the W*-module tensor products to the natural identification $L^2(A) \otimes_A L^2(A) \cong L^2(A)$ of regular representation spaces.

Anyway, in an algebraically legitimate way, we have tensor products of W*-bimodules, which satisfy the associativity in a categorical manner. In particular, given a W*-algebra A, A-A bimodules constitute a C*-tensor category with the unit object given by ${}_AL^2(A)_A$. Note that the unit object is simple (or irreducible); i.e., $\mathrm{End}({}_AL^2(A)_A) = \mathbb{C}1_A$ if and only if A has a trivial center (so-called a factor). In view of this fact, one is much interested in W*-bimodules for which the acting algebras are factors. Among factors, the simplest would be the full operator algebra

$\mathcal{B}(\mathcal{H})$ on a Hilbert space $\mathcal{H}$, which, however, gives rise to a trivial tensor category; the tensor category of $\mathcal{B}(\mathcal{H})$-$\mathcal{B}(\mathcal{H})$ bimodules is isomorphic to the tensor category of Hilbert spaces.

Thus, to get a nontrivial tensor category, it is necessary to deal with other factors than the full operator algebras. Among such factors, well-studied are the ones which admit (finite) traces (called finite factors).

DEFINITION 5.1. A (unitary) tensor functor F from a C*-tensor category $\mathcal{C}$ into another C*-tensor category $\mathcal{C}'$ is called a fully faithful embedding if $F : \mathrm{Hom}(X,Y) \to \mathrm{Hom}(F(X),F(Y))$ is isomorphic for any choice of objects X, Y in $\mathcal{C}$.

THEOREM 5.2 (Yamagami [**81**]). *Let $\mathcal{C}$ be a rigid C*-tensor category. If the isomorphism classes of simple objects in $\mathcal{C}$ form a countable set, then we can find a finite factor A and a fully faithful embedding of $\mathcal{C}$ into the C*-tensor category of A-A bimodules.*

Among finite factors, fundamental is a factor called AFD (approximately finite-dimensional); a factor A is said to be AFD if we can find an increasing sequence of finite-dimensional *-subalgebras which approximate the whole factor with respect to the weak operator topology. In spite of their various ways of appearance, all the AFD finite factors have been known to be isomorphic since the advent of operator algebras by Murrray and von Neumann.

THEOREM 5.3 (Hayashi-Yamagami [**25**]). *Adding to the previous assumption, if the associated dimension function is supposed to be amenable, we can find a fully faithful embedding into the C*-tensor category A-A bimodules with A an AFD finite factor.*

These embedding results are analogous to S. Popa's existence theorems in subfactor theory ([**57, 58**]), but we need a nontrivial device to incorporate the structure of tensor categories. In connection with the last theorem, it is an open question to determine whether the amenability is actually necessary or not. In particular, the realization of the Tannaka dual of the quantum group $\mathrm{SU}_q(2)$ as AFD bimodules has been an outstanding open problem since the very beginning of Jones' index theory.

6. Reference guide

We shall here give a brief guide to the references, which should have been touched upon in the text but have been omitted owing to the author's disacquaintance as well as lack of space.

(i) Fundamentals

Mac Lane [**48**] is the bible on category theory which contains explanations on tensor categories. For users, however, the books of Borceux [**5, 6**] might be more convenient. Today tensor categories are discussed mostly in connection with representations of quantum groups, and some of the books on the subject include discussions on tensor categories (Chari-Pressley [**8**], Etingof-Schiffman [**20**], Kassel [**38**], Shnider-Sternberg [**64**]).

Basic results are also included in Saavedra Rivano [**61**]. As review articles, we refer to Calaque-Etingof [**7**] and Davydov [**9**]. For physical backgrounds, see Fröhlich-Kerler [**23**].

(ii) Finite tensor category

As for the algebraic theory of tensor categories, especially for finite tensor categories, remarkable developments are in progress ([**17**], [**18**], [**55**]). Here mathematical physics (especially conformal field theory) is influencial and the notion of module category (not the category of modules), which is a kind of action of a tensor category on a linear category, seems to deserves serious investigations.

Looking into the inside of tensor categories themselves, it is meaningful to think of formal definitions of algebraic systems such as group, ring and so on inside tensor categories. For example, orbifold constructions can be dealt with in this context.

(iii) Modular tensor category

Related with applications to geometric invariants, we should be aware of finite tensor categories such as modular tensor categories and ribbon categories, which are special classes of braided tensor categories and have been most extensively studied (Bakalov-Kirillov [**2**], Turaev [**70**]).

(iv) Conformal field theory

This branch of quantum physics has certainly been having much effect on tensor categories. Lots of important mathematical notions such as vertex operator algebra, ADE classification, orbifold construction, fusion rule algebras and so on, have their origins in conformal field theory. Although there are plenty of monographs available from the physics side, we mention Yamada [**78**] as a reference to mathematical frameworks of this important subject. In relation with operator algebras, we refer to Evans-Kawahigashi [**21**] and Kawahigashi [**40**] for recent developments.

(v) Quantum groups and Tannaka duals

Quantum groups are good sources of tensor categories and constitute an important class of them. There are a certain number of results on quantum groups which more or less concern tensor categories. In particular, for the most basic quantum groups $\mathrm{SL}_q(2,\mathbb{C})$ or $\mathrm{SU}_q(2)$, all the details seem to be available. However, focusing on tensor categorical aspects, there still remain things to be exploited and we refer to Kirillov-Ostrik [**45**] and Ocneanu [**53**] for quantum McKay correspondence.

For a monograph on Hopf algebraic Tannaka duality, we refer to Schauenburg [**62**].

(vi) Tannakian category

When one sticks to symmetric tensor categories, the relation with group theory will be more dominant. For studies on tensor categories of representations of algebraic groups, consult Saavedra Rivano-Deligne-Milne [**61, 12, 10**]. Deligne's characterization of Tannaka duals in terms of Young's symmetrizers is afterwards generalized to the existence theorem of super fiber functors ([**11**], see also Müger [**51**], Ostrik [**54**]).

(vii) C*-tensor category

Basic results on C*-tensor categories are stated in the text. For researchers not specialized in operator algebras, the algebraic or combinatorial structures behind them would be more useful than truly analytical results (Müger [**50**]). Related to this, combinatorial information based on subfactor theory such as paragroups can be found in Evans-Kawahigashi [**21**].

(viii) Hypergroups and fusion algebras

The notion of a fusion rule algebra is a way of generalization of representation rings of groups, which can be seen as a discrete hypergroup, where hypergroup is another way to generalize representation rings. See Bloom-Heyer [4], Heyer [**26**] and Wildberger [**76**] for more information on hypergroups.

(ix) Operad and loop spaces

Geometrically the coherence theorem on associativity in tensor categories can be equivalently interpreted as the triviality of Stasheff polyhedra (Etingof-Schiffman[**20**], Shnider-Sternberg[**64**]) and we find connections between homotopy theory and tensor categories. Among them, the notion of operad is noteworthy, which is extracted from composition operations in multiple loop spaces (this is closely related to the diagrammatic tensor calculus) and supplies examples of tensor or monoidal categories (Jones [**32**], Voronov [**74**]).

References

[1] Z. Arad and H.I. Blau, On table algebras and applications to finite group theory, J. Algebra, 138(1991), 137–185. MR1102572 (92f:20007)

[2] B. Bakalov and A. Kirillov, Jr., Lectures on tensor categories and modular functors. University Lecture Series, 21. American Mathematical Society, Providence, RI, 2001. MR1797619 (2002d:18003)

[3] J.W. Barrett and B.W. Westbury, Spherical categories, *Adv. Math.*, 143(1999), 357–375. MR1686423 (2000c:18007)

[4] W.R. Bloom and H. Heyer, Harmonic analysis of probability measures on hypergroups. Walter de Gruyter, Berlin, 1995. MR1312826 (96a:43001)

[5] F. Borceux, Handbook of categorical algebra. 2, volume 51 of Encyclopedia of Mathematics and its Applications. Cambridge University Press, Cambridge, 1994. MR1313497 (96g:18001b)

[6] F. Borceux, Handbook of categorical algebra. 3, volume 52 of Encyclopedia of Mathematics and its Applications. Cambridge University Press, Cambridge, 1994. MR1315049 (96g:18001c)

[7] D. Calaque and P. Etingof, Lectures on tensor categories, math.QA/0401246.

[8] V. Chari and A. Pressley, A Guide to Quantum Groups, Cambridge Univ. Press, 1995. MR1358358 (96h:17014)

[9] A.A. Davydov, Monoidal categories, J. Math. Sci. (New York), 88(1998), 457–519. MR1613071 (2000c:18008)

[10] P. Deligne, Catégories Tannakiennes, Grothendieck Festschrift, Vol. 2, P. Cartier et al. (eds.), 111–195, Birkhäuser, Boston, 1991. MR1106898 (92d:14002)

[11] P. Deligne, Catégories tensorielles, Moscow Math. J., 2(2002), 227–248. MR1944506 (2003k:18010)

[12] P. Deligne and J. Milne, Tannakian categories, Hodge Cycles, Motives and Shimura Varieties, Lecture Notes in Math. 900, 101–228, Springer, Berlin, 1982. MR654325 (84m:14046)

[13] S. Doplicher and J.E. Roberts, A new duality theory for compact groups. Invent. Math., 98(1989), 157–218. MR1010160 (90k:22005)

[14] S. Eilenberg and G.M. Kelly, Closed categories, Proc. Conf. Categorical Algebra (La Jolla, Calif., 1965), 421–562, Springer, New York 1996. MR0225841 (37:1432)

[15] P. Etingof and S. Gelaki, Isocategorical groups, Internat. Math. Res. Notices 2001, no. 2, 59–76. MR1810480 (2001j:20011)

[16] P. Etingof and M. Khovanov, Representations of tensor categories and Dynkin diagrams. Internat. Math. Res. Notices 1995, no. 5, 235–247. MR1333750 (96m:17026)

[17] P. Etingof, D. Nikshych and V. Ostrik, On fusion categories, Ann. of Math. (2), 162(2005), 581–642. MR2183279 (2006m:16051)

[18] P. Etingof and V. Ostrik, Finite tensor categories, Mosc. Math. J., 4(2004), 627–654, 782–783. MR2119143 (2005j:18006)

[19] P. Etingof and V. Ostrik, Module categories over representations of $\mathrm{SL}_q(2)$ and graphs, Math. Res. Lett., 11(2004), 103–114. MR2046203 (2005d:20088)

[20] P. Etingof and O. Schiffman, Lectures on Quantum Groups, International Press, 1998. MR1698405 (2000e:17016)

[21] D.E. Evans and Y. Kawahigashi, Quantum symmetries on operator algebras. Oxford Mathematical Monographs. Oxford Science Publications. Oxford University Press, New York, 1998. MR1642584 (99m:46148)

[22] P.J. Freyd and D.N. Yetter, Braided compact closed categories with applications to low-dimensional topology, Adv. Math., 77(1989), 156–182. MR1020583 (91c:57019)

[23] J. Fröhlich and T. Kerler, Quantum Groups, Quantum Categories and Quantum Field Theory, Lec. Notes in Math. 1542, Springer-Verlag, 1993. MR1239440 (95f:81042)

[24] P. Ghez, R. Lima and J.E. Roberts, W*-categories, Pacific J. Math., 120(1985), 79–109. MR808930 (87g:46091)

[25] T. Hayashi and S. Yamagami, Amenable tensor categories and their realizations as AFD bimodules, J. Funct. Anal., 172(2000), 19–75. MR1749868 (2001d:46092)

[26] H. Heyer, Probability theory on hypergroups: a survey. Probability measures on groups, VII (Oberwolfach, 1983), 481–550, Lecture Notes in Math., 1064, Springer, 1984. MR772428 (86c:60016)

[27] F. Hiai and M. Izumi, Amenability and strong amenability for fusion algebras with applications to subfactor theory, Internat. J. Math., 9(1998), 669–722. MR1644299 (99h:46116)

[28] Y.Z. Huang and J. Lepowsky, A theory of tensor products for module categories for a vertex operator algebra. I, II, Selecta Math., 1(1995), 699–756, 757–786. MR1383584 (98a:17047)

[29] M. Izumi, Classification theory in C*-algebras (in Japanese), Sugaku, 57(2005), 282–301. MR2163673

[30] M. Jimbo, Quantum Groups and Yang-Baxter Equations (in Japanese), Springer-Verlag, Tokyo, 1990.

[31] V.F.R. Jones, Index for subfactors, Invent. Math., 72(1983), 1–15. MR696688 (84d:46097)

[32] V.F.R. Jones, Planar algebras I, math.QA/9909027.

[33] A. Joyal and R. Street, Braided tensor categories. Adv. Math., 102(1993), 20–78. MR1250465 (94m:18008)

[34] A. Joyal and R. Street, Traced monoidal categories. Math. Proc. Cambridge Philos. Soc., 119(1996), 447–468. MR1357057 (96m:18014)

[35] T. Kajiwara and Y. Watatani, Jones index theory by Hilbert C^*-bimodules and K-theory, Trans. Amer. Math. Soc., 352(2000), 3429–3472. MR1624182 (2000m:46128)

[36] T. Kajiwara and Y. Watatani, Hilbert C^*-bimodules and continuous Cuntz-Krieger algebras, J. Math. Soc. Japan, 54(2002), 35–59. MR1864927 (2002k:46154)

[37] T. Kajiwara, C. Pinzari and Y. Watatani, Jones index theory for Hilbert C*-bimodules and its equivalence with conjugation theory, J. Funct. Anal., 215(2004), 1–49. MR2085108 (2005f:46114)

[38] C. Kassel, Quantum Groups, Springer-Verlag, Berlin-New York, 1995. MR1321145 (96e:17041)

[39] Y. Kawahigashi, Paragroups as quantum Galois groups of subfactors (in Japanese), Sugaku, 45(1993), 346–358. MR1256473 (95h:46095)

[40] Y. Kawahigashi, Subfactor theory and its applications, Sugaku, 54(2002), 337–347. MR1947768 (2003j:46095)

[41] Y. Kawahigashi and R. Longo, Classification of local conformal nets. Case $c < 1$, Ann. of Math. (2), 160(2004), 493–522. MR2123931 (2006i:81119)

[42] Y. Kawahigashi and R. Longo, Classification of two-dimensional local conformal nets with $c < 1$ and 2-cohomology vanishing for tensor categories, Comm. Math. Phys., 244(2004), 63–97. MR2029950 (2005d:81228)

[43] D. Kazhdan and H. Wenzl, Reconstructing monoidal categories, Adv. Soviet Math., 16(1993), 111–136. MR1237835 (95e:18007)

[44] G.M. Kelley and M.I. Laplaza, Coherence for compact closed categories, J. Pure Appl. Algebra, 19(1980), 193–213. MR593254 (81m:18008)

[45] A.A. Kirillov and V. Ostrik, On a q-analogue of the McKay correspondence and the ADE classification of sl_2 conformal field theories, Adv. Math., 171(2002), 183–227. MR1936496 (2003j:17019)

[46] R. Longo and J.E. Roberts, A theory of dimension, K-Theory, 11(1997), 103–159. MR1444286 (98i:46065)

[47] V. Lyubashenko, Modular transformations for tensor categories. J. Pure Appl. Algebra, 98(1995), 279–327. MR1324034 (96f:18010)

[48] S. Mac Lane, Categories for the Working Mathematician, Second Edition, Springer-Verlag, Berlin-New York, 1998. MR1712872 (2001j:18001)

[49] G. Maltsiniotis, Traces dans les catégories monoïdales, dualité et catégories monoïdales fibrées, Cahiers Topologie Géom. Différentielle Catég., 36(1995), 195–288. MR1352535 (97d:18007)

[50] M. Müger, From subfactors to categories and topology. I. II., J. Pure Appl. Algebra, 180(2003), 81–157, 159–219. MR1966524 (2004f:18013); MR1966525 (2004f:18014)

[51] M. Müger, Abstract duality theory for symmetric tensor *-categories, in an appendix of "Algebraic Quantum Field Theory" by H. Halvorson.

[52] J. Murakami, Knots and Quantum Groups (in Japanese), Asakurashoten, 2000.

[53] A. Ocneanu, Paths on Coxeter diagrams: From platonic solids and singularities to minimal models and subfactors (Notes recorded by S. Goto), Lectures on operator theory, The Fields Institute Monographs. vol. 13, AMS Publications 1999, 243–323. MR1743202 (2001j:46077)

[54] V. Ostrik, Tensor categories (after P. Deligne), math.CT/0401347.

[55] V. Ostrik, Module categories, weak Hopf algebras and modular invariants, Transform. Groups, 8(2003), 177–206. MR1976459 (2004h:18006)

[56] N. Ozawa, Operator spaces and their applications, Sugaku, 56(2004), 297–307. MR2086117

[57] S. Popa, Markov traces on universal Jones algebras and subfactors of finite index, Invent. Math., 111(1993), 375–405. MR1198815 (94c:46128)

[58] S. Popa, Classification of amenable subfactors of type II, Acta Math., 172(1994), 352–445. MR1278111 (95f:46105)

[59] M. Rosso, Finite-dimensional representations of the quantum analog of the enveloping algebra of a complex simple Lie algebra, Comm. Math. Phys., 117(1988), 581–593. MR953821 (90c:17019)

[60] E.C. Rowell, From Quantum Groups to Unitary Modular Tensor Categories, arXiv:math. QA/0503226. Contemp. Math. vol. 413, Amer. Math. Soc., Providence, RI, 2006. MR2263097 (2007i:18007)

[61] N. Saavedra Rivano, Catégories Tannakiennes, Lecture Notes in Math. 265, Springer, Berlin, 1972. MR0338002 (49:2769)

[62] P. Schauenburg, Tannaka duality for arbitrary Hopf algebras, Algebra Berichte 66, Verlag Reinhard Fischer, Munich, 1992. MR1623637 (99c:18005)

[63] K. Shimakawa, Mackey systems of monoidal categories and G-spectra, K-Theory, 5(1991/92), 373–394. MR1162448 (94b:55016)

[64] S. Shnider and S. Sternberg, Quantum Groups, International Press, 1993. MR1287162 (95e:17022)

[65] M.-C. Shum, Tortile tensor categories. J. Pure Appl. Algebra, 93(1994), 57–110. MR1268782 (95a:18008)

[66] V.S. Sunder, II_1 factors, their bimodules and hypergroups, Trans. Amer. Math. Soc., 330(1992), 227–256. MR1049618 (92f:46076)

[67] M. Takesaki, Theory of Operator Algebras II, Springer, 2003. MR1943006 (2004g:46079)

[68] D. Tambara, Representations of tensor categories with fusion rules of self-duality for abelian groups, Israel J. Math., 118(2000), 29–60. MR1776075 (2001j:18015)

[69] D. Tambara and S. Yamagami, Tensor categories with fusion rules of self-duality for finite abelian groups, J. Algebra, 209(1998), 692–707. MR1659954 (2000b:18013)

[70] V.G. Turaev, Quantum Invariants of Knots and 3-Manifolds, Walter de Gruyter, Berlin-New York, 1994. MR1292673 (95k:57014)

[71] K.-H. Ulbrich, On Hopf algebras and rigid monoidal categories, Israel J. Math., 72(1990), 252–256. MR1098991 (92e:16031)

[72] E.M. Vitale, Monoidal categories for Morita theory. Cahiers Topologie Géom. Différentielle Catégoriques, 33(1992), 331–343. MR1197429 (94d:18014)

[73] M. Voit, Positive characters on commutative hypergroups and some applications, Math. Z., 198(1988), 405–421. MR946612 (89j:43005)

[74] A. Voronov, Lecture notes on operads, http://www.math.umn.edu/~voronov/8390/

[75] H. Wenzl, C^* tensor categories from quantum groups. J. Amer. Math. Soc., 11(1998), 261–282. MR1470857 (98k:46123)

[76] N. J. Wildberger, Finite commutative hypergroups and applications from group theory to conformal field theory. Applications of hypergroups and related measure algebras (Seattle, WA, 1993), 413–434, Contemp. Math., 183, Amer. Math. Soc., Providence, RI, 1995. MR1334792 (96e:20102)

[77] S.L. Woronowicz, Tannaka-Krein duality for compact matrix pseudogroups. Twisted $SU(N)$ groups, Invent. Math., 93(1988), 35–76. MR943923 (90e:22033)

[78] Y. Yamada, Introduction to Conformal Field Theory, Baifukan, 2006.

[79] S. Yamagami, A note on Ocneanu's approach to Jones index theory, Intern. J. Math., 4(1993), 859–871. MR1245354 (95f:46114)

[80] S. Yamagami, Notes on amenability of commutative fusion algebras, Positivity, 3(1999), 377–388. MR1721584 (2000m:46130)

[81] S. Yamagami, C^*-tensor categories and free product bimodules, J. Funct. Anal., 197(2003), 323–346. MR1960417 (2004a:46062)

[82] S. Yamagami, Frobenius duality in C^*-tensor categories, J. Operator Theory, 52(2004), 3–20. MR2091457 (2005f:46109)

[83] D.N. Yetter, Quantum groups and representations of monoidal categories. Math. Proc. Cambridge Philos. Soc., 108(1990), 261–290. MR1074714 (91k:16028)

Department of Mathematical Sciences, Ibaraki University, MITO 310-8512, Japan

Translated by SHIGERU YAMAGAMI